011858191

519

D0302950

. A fine
m y be r
ader

30

DUE
3 0 JU

DUE
3 0
2 y

Fuzzy Sets, Uncertainty, and Information

GEORGE J. KLIR AND TINA A. FOLGER

State University of New York, Binghamton

Prentice Hall, Englewood Cliffs, New Jersey 07632

Library of Congress Cataloging-in-Publicaiton Data

KLIR, GEORGE J. (DATE)
    Fuzzy sets, uncertainty, and information.

    Bibliography: p.
    1. Fuzzy sets.   2. System analysis.
3. Fuzzy systems.   I. Folger, Tina A.   II. Title.
QA248.K49 1988   511.3'2   87-6907
ISBN 0-13-345984-5

Editorial/production supervision
  and interior design: *Gloria L. Jordan*
Cover design: *Photo Plus Art*
Manufacturing buyer: *S. Gordon Osbourne*

Printed in the United States of America

10   9   8   7   6   5   4   3   2   1

ISBN   0-13-345984-5   025

Prentice-Hall International (UK) Limited, *London*
Prentice-Hall of Australia Pty. Limited, *Sydney*
Prentice-Hall Canada Inc., *Toronto*
Prentice-Hall Hispanoamericana, S.A., *Mexico*
Prentice-Hall of India Private Limited, *New Delhi*
Prentice-Hall of Japan, Inc., *Tokyo*
Prentice-Hall of Southeast Asia Pte. Ltd., *Singapore*
Editora Prentice-Hall do Brasil, Ltda., *Rio de Janeiro*

CONTENTS

PREFACE

It has increasingly been recognized that our society is undergoing a significant transformation, usually described as a transition from an industrial to an information society. There is little doubt that this transition is strongly connected with the emergence and development of computer technology and with the associated intellectual activities resulting in new fields of inquiry such as systems science, information science, decision analysis, or artificial intelligence.

Advances in computer technology have been steadily extending our capabilities for coping with systems of an increasingly broad range, including systems that were previously intractable to us by virtue of their nature and complexity. While the level of complexity we can manage continues to increase, we begin to realize that there are fundamental limits in this respect. As a consequence, we begin to understand that the necessity for simplification of systems, many of which have become essential for characterizing certain currently relevant problem situations, is often unavoidable. In general, a good simplification should minimize the loss of information relevant to the problem of concern. Information and complexity are thus closely interrelated.

One way of simplifying a very complex system—perhaps the most significant one—is to allow some degree of uncertainty in its description. This entails an appropriate aggregation or summary of the various entities within the system. Statements obtained from this simplified system are less precise (certain), but their relevance to the original system is fully maintained. That is, the information loss that is necessary for reducing the complexity of the system to a manageable level is expressed in uncertainty. The concept of uncertainty is thus connected with both complexity and information.

It is now realized that there are several fundamentally different types of uncertainty and that each of them plays a distinct role in the simplification problem. A mathematical formulation within which these various types of uncertainty can be properly characterized and investigated is now available in terms of the theory of fuzzy sets and fuzzy measures.

The primary purpose of this book is to bring this new mathematical formalism into the education system, not merely for its own sake, but as a basic framework for characterizing the full scope of the concept of uncertainty and its relationship to the increasingly important concepts of information and complexity. It should be stressed that these concepts arise in virtually all fields of inquiry; the usefulness of the mathematical framework presented in this book thus transcends the artificial boundaries of the various areas and specializations in the sciences and professions. This book is intended, therefore, to make an understanding of this mathematical formalism accessible to students and professionals in a broad range of disciplines. It is written specifically as a text for a one-semester course at the graduate or upper division undergraduate level that covers the various issues of uncertainty, information, and complexity from a broad perspective based on the formalism of fuzzy set theory. It is our hope that this book will encourage the initiation of new courses of this type in the various programs of higher education as well as in programs of industrial and continuing education. The book is, in fact, a by-product of one such graduate level course, which has been taught at the State University of New York at Binghamton for the last three years.

No previous knowledge of fuzzy set theory or information theory is required for an understanding of the material in this book, thus making it a virtually self-contained text. Although we assume that the reader is familiar with the basic notions of classical (nonfuzzy) set theory, classical (two-valued) logic, and probability theory, the fundamentals of these subject areas are briefly overviewed in the book. In addition, the basic ideas of classical information theory (based on the Hartley and Shannon information measures) are also introduced. For the convenience of the reader, we have included in Appendix B a glossary of the symbols most frequently used in the text.

Chapters 1–3 cover the fundamentals of fuzzy set theory and its connection with fuzzy logic. Particular emphasis is given to a comprehensive coverage of operations on fuzzy sets (Chap. 2) and to various aspects of fuzzy relations (Chap. 3). The concept of general fuzzy measures is introduced in Chap. 4, but the main focus of this chapter is on the dual classes of belief and plausibility measures along with some of their special subclasses (possibility, necessity, and probability measures); this chapter does not require a previous reading of Chapters 2 and 3. Chapter 5 introduces the various types of uncertainty and discusses their relation to information and complexity. Measures of the individual types of uncertainty are investigated in detail and proofs of the uniqueness of some of these are included in Appendix A. The classical information theory (based on the Hartley and Shannon measures of uncertainty) is overviewed, but the major emphasis is given to the new measures of uncertainty and information that have emerged from fuzzy set theory. While Chapters 1–5 focus on theoretical developments, Chap. 6 offers a brief look at some of the areas in which successful applications of this mathematical formalism have been made. Each section of Chap. 6 gives a brief overview of a major area of application along with some specific illustrative examples. We

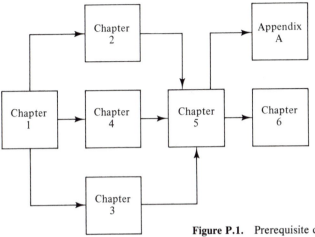

Figure P.1. Prerequisite dependencies among chapters of this book.

have attempted to provide the reader with a flavor of the numerous and diverse areas of application of fuzzy set theory and information theory without attempting an exhaustive study of each one. Ample references are included, however, which will allow the interested reader to pursue further study in the application area of concern.

The prerequisite dependencies among the individual chapters are expressed by the diagram in Fig. P.1. It is clear that the reader has some flexibility in studying the material; for instance, the chapters may be read in order, or the study of Chap. 4 may preceed that of Chaps. 2 and 3.

In order to avoid interruptions in the main text, virtually all bibliographical, historical, and other side remarks are incorporated into the notes that follow each individual chapter. These notes are uniquely numbered and are only occasionally referred to in the text.

When the book is used at the undergraduate level, coverage of some or all of the proofs of the various mathematical theorems may be omitted, depending on the background of the students. At the graduate level, on the other hand, we encourage coverage of most of these proofs in order to effect a deeper understanding of the material. In all cases, the relevance of the material to the specific area of student interest or study can be emphasized with additional application-oriented readings; the notes to Chap. 6 contain annotated references to guide in the selection of such readings from the literature.

Chapters 1–5 are each followed by a set of exercises, which are intended to enhance an understanding of the material presented in the chapter. The solutions to a selected subset of these exercises are provided in the instructor's manual; the remaining exercises are left unanswered so as to be suitable for examination use. Further suggestions for the use of this book in the teaching context can be found in the instructor's manual.

George J. Klir and Tina A. Folger
Binghamton, New York

ACKNOWLEDGMENTS

The emergence of this book was greatly influenced by contacts with many colleagues all over the world. In the area of fuzzy sets, the main influence was Lotfi A. Zadeh, the founder of fuzzy set theory as well as some of the key contributors to the theory, particularly Wyllis Bandler, Didier Dubois, Brian R. Gaines, Ladislav J. Kohout, Maria Nowakowska, Henri Prade, Ronald R. Yager, and Hans-J. Zimmermann. As far as the concepts of uncertainty, information, and complexity are concerned, the main influence was the late W. Ross Ashby, a pioneer of the use of information theory in the investigation of complex systems, and Ronald Christensen, who achieved a great mastery in the use of the minimum and maximum entropy principles. In addition, some material in Chapter 5 was influenced by contacts with Gerrit Broekstra, Roger Conant, and Klaus Krippendorff.

The book is based upon class notes prepared for a course offered by the Systems Science Department of the Thomas J. Watson School of Engineering, Applied Science, and Technology, State University of New York at Binghamton during the last few years. We are grateful to the faculty of the department for supporting this innovative course.

A draft of the book manuscript was critically evaluated by students in the Spring 1986 class. In particular, we received useful comments from Angela Bartlett, Kevin Hufford, Marlene Kaye, Efiok Otudor, Kathy Pendleton, and Dennis Rookwood. We are especially grateful to Mark Wierman; in addition to his many constructive comments, he solved almost all exercises in the book and prepared a few useful APL programs for solving some of the problems involved in the exercises. These solutions as well as programs are included in the Instructor's Guide. Our special thanks also go to Marlene Kaye; she was extremely helpful in proofreading the manuscript and Instructor's Guide and in making creative suggestions for the cover design of this book. Finally, we are grateful to Bonnie Cornick for her excellent typing of the Instructor's Guide.

The book contains some original results, particularly in the areas of uncertainty and information measures. Some of the research that led to these results was supported by the National Science Foundation under Research Grants IST-8401220, IST-8544191, and IST-8644676.

The book contains several excellent quotes and we are grateful to the copyright owners for permitting us to use the material. They are Chapman & Hall, London (see p. 193), the Institute of Electrical and Electronic Engineers, New York (p. 202), Gordon and Breach, New York (p. 211), Academic Press, London (p. 271), and Elsevier Science Publishers, Amsterdam, (p. 235).

1

CRISP SETS AND FUZZY SETS

1.1 INTRODUCTION

The process and progress of knowledge unfolds into two stages: an attempt to know the character of the world and a subsequent attempt to know the character of knowledge itself. The second reflective stage arises from the failures of the first; it generates an inquiry into the possibility of knowledge and into the limits of that possibility. It is in this second stage of inquiry that we find ourselves today. As a result, our concerns with knowledge, perceptions of problems and attempts at solutions are of a different order than in the past. We want to know not only specific facts or truths but what we can and cannot know, what we do and do not know, and how we know at all. Our problems have shifted from questions of how to cope with the world (how to provide ourselves with food, shelter, and so on), to questions of how to cope with knowledge (and ignorance) itself. Ours has been called an "information society," and a major portion of our economy is devoted to the handling, processing, selecting, storing, disseminating, protecting, collecting, analyzing, and sorting of information, our best tool for this being, of course, the computer.

Our problems are seen in terms of decision, management, and prediction; solutions are seen in terms of faster access to more information and of increased aid in analyzing, understanding and utilizing the information that is available and in coping with the information that is not. These two elements, large amounts of information coupled with large amounts of uncertainty, taken together constitute the ground of many of our problems today: complexity. As we become aware of how much we know and of how much we do not know, as information and uncertainty themselves become the focus of our concern, we begin to see our problems as centering around the issue of complexity.

The fact that complexity itself includes both the element of how much we know, or how well we can describe, and the element of how much we do not know, or how uncertain we are, can be illustrated with the simple example of driving a car. We can probably agree that driving a car is (at least relatively) complex. Further, driving a standard transmission or stick-shift car is more complex than driving a car with an automatic transmission, one index of this being that more description is needed to cover adequately our knowledge of driving in the former case than in the latter. Thus, because more knowledge is involved in the driving of a standard-transmission car (we must know, for instance, the revolutions per minute of the engine and how to use the clutch), it is more complex. However, the complexity of driving also involves the degree of our uncertainty; for example, we do not know precisely when we will have to stop or swerve to avoid an obstacle. As our uncertainty increases—for instance, in heavy traffic or on unfamiliar roads—so does the complexity of the task. Thus, our perception of complexity increases both when we realize how much we know and when we realize how much we do not know.

How do we manage to cope with complexity as well as we do, and how could we manage to cope better? The answer seems to lie in the notion of simplifying complexity by making a satisfactory trade-off or compromise between the information available to us and the amount of uncertainty we allow. One option is to increase the amount of allowable uncertainty by sacrificing some of the precise information in favor of a vague but more robust summary. For instance, instead of describing the weather today in terms of the exact percentage of cloud cover (which would be much too complex), we could just say that it is sunny, which is more uncertain and less precise but more useful. In fact, it is important to realize that the imprecision or vagueness that is characteristic of natural language does not necessarily imply a loss of accuracy or meaningfulness. It is, for instance, generally more meaningful to give travel directions in terms of city blocks than in terms of inches, although the former is much less precise than the latter. It is also more accurate to say that it is usually warm in the summer than to say that it is usually 72° in the summer. In order for a term such as *sunny* to accomplish the desired introduction of vagueness, however, we cannot use it to mean precisely 0 percent cloud cover. Its meaning is not totally arbitrary, however; a cloud cover of 100 percent is not sunny and neither, in fact, is a cloud cover of 80 percent. We can accept certain intermediate states, such as 10 or 20 percent cloud cover, as sunny. But where do we draw the line? If, for instance, any cloud cover of 25 percent or less is considered sunny, does this mean that a cloud cover of 26 percent is not? This is clearly unacceptable since 1 percent of cloud cover hardly seems like a distinguishing characteristic between sunny and not sunny. We could, therefore, add a qualification that any amount of cloud cover 1 percent greater than a cloud cover already considered to be sunny (that is, 25 percent or less) will also be labeled as sunny. We can see, however, that this definition eventually leads us to accept all degrees of cloud cover as sunny, no matter how gloomy the weather looks! In order to resolve this paradox, the term *sunny* may introduce vagueness by allowing some sort of gradual transition from degrees of cloud cover that are considered to be sunny and those that are not. This is, in fact, precisely the basic concept of the *fuzzy set*, a concept that is both simple and intuitively pleasing and that forms, in essence, a generalization of the classical or *crisp set*.

The crisp set is defined in such a way as to dichotomize the individuals in some given universe of discourse into two groups: members (those that certainly belong in the set) and nonmembers (those that certainly do not). A sharp, unambiguous distinction exists between the members and nonmembers of the class or category represented by the crisp set. Many of the collections and categories we commonly employ, however (for instance, in natural language), such as the classes of tall people, expensive cars, highly contagious diseases, numbers much greater than 1, or sunny days, do not exhibit this characteristic. Instead, their boundaries seem vague, and the transition from member to nonmember appears gradual rather than abrupt. Thus, the fuzzy set introduces vagueness (with the aim of reducing complexity) by eliminating the sharp boundary dividing members of the class from nonmembers. A fuzzy set can be defined mathematically by assigning to each possible individual in the universe of discourse a value representing its grade of membership in the fuzzy set. This grade corresponds to the degree to which that individual is similar or compatible with the concept represented by the fuzzy set. Thus, individuals may belong in the fuzzy set to a greater or lesser degree as indicated by a larger or smaller membership grade. These membership grades are very often represented by real-number values ranging in the closed interval between 0 and 1. Thus, a fuzzy set representing our concept of sunny might assign a degree of membership of 1 to a cloud cover of 0 percent, .8 to a cloud cover of 20 percent, .4 to a cloud cover of 30 percent and 0 to a cloud cover of 75 percent. These grades signify the degree to which each percentage of cloud cover approximates our subjective concept of *sunny*, and the set itself models the semantic flexibility inherent in such a common linguistic term. Because full membership and full nonmembership in the fuzzy set can still be indicated by the values of 1 and 0, respectively, we can consider the crisp set to be a restricted case of the more general fuzzy set for which only these two grades of membership are allowed.

Research on the theory of fuzzy sets has been abundant, and in this book we present an introduction to the major developments of the theory. There are, however, several types of uncertainty other than the type represented by the fuzzy set. The classical probability theory, in fact, represents one of these alternative and distinct forms of uncertainty. Understanding these various types of uncertainty and their relationships with information and complexity is currently an area of active and promising research. Therefore, in addition to offering a thorough introduction to the fuzzy set theory, this book provides an overview of the larger framework of issues of uncertainty, information, and complexity and places the fuzzy set theory within this framework of mathematical explorations.

In addition to presenting the theoretical foundations of fuzzy set theory and associated measures of uncertainty and information, the last chapter of this book offers a glimpse at some of the successful applications of this new conceptual framework to real-world problems. As general tools for dealing with complexity independent of the particular content of concern, the theory of fuzzy sets and the various mathematical representations and measurements of uncertainty and information have a virtually unrestricted applicability. Indeed, possibilities for application include any field that examines how we process or act on information, make decisions, recognize patterns, or diagnose problems or any field in which the complexity of the necessary knowledge requires some form of simplification. Successful applications have, in fact, been made in fields as numerous and diverse as engineering, psychology, artificial intelligence, medicine, ecology, decision theory, pattern recognition, information retrieval, sociology, and meteorology. Few fields remain, in fact, in which conceptions of the major problems and obstacles have not been reformulated in terms of the handling of information and uncertainty. While the diversity of successful applications has thus been expanding rapidly, the theory of fuzzy sets in particular and the mathematics of uncertainty and information in general have been achieving a secure identity as valid and useful extensions of classical mathematics. They will undoubtedly continue to constitute an important framework for further investigations into rigorous representations of uncertainty, information, and complexity.

1.2 CRISP SETS: AN OVERVIEW

This text is devoted to an examination of fuzzy sets as a broad conceptual framework for dealing with uncertainty and information. The reader's familiarity with the basic theory of crisp sets is assumed. Therefore, this section is intended to serve simply to refresh the basic concepts of crisp sets and to introduce notation and terminology useful for our discussion of fuzzy sets.

Throughout this book, sets are denoted by capital letters and their members by lower-case letters. The letter X denotes the universe of discourse, or *universal set*. This set contains all the possible elements of concern in each particular context or application from which sets can be formed. Unless otherwise stated, X is assumed in this text to contain a finite number of elements.

To indicate that an individual object x is a *member* or *element* of a set A, we write

$$x \in A.$$

Whenever x is not an element of a set A, we write

$$x \notin A.$$

A set can be described either by naming all its members (the *list method*) or by specifying some well-defined properties satisfied by the members of the set (the *rule method*). The list method, however, can be used only for finite sets. The set A whose members are $a_1, a_2, \ldots, a_n$ is usually written as

$$A = \{a_1, a_2, \ldots, a_n\},$$

and the set B whose members satisfy the properties $P_1, P_2, \ldots, P_n$ is usually

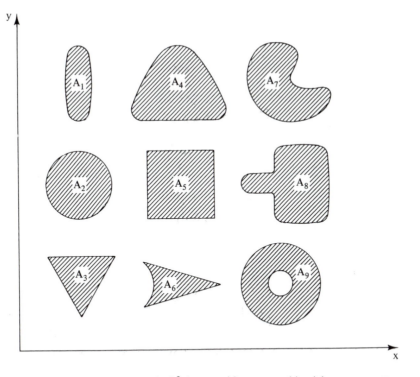

Figure 1.1.   Example of sets in $\mathbb{R}^2$ that are either convex (A_1–A_5) or nonconvex (A_6–A_9).

written as

$$B = \{b \mid b \text{ has properties } P_1, P_2, \ldots, P_n\},$$

where the symbol | denotes the phrase "such that."

An important and frequently used universal set is the set of all points in the n-dimensional Euclidean vector space $\mathbb{R}^n$ (i.e., all n-tuples of real numbers). Sets defined in terms of $\mathbb{R}^n$ are often required to possess a property referred to as convexity. A set A in $\mathbb{R}^n$ is called *convex* if, for every pair of points*

$$\mathbf{r} = (r_i \mid i \in \mathbb{N}_n) \quad \text{and} \quad \mathbf{s} = (s_i \mid i \in \mathbb{N}_n)$$

in A and every real number λ between 0 and 1, exclusively, the point

$$\mathbf{t} = (\lambda r_i + (1 - \lambda)s_i \mid i \in \mathbb{N}_n)$$

is also in A. In other words, a set A in $\mathbb{R}^n$ is convex if, for every pair of points $\mathbf{r}$ and $\mathbf{s}$ in A, all points located on the straight line segment connecting $\mathbf{r}$ and $\mathbf{s}$ are also in A. Examples of convex and nonconvex sets in $\mathbb{R}^2$ are given in Fig. 1.1.

---

* $\mathbb{N}$ subscripted by a positive integer is used in this text to denote the set of all integers from 1 through the value of the subscript; that is, $\mathbb{N}_n = \{1, 2, \ldots, n\}$.

A set whose elements are themselves sets is often referred to as a *family of sets*. It can be defined in the form

$$\{A_i \mid i \in I\},$$

where i and I are called the *set identifier* and the *identification set,* respectively. Because the index i is used to reference the sets A_i, the family of sets is also called an *indexed set*.

If every member of set A is also a member of set B—that is, if $x \in A$ implies $x \in B$—then A is called a *subset* of B, and this is written as

$$A \subseteq B.$$

Every set is a subset of itself and every set is a subset of the universal set. If $A \subseteq B$ and $B \subseteq A$, then A and B contain the same members. They are then called *equal* sets; this is denoted by

$$A = B.$$

To indicate that A and B are not equal, we write

$$A \neq B.$$

If both $A \subseteq B$ and $A \neq B$, then B contains at least one individual that is not a member of A. In this case, A is called a *proper subset* of B, which is denoted by

$$A \subset B.$$

The set that contains no members is called the *empty set* and is denoted by $\emptyset$. The empty set is a subset of every set and is a proper subset of every set except itself.

The process by which individuals from the universal set X are determined to be either members or nonmembers of a set can be defined by a *characteristic,* or *discrimination, function*. For a given set A, this function assigns a value $\mu_A(x)$ to every $x \in X$ such that

$$\mu_A(x) = \begin{cases} 1 & \text{if and only if } x \in A, \\ 0 & \text{if and only if } x \notin A. \end{cases}$$

Thus, the function maps elements of the universal set to the set containing 0 and 1. This can be indicated by

$$\mu_A : X \to \{0, 1\}.$$

The number of elements that belong to a set A is called the *cardinality* of the set and is denoted by $|A|$. A set that is defined by the rule method may contain an infinite number of elements.

The family of sets consisting of all the subsets of a particular set A is referred to as the *power set* of A and is indicated by $\mathscr{P}(A)$. It is always the case that

$$|\mathscr{P}(A)| = 2^{|A|}.$$

The *relative complement* of a set A with respect to set B is the set containing

all the members of B that are not also members of A. This can be written $B -$
A. Thus,

$$B - A = \{x \mid x \in B \quad \text{and} \quad x \notin A\}.$$

If the set B is the universal set, the complement is absolute and is usually denoted
by $\overline{A}$. Complementation is always *involutive*; that is, taking the complement of a
complement yields the original set, or

$$\overline{\overline{A}} = A.$$

The absolute complement of the empty set equals the universal set, and the absolute complement of the universal set equals the empty set. That is,

$$\overline{\varnothing} = X,$$

and

$$\overline{X} = \varnothing.$$

The *union* of sets A and B is the set containing all the elements that belong
either to set A alone, to set B alone, or to both set A and set B. This is denoted
by $A \cup B$. Thus,

$$A \cup B = \{x \mid x \in A \quad \text{or} \quad x \in B\}.$$

The union operation can be generalized for any number of sets. For a family of
sets $\{A_i \mid i \in I\}$, this is defined as

$$\bigcup_{i \in I} A_i = \{x \mid x \in A_i \text{ for some } i \in I\}.$$

The union of any set with the universal set yields the universal set, whereas the
union of any set with the empty set yields the set itself. We can write this as

$$A \cup X = X$$

and

$$A \cup \varnothing = A.$$

Because all the elements of the universal set necessarily belong either to a set A
or to its absolute complement, $\overline{A}$, the union of A and $\overline{A}$ yields the universal set.
Thus,

$$A \cup \overline{A} = X.$$

This property is usually called the *law of excluded middle*.

The *intersection* of sets A and B is the set containing all the elements belonging to both set A and set B. It is denoted by $A \cap B$. Thus,

$$A \cap B = \{x \mid x \in A \text{ and } x \in B\}.$$

The generalization of the intersection for a family of sets $\{A_i \mid i \in I\}$ is defined as

$$\bigcap_{i \in I} A_i = \{x \mid x \in A_i \text{ for all } i \in I\}.$$

The intersection of any set with the universal set yields the set itself, and the intersection of any set with the empty set yields the empty set. This can be indicated by writing

$$A \cap X = A$$

and

$$A \cap \emptyset = \emptyset.$$

Since a set and its absolute complement by definition share no elements, their intersection yields the empty set. Thus,

$$A \cap \overline{A} = \emptyset.$$

This property is usually called the *law of contradiction.*

Any two sets A and B are *disjoint* if they have no elements in common, that is, if

$$A \cap B = \emptyset.$$

It follows directly from the law of contradiction that a set and its absolute complement are always disjoint.

A collection of pairwise disjoint nonempty subsets of a set A is called a *partition* on A if the union of these subsets yields the original set A. We denote a partition on A by the symbol $\pi(A)$. Formally,

$$\pi(A) = \{A_i \mid i \in I, \quad A_i \subseteq A\},$$

where $A_i \neq \emptyset$, is a partition on A if and only if

$$A_i \cap A_j = \emptyset.$$

for each pair $i \neq j$, $i, j \in I$, and

$$\bigcup_{i \in I} A_i = A.$$

Thus, each element of A belongs to one and only one of the subsets forming the partition.

There are several important properties that are satisfied by the operations of union, intersection and complement. Both union and intersection are *commutative*, that is, the result they yield is not affected by the order of their operands. Thus,

$$A \cup B = B \cup A,$$

$$A \cap B = B \cap A.$$

Union and intersection can also be applied pairwise in any order without altering the result. We call this property *associativity* and express it by the equations

$$A \cup B \cup C = (A \cup B) \cup C = A \cup (B \cup C),$$

$$A \cap B \cap C = (A \cap B) \cap C = A \cap (B \cap C),$$

where the operations in parentheses are performed first.

Because the union and intersection of any set with itself yields that same set, we say that these two operations are *idempotent*. Thus,

$$A \cup A = A,$$

$$A \cap A = A.$$

The *distributive law* is also satisfied by union and intersection in the following ways:

$$A \cap (B \cup C) = (A \cap B) \cup (A \cap C),$$

$$A \cup (B \cap C) = (A \cup B) \cap (A \cup C).$$

Finally, *DeMorgan's law* for union, intersection, and complement states that the complement of the intersection of any two sets equals the union of their complements. Likewise, the complement of the union of two sets equals the intersection of their complements. This can be written as

$$\overline{A \cap B} = \overline{A} \cup \overline{B},$$

$$\overline{A \cup B} = \overline{A} \cap \overline{B}.$$

These and some additional properties are summarized in Table 1.1. Note that all the equations in this table that involve the set union and intersection are arranged in pairs. The second equation in each pair can be obtained from the first by replacing $\varnothing$, $\cup$, and $\cap$ with X, $\cap$, and $\cup$, respectively, and vice versa. We

TABLE 1.1.   PROPERTIES OF CRISP SET OPERATIONS.

Involution	$\overline{\overline{A}} = A$
Commutativity	$A \cup B = B \cup A$
	$A \cap B = B \cap A$
Associativity	$(A \cup B) \cup C = A \cup (B \cup C)$
	$(A \cap B) \cap C = A \cap (B \cap C)$
Distributivity	$A \cap (B \cup C) = (A \cap B) \cup (A \cap C)$
	$A \cup (B \cap C) = (A \cup B) \cap (A \cup C)$
Idempotence	$A \cup A = A$
	$A \cap A = A$
Absorption	$A \cup (A \cap B) = A$
	$A \cap (A \cup B) = A$
Absorption of complement	$A \cup (\overline{A} \cap B) = A \cup B$
	$A \cap (\overline{A} \cup B) = A \cap B$
Absorption by X and $\varnothing$	$A \cup X = X$
	$A \cap \varnothing = \varnothing$
Identity	$A \cup \varnothing = A$
	$A \cap X = A$
Law of contradiction	$A \cap \overline{A} = \varnothing$
Law of excluded middle	$A \cup \overline{A} = X$
DeMorgan's laws	$\overline{A \cap B} = \overline{A} \cup \overline{B}$
	$\overline{A \cup B} = \overline{A} \cap \overline{B}$

are thus concerned with pairs of dual equations. They exemplify a *general principle of duality*: for each valid equation in set theory that is based on the union and intersection operations, there corresponds a dual equation, also valid, that is obtained by the above specified replacement.

1.3 THE NOTION OF FUZZY SETS

As defined in the previous section, the characteristic function of a crisp set assigns a value of either 1 or 0 to each individual in the universal set, thereby discriminating between members and nonmembers of the crisp set under consideration. This function can be generalized such that the values assigned to the elements of the universal set fall within a specified range and indicate the membership grade of these elements in the set in question. Larger values denote higher degrees of set membership. Such a function is called a *membership function* and the set defined by it a *fuzzy set*.

Let X denote a universal set. Then, the membership function μ_A by which a fuzzy set A is usually defined has the form

$$\mu_A : X \to [0, 1],$$

where [0, 1] denotes the interval of real numbers from 0 to 1, inclusive.

For example, we can define a possible membership function for the fuzzy set of real numbers close to 0 as follows:

$$\mu_A(x) = \frac{1}{1 + 10x^2}.$$

The graph of this function is pictured in Fig. 1.2. Using this function, we can determine the membership grade of each real number in this fuzzy set, which

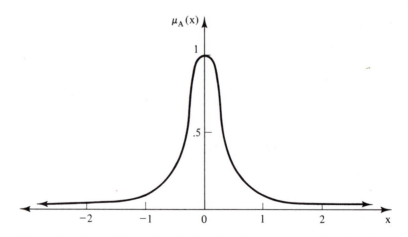

Figure 1.2. A possible membership function of the fuzzy set of real numbers close to zero.

signifies the degree to which that number is close to 0. For instance, the number 3 is assigned a grade of .01, the number 1 a grade of .09, the number .25 a grade of .62, and the number 0 a grade of 1. We might intuitively expect that by performing some operation on the function corresponding to the set of numbers close to 0, we could obtain a function representing the set of numbers very close to 0. One possible way of accomplishing this is to square the function, that is,

$$\mu_A(x) = \left(\frac{1}{1 + 10x^2}\right)^2 .$$

We could also generalize this function to a family of functions representing the set of real numbers close to any given number a as follows:

$$\mu_A(x) = \frac{1}{1 + 10(x - a)^2} .$$

Although the range of values between 0 and 1, inclusive, is the one most commonly used for representing membership grades, any arbitrary set with some natural full or partial ordering can in fact be used. Elements of this set are not required to be numbers as long as the ordering among them can be interpreted as representing various strengths of membership degree. This generalized membership function has the form

$$\mu_A : X \rightarrow L,$$

where L denotes any set that is at least partially ordered. Since L is most frequently a lattice, fuzzy sets defined by this generalized membership grade function are called *L-fuzzy sets,* where L is intended as an abbreviation for *lattice.* (The full definitions of partial ordering, total ordering, and lattice are given in Sec. 3.6.) *L*-fuzzy sets are important in certain applications, perhaps the most important being those in which $L = [0, 1]^n$. The symbol $[0, 1]^n$ is a shorthand notation of the Cartesian product

$$\underbrace{[0, 1] \times [0, 1] \times \cdots \times [0, 1]}_{n \text{ times}}$$

(see Sec. 3.1). Although the set $[0, 1]$ is totally ordered, sets $[0, 1]^n$ for any $n \geq 2$ are ordered only partially. For example, any two pairs $(a_1, b_1) \in [0, 1]^2$ and $(a_2, b_2) \in [0, 1]^2$ are not comparable (ordered) whenever $a_1 < a_2$ and $b_1 > b_2$.

A few examples in this book demonstrate the utility of *L*-fuzzy sets. For the most part, however, our discussions and examples focus on the classical representation of membership grades using real-number values in the interval $[0, 1]$.

Fuzzy sets are often incorrectly assumed to indicate some form of probability. Despite the fact that they can take on similar values, it is important to realize that membership grades are *not* probabilities. One immediately apparent difference is that the summation of probabilities on a finite universal set must equal 1, while there is no such requirement for membership grades. A more thorough discussion of the distinction between these two expressions of uncertainty is made in Chap. 4.

A further distinction must be drawn between the concept of a fuzzy set and another representation of uncertainty known as the *fuzzy measure*. Given a particular element of a universal set of concern whose membership in the various *crisp subsets* of this universal set is not known with certainty, a fuzzy measure g assigns a graded value to each of these crisp subsets, which indicates the degree of evidence or subjective certainty that the element belongs in the subset. Thus, the fuzzy measure is defined by the function

$$g : \mathcal{P}(X) \to [0, 1],$$

which satisfies certain properties. Fuzzy measures are covered in Chap. 4.

The difference between fuzzy sets and fuzzy measures can be briefly illustrated by an example. For any particular person under consideration, the evidence of age that would be necessary to place that person with certainty into the group of people in their twenties, thirties, forties, or fifties may be lacking. Note that these sets are crisp; there is no fuzziness associated with their boundaries. The set assigned the highest value in this particular fuzzy measure is our best guess of the person's age; the next highest value indicates the degree of certainty associated with our next best guess, and so on. Better evidence would result in a higher value for the best guess until absolute proof would allow us to assign a grade of 1 to a single crisp set and 0 to all the others. This can be contrasted with a problem formulated in terms of fuzzy sets in which we know the person's age but must determine to what degree he or she is considered, for instance, "old" or "young." Thus, the type of uncertainty represented by the fuzzy measure should not be confused with that represented by fuzzy sets. Chapter 4 contains a further elaboration of this distinction.

Obviously, the usefulness of a fuzzy set for modeling a conceptual class or a linguistic label depends on the appropriateness of its membership function. Therefore, the practical determination of an accurate and justifiable function for any particular situation is of major concern. The methods proposed for accomplishing this have been largely empirical and usually involve the design of experiments on a test population to measure subjective perceptions of membership degrees for some particular conceptual class. There are various means for implementing such measurements. Subjects may assign actual membership grades, the statistical response pattern for the true or false question of set membership may be sampled, or the time of response to this question may be measured, where shorter response times are taken to indicate higher subjective degrees of membership. Once these data are collected, there are several ways in which a membership function reflecting the results can be derived. Since many applications for fuzzy sets involve modeling the perceptions of a limited population for specified concepts, these methods of devising membership functions are, on the whole, quite useful. More detailed examples of some applied derivation methods are discussed in Chap. 6.

The accuracy of any membership function is necessarily limited. In addition, it may seem problematical, if not paradoxical, that a representation of fuzziness is made using membership grades that are themselves precise real numbers. Although this does not pose a serious problem for many applications, it is nevertheless possible to extend the concept of the fuzzy set to allow the distinction between grades of membership to become blurred. Sets described in this way are known as *type 2 fuzzy sets*. By definition, a type 1 fuzzy set is an ordinary fuzzy set and the elements of a type 2 fuzzy set have membership grades that are themselves type 1 (i.e., ordinary) fuzzy sets defined on some universal set Y. For example, if we define a type 2 fuzzy set "*intelligent*," membership grades assigned to elements of X (a population of human beings) might be type 1 fuzzy sets such as *average, below average, superior, genius,* and so on. Note that every fuzzy set of type 2 is an L-fuzzy set. When the membership grades employed in the definition of a type 2 fuzzy set are themselves type 2 fuzzy sets, the set is viewed as a type 3 fuzzy set. In the same way, higher types of fuzzy sets are defined.

A different extension of the fuzzy set concept involves creating fuzzy subsets of a universal set whose elements are fuzzy sets. These fuzzy sets are known as *level k fuzzy sets*, where k indicates the depth of nesting. For instance, the elements of a level 3 fuzzy set are level 2 fuzzy sets whose elements are in turn level 1 fuzzy sets. One example of a level 2 fuzzy set is the collection of desired attributes for a new car, where elements from the universe of discourse are ordinary (level 1) fuzzy sets such as *inexpensive, reliable, sporty,* and so on.

Given a crisp universal set X, let $\tilde{\mathcal{P}}(X)$ denote the set of all fuzzy subsets of X and let $\tilde{\mathcal{P}}^k(X)$ be defined recursively by the equation

$$\tilde{\mathcal{P}}^k(X) = \tilde{\mathcal{P}}(\tilde{\mathcal{P}}^{k-1}(X)),$$

for all integers $k \geq 2$. Then, fuzzy sets of level k are formally defined by membership functions of the form

$$\mu_A : \tilde{\mathcal{P}}^{k-1}(X) \to [0, 1],$$

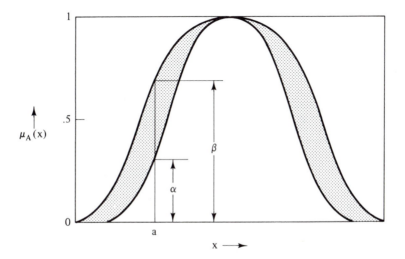

Figure 1.3.   An example of an interval-valued fuzzy set ($\mu_A(a) = [\alpha, \beta]$).

or, when extended to L-fuzzy sets, by functions

$$\mu_A : \tilde{\mathcal{P}}^{k-1}(X) \to L.$$

The requirement for a precise membership function can also be relaxed by allowing values $\mu_A(x)$ to be intervals of real numbers in [0, 1] rather than single numbers. Fuzzy sets of this sort are called *interval-valued fuzzy sets*. They are formally defined by membership functions of the form

$$\mu_A : X \to \mathcal{P}([0, 1]).$$

where $\mu_A(x)$ is a closed interval in [0, 1] for each $x \in X$. An example of this kind of membership function is given in Fig. 1.3; for each x, $\mu_A(x)$ is represented by the segment between the two curves. It is clear that the concept of interval-valued fuzzy sets can be extended to L-fuzzy sets by replacing [0, 1] with a partially ordered set L and requiring that, for each $x \in X$, $\mu_A(x)$ be a segment of totally ordered elements in L.

1.4 BASIC CONCEPTS OF FUZZY SETS

This section introduces some of the basic concepts and terminology of fuzzy sets. Many of these are extensions and generalizations of the basic concepts of crisp sets, but others are unique to the fuzzy set framework. To illustrate some of the concepts, we consider the membership grades of the elements of a small universal set in four different fuzzy sets as listed in Table 1.2 and graphically expressed in Fig. 1.4. Here the crisp universal set X of ages that we have selected is

$$X = \{5, 10, 20, 30, 40, 50, 60, 70, 80\},$$

and the fuzzy sets labeled as *infant, adult, young,* and *old* are four of the elements of the power set containing all possible fuzzy subsets of X, which is denoted by $\tilde{\mathcal{P}}(X)$.

The *support* of a fuzzy set A in the universal set X is the crisp set that contains all the elements of X that have a nonzero membership grade in A. That

TABLE 1.2. EXAMPLES OF FUZZY SETS.

Elements (ages)	Infant	Adult	Young	Old
5	0	0	1	0
10	0	0	1	0
20	0	.8	.8	.1
30	0	1	.5	.2
40	0	1	.2	.4
50	0	1	.1	.6
60	0	1	0	.8
70	0	1	0	1
80	0	1	0	1

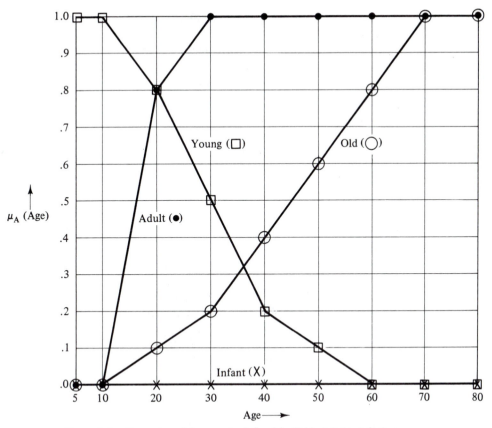

Figure 1.4.   Examples of fuzzy sets defined in Table 1.2 ($A \in \{infant, young, adult, old\}$).

is, supports of fuzzy sets in X are obtained by the function

$$\text{supp: } \tilde{\mathcal{P}}(X) \rightarrow \mathcal{P}(X),$$

where

$$\text{supp } A = \{x \in X \mid \mu_A(x) > 0\}.$$

For instance, the support of the fuzzy set *young* from Table 1.2 is the crisp set

$$\text{supp}(young) = \{5, 10, 20, 30, 40, 50\}.$$

An *empty fuzzy set* has an empty support; that is, the membership function assigns 0 to all elements of the universal set. The fuzzy set *infant* as defined in Table 1.2 is one example of an empty fuzzy set within the chosen universe.

Let us introduce a special notation that is often used in the literature for defining fuzzy sets with a finite support. Assume that x_i is an element of the support of fuzzy set A and that μ_i is its grade of membership in A. Then A is written as

$$A = \mu_1/x_1 + \mu_2/x_2 + \cdots + \mu_n/x_n,$$

where the slash is employed to link the elements of the support with their grades of membership in A and the plus sign indicates, rather than any sort of algebraic addition, that the listed pairs of elements and membership grades collectively form the definition of the set A. For the case in which a fuzzy set A is defined on a universal set that is finite and countable, we may write

$$A = \sum_{i=1}^{n} \mu_i/x_i.$$

Similarly, when X is an interval of real numbers, a fuzzy set A is often written in the form

$$A = \int_X \mu_A(x)/x.$$

The *height* of a fuzzy set is the largest membership grade attained by any element in that set. A fuzzy set is called *normalized* when at least one of its elements attains the maximum possible membership grade. If membership grades range in the closed interval between 0 and 1, for instance, then at least one element must have a membership grade of 1 for the fuzzy set to be considered normalized. Clearly, this will also imply that the height of the fuzzy set is equal to 1. The three fuzzy sets *adult, young,* and *old* from Table 1.2 as well as those defined by Figs. 1.2 and 1.3 are all normalized, and thus the height of each is equal to 1. Figure 1.5 illustrates a fuzzy set that is not normalized.

An α-cut of a fuzzy set A is a crisp set A_α that contains all the elements of the universal set X that have a membership grade in A greater than or equal to

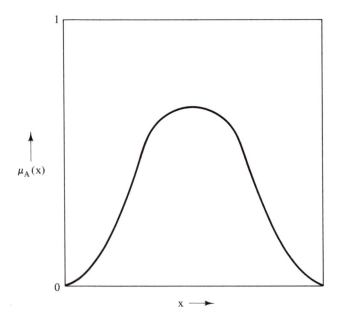

Figure 1.5. Nonnormalized fuzzy set that is convex.

the specified value of α. This definition can be written as

$$A_\alpha = \{x \in X \mid \mu_A(x) \geq \alpha\}.$$

The value α can be chosen arbitrarily but is often designated at the values of the membership grades appearing in the fuzzy set under consideration. For instance, for $\alpha = .2$, the α-cut of the fuzzy set *young* from Table 1.2 is the crisp set

$$young_{.2} = \{5, 10, 20, 30, 40\}.$$

For $\alpha = .8$,

$$young_{.8} = \{5, 10, 20\},$$

and for $\alpha = 1$,

$$young_1 = \{5, 10\}.$$

Observe that the set of all α-cuts of any fuzzy set on X is a family of nested crisp subsets of X.

The set of all levels $\alpha \in [0, 1]$ that represent distinct α-cuts of a given fuzzy set A is called a *level set* of A. Formally,

$$\Lambda_A = \{\alpha \mid \mu_A(x) = \alpha \text{ for some } x \in X\},$$

where Λ_A denotes the level set of fuzzy set A defined on X.

When the universal set is the set of all *n*-tuples of real numbers in the *n*-dimensional Euclidean vector space $\mathbb{R}^n$, the concept of set convexity can be generalized to fuzzy sets. A fuzzy set is *convex* if and only if each of its α-cuts is a convex set. Equivalently we may say that a fuzzy set A is convex if and only if

$$\mu_A(\lambda \mathbf{r} + (1 - \lambda)\mathbf{s}) \geq \min[\mu_A(\mathbf{r}), \mu_A(\mathbf{s})],$$

for all $\mathbf{r}, \mathbf{s} \in \mathbb{R}^n$ and all $\lambda \in [0, 1]$. Figures 1.2, 1.4, and 1.5 illustrate convex fuzzy sets, whereas Fig. 1.6 illustrates a nonconvex fuzzy set on $\mathbb{R}$. Figure 1.7 illustrates a convex fuzzy set on $\mathbb{R}^2$ expressed by the α-cuts for all α in its level set. Note that the definition of convexity for fuzzy sets does not mean that the membership function of a convex fuzzy set is necessarily a convex function.

A convex and normalized fuzzy set defined on $\mathbb{R}$ whose membership function is piecewise continuous is called a *fuzzy number*. Thus, a fuzzy number can be thought of as containing the real numbers within some interval to varying degrees. For example, the membership function given in Fig. 1.2 can be viewed as a representation of a fuzzy number.

The *scalar cardinality* of a fuzzy set A defined on a finite universal set X is the summation of the membership grades of all the elements of X in A. Thus,

$$|A| = \sum_{x \in X} \mu_A(x).$$

The scalar cardinality of the fuzzy set *old* from Table 1.2 is

$$|old| = 0 + 0 + .1 + .2 + .4 + .6 + .8 + 1 + 1 = 4.1$$

The scalar cardinality of the fuzzy set *infant* is 0.

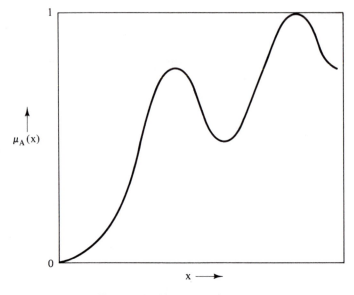

Figure 1.6.   Nonconvex fuzzy set.

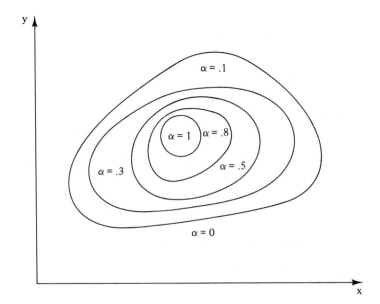

Figure 1.7. α-cuts of a convex fuzzy set defined on $\mathbb{R}^2$.

Other forms of cardinality have been proposed for fuzzy sets. One of these, which is called the *fuzzy cardinality,* is defined as a fuzzy number rather than as a real number, as is the case for the scalar cardinality. When a fuzzy set A has a finite support, its fuzzy cardinality $|\tilde{A}|$ is a fuzzy set (fuzzy number) defined on $\mathbb{N}$ whose membership function is defined by

$$\mu_{|\tilde{A}|}(|A_{\alpha}|) = \alpha,$$

for all α in the level set of A ($\alpha \in \Lambda_A$). The fuzzy cardinality of the fuzzy set *old* from Table 1.2 is

$$|\widetilde{old}| = .1/7 + .2/6 + .4/5 + .6/4 + .8/3 + 1/2.$$

There are many ways of extending the set inclusion as well as the basic crisp set operations for application to fuzzy sets. Several of these are examined in detail in Chap. 2. The discussion here is a brief introduction to the simple definitions of set inclusion and complement and to the union and intersection operations that were first proposed for fuzzy sets.

If the membership grade of each element of the universal set X in fuzzy set A is less than or equal its membership grade in fuzzy set B, then A is called a *subset* of B. Thus, if

$$\mu_A(x) \leq \mu_B(x),$$

for every $x \in X$, then

$$A \subseteq B.$$

The fuzzy set *old* from Table 1.2 is a subset of the fuzzy set *adult* since for each element in our universal set

$$\mu_{old}(x) \leq \mu_{adult}(x).$$

Fuzzy sets A and B are called *equal* if $\mu_A(x) = \mu_B(x)$ for every element $x \in X$. This is denoted by

$$A = B.$$

Clearly, if $A = B$, then $A \subseteq B$ and $B \subseteq A$.

If fuzzy sets A and B are not equal ($\mu_A(x) \neq \mu_B(x)$ for at least one $x \in X$), we write

$$A \neq B.$$

None of the four fuzzy sets defined in Table 1.2 is equal to any of the others.

Fuzzy set A is called a *proper subset* of fuzzy set B when A is a subset of B and the two sets are not equal; that is, $\mu_A(x) \leq \mu_B(x)$ for every $x \in X$ and $\mu_A(x) < \mu_B(x)$ for at least one $x \in X$. We can denote this by writing

$$A \subset B \quad \text{if and only if} \quad A \subseteq B \text{ and } A \neq B.$$

It was mentioned that the fuzzy set *old* from Table 1.2 is a subset of the fuzzy set *adult* and that these two fuzzy sets are not equal. Therefore, *old* can be said to be a proper subset of *adult*.

When membership grades range in the closed interval between 0 and 1, we denote the *complement* of a fuzzy set with respect to the universal set X by $\overline{A}$ and define it by

$$\mu_{\overline{A}}(x) = 1 - \mu_A(x),$$

for every $x \in X$. Thus, if an element has a membership grade of .8 in a fuzzy set A, its membership grade in the complement of A will be .2. For instance, taking the complement of the fuzzy set *old* from Table 1.2 produces the fuzzy set *not old* defined as

$$not\ old = 1/5 + 1/10 + .9/20 + .8/30 + .6/40 + .4/50 + .2/60.$$

Note that in this particular case *not old* is not equal to the fuzzy set *young*.

The *union* of two fuzzy sets A and B is a fuzzy set $A \cup B$ such that

$$\mu_{A \cup B}(x) = \max[\mu_A(x), \mu_B(x)],$$

for every $x \in X$. Thus, the membership grade of each element of the universal set in $A \cup B$ is either its membership grade in A or its membership grade in B, whichever is the larger value. From this definition it can be seen that fuzzy sets A and B are both subsets of the fuzzy set $A \cup B$, a property we would in fact expect from a union operation. When we take the union of the fuzzy sets *young* and *old* from Table 1.2, the following fuzzy set is created:

$$young \cup old = 1/5 + 1/10 + .8/20 + .5/30 + .4/40$$
$$+ .6/50 + .8/60 + 1/70 + 1/80.$$

The *intersection* of fuzzy sets A and B is a fuzzy set $A \cap B$ such that

$$\mu_{A \cap B}(x) = \min[\mu_A(x), \mu_B(x)],$$

for every $x \in X$. Here, the membership grade of an element x in fuzzy set $A \cap B$ is the smaller of its membership grades in set A and set B. As is desirable for an intersection operation, the fuzzy set $A \cap B$ is a subset of both A and B. The intersection of fuzzy sets *young* and *old* from Table 1.2 is a fuzzy set defined as

$$young \cap old = .1/20 + .2/30 + .2/40 + .1/50.$$

These original formulations of fuzzy complement, union, and intersection perform identically to the corresponding crisp set operators when membership grades are restricted to the values 0 and 1. They are, therefore, good generalizations of the classical crisp set operators. Chapter 2 contains a further discussion of the properties of these original operators and of their relation to the other classes of operators subsequently proposed.

A basic principle that allows the generalization of crisp mathematical concepts to the fuzzy framework is known as the *extension principle*. It provides the means for any function f that maps points $x_1, x_2, \ldots, x_n$ in the crisp set X to the crisp set Y to be generalized such that it maps fuzzy subsets of X to Y. Formally, given a function f mapping points in set X to points in set Y and any fuzzy set $A \in \tilde{P}(X)$, where

$$A = \mu_1/x_1 + \mu_2/x_2 + \cdots + \mu_n/x_n,$$

the extension principle states that

$$f(A) = f(\mu_1/x_1 + \mu_2/x_2 + \cdots + \mu_n/x_n)$$

$$= \mu_1/f(x_1) + \mu_2/f(x_2) + \cdots + \mu_n/f(x_n).$$

If more than one element of X is mapped by f to the same element $y \in Y$, then the maximum of the membership grades of these elements in the fuzzy set A is chosen as the membership grade for y in $f(A)$. If no element $x \in X$ is mapped to y, then the membership grade of y in $f(A)$ is zero. Often a function f maps ordered tuples of elements of several different sets $X_1, X_2, \ldots, X_n$ such that $f(x_1, x_2, \ldots, x_n) = y, y \in Y$. In this case, for any arbitrary fuzzy sets $A_1, A_2, \ldots, A_n$ defined on $X_1, X_2, \ldots, X_n$, respectively, the membership grade of element y in $f(A_1, A_2, \ldots, A_n)$ is equal to the minimum of the membership grades of $x_1, x_2, \ldots, x_n$ in $A_1, A_2, \ldots, A_n$, respectively.

As a simple illustration of the use of this principle, suppose that f is a function mapping ordered pairs from $X_1 = \{a, b, c\}$ and $X_2 = \{x, y\}$ to $Y = \{p, q, r\}$. Let f be specified by the following matrix:

$$\begin{array}{c}
 \\
a \\
b \\
c
\end{array}
\begin{array}{cc}
x & y \\
\left[\begin{array}{cc} p & p \\ q & r \\ r & p \end{array}\right]
\end{array}$$

Let A_1 be a fuzzy set defined on X_1 and let A_2 be a fuzzy set defined on X_2 such that

$$A_1 = .3/a + .9/b + .5/c$$

and

$$A_2 = .5/x + 1/y.$$

The membership grades of p, q, and r in the fuzzy set $B = f(A_1, A_2) \in \tilde{P}(Y)$ can be calculated from the extension principle as follows:

$$\mu_B(p) = \max[\min(.3, .5), \min(.3, 1), \min(.5, 1)] = .5;$$

$$\mu_B(q) = \max[\min(.9, .5)] = .5;$$

$$\mu_B(r) = \max[\min(.5, .5), \min(.9, 1)] = .9.$$

Thus, by the extension principle

$$f(A_1, A_2) = .5/p + .5/q + .9/r.$$

1.5 CLASSICAL LOGIC: AN OVERVIEW

We assume that the reader of this book is familiar with the fundamentals of classical logic. Therefore, this section is solely intended to provide a brief overview of the basic concepts of classical logic and to introduce terminology and notation employed in our discussion of fuzzy logic.

Logic is the study of the methods and principles of *reasoning* in all its possible forms. Classical logic deals with *propositions* that are required to be either *true* or *false*. Each proposition has an opposite, which is usually called a *negation* of the proposition. A proposition and its negation are required to assume opposite truth values.

One area of logic, referred to as *propositional logic*, deals with combinations of variables that stand for arbitrary propositions. These variables are usually called *logic variables* (or propositional variables). As each variable stands for a hypothetical proposition, it may assume either of the two truth values; the variable is not committed to either truth value unless a particular proposition is substituted for it.

One of the main concerns of propositional logic is the study of rules by which new logic variables can be produced as functions of some given logic variables. It is not concerned with the internal structure of the propositions represented by the logic variables.

Assume that n logic variables $v_1, v_2, \ldots, v_n$ are given. A new logic variable can then be defined by a function that assigns a particular truth value to the new variable for each combination of truth values of the given variables. This function is usually called a *logic function*. Since n logic variables may assume 2^n prospective truth values, there are 2^{2^n} possible logic functions defining these variables. For example, all the logic functions of two variables are listed in Table 1.3, where falsity and truth are denoted by 0 and 1, respectively, and the resulting 16 logic variables are denoted by $w_1, w_2, \ldots, w_{16}$. Logic functions of one or two variables are usually called *logic operations*.

TABLE 1.3. LOGIC FUNCTIONS OF TWO VARIABLES.

v_2	1 1 0 0	Adopted name	Adopted	Other names used	Other symbols used
v_1	1 0 1 0	of function	Symbol	in the literature	in the literature
w_1	0 0 0 0	Zero function	0	Falsum	$F, \perp$
w_2	0 0 0 1	Nor function	$v_1 \barvee v_2$	Pierce function	$v_1 \downarrow v_2$, NOR(v_1, v_2)
w_3	0 0 1 0	Inhibition	$v_1 \nLeftarrow v_2$	Proper inequality	$v_1 > v_2$
w_4	0 0 1 1	Negation	$\bar{v}_2$	Complement	$\neg v_2$, $\sim v_2$, v_2^0
w_5	0 1 0 0	Inhibition	$v_1 \nRightarrow v_2$	Proper inequality	$v_1 < v_2$
w_6	0 1 0 1	Negation	$\bar{v}_1$	Complement	$\neg v_1$, $\sim v_1$, v_1^0
w_7	0 1 1 0	Exclusive-or function	$v_1 \ovee v_2$	Nonequivalence	$v_1 \neq v_2$, $v_1 \oplus v_2$
w_8	0 1 1 1	Nand function	$v_1 \barwedge v_2$	Sheffer stroke	$v_1 \mid v_2$, NAND(v_1, v_2)
w_9	1 0 0 0	Conjunction	$v_1 \wedge v_2$	And function	$v_1 \& v_2$, $v_1 v_2$
w_{10}	1 0 0 1	Biconditional	$v_1 \Leftrightarrow v_2$	Equivalence	$v_1 \equiv v_2$
w_{11}	1 0 1 0	Assertion	v_1	Identity	v_1^1
w_{12}	1 0 1 1	Implication	$v_1 \Leftarrow v_2$	Conditional, inequality	$v_1 \subset v_2$, $v_1 \geq v_2$
w_{13}	1 1 0 0	Assertion	v_2	Identity	v_2^1
w_{14}	1 1 0 1	Implication	$v_1 \Rightarrow v_2$	Conditional, inequality	$v_1 \supset v_2$, $v_1 \leq v_2$
w_{15}	1 1 1 0	Disjunction	$v_1 \vee v_2$	Or function	$v_1 + v_2$
w_{16}	1 1 1 1	One function	1	Verum	T, I

The key issue of propositional logic is the expression of all the logic functions of n variables ($n \in N$), the number of which grows extremely rapidly with increasing values of n, with the aid of a small number of simple logic functions. These simple functions are preferably logic operations of one or two variables, which are called *logic primitives*. It is known that this can be accomplished only with some sets of logic primitives. We say that a set of primitives is *complete* if and only if any logic function of variables $v_1, v_2, \ldots, v_n$ (for any finite n) can be composed by a finite number of these primitives.

Two of the many complete sets of primitives have been predominant in propositional logic: (1) negation, conjunction, and disjunction, and (2) negation and implication. By combining, for example, negations, conjunctions, and disjunctions (employed as primitives) in appropriate algebraic expressions, referred to as *logic formulas*, we can form any other logic function. Logic formulas are then defined recursively as follows:

1. The truth values 0 and 1 are logic formulas.
2. If v denotes a logic variable, then v and $\bar{v}$ are logic formulas.
3. If a and b denote logic formulas, then $a \wedge b$ and $a \vee b$ are also logic formulas.
4. The only logic formulas are those defined by statements 1 through 3.

Every logic formula of this type defines a logic function by composing it from the three primary functions. To define a unique function, the order in which the individual compositions are to be performed must be specified in some way. There are various ways in which this order can be specified. The most common is the usual use of parentheses, as in any other algebraic expression.

Other types of logic formulas can be defined by replacing some of the three operations in this definition with other operations or by including some additional operations. We may replace, for example, $a \wedge b$ and $a \vee b$ in the definition with $a \Rightarrow b$, or we may simply add $a \Rightarrow b$ to the definition.

While each proper logic formula represents a single logic function and the associated logic variable, different formulas may represent the same function and variable. If they do, we consider them equivalent. When logic formulas a and b are equivalent, we write $a = b$. For example,

$$(\bar{v}_1 \wedge \bar{v}_2) \vee (v_1 \wedge \bar{v}_3) \vee (v_2 \wedge v_3) = (\bar{v}_2 \wedge \bar{v}_3) \vee (\bar{v}_1 \wedge v_3) \vee (v_1 \wedge v_2),$$

as can easily be verified by evaluating each of the formulas for all eight combinations of truth values of the logic variables v_1, v_2, and v_3.

When the variable represented by a logic formula is always true regardless of the truth values assigned to the variables participating in the formula, it is called a *tautology*; when it is always false, it is called a *contradiction*. For example, when two logic formulas a and b are equivalent, then $a \Leftrightarrow b$ is a tautology, whereas the formula $a \,\textcircled{v}\, b$ is a contradiction. Tautologies are important for deductive reasoning, since they represent logic formulas that, due to their form, are true on logical grounds alone.

Various forms of tautologies can be used for making deductive inferences.

TABLE 1.4. PROPERTIES OF BOOLEAN ALGEBRAS.

(B1)	Idempotence	$a + a = a$
		$a \cdot a = a$
(B2)	Commutativity	$a + b = b + a$
		$a \cdot b = b \cdot a$
(B3)	Associativity	$(a + b) + c = a + (b + c)$
		$(a \cdot b) \cdot c = a \cdot (b \cdot c)$
(B4)	Absorption	$a + (a \cdot b) = a$
		$a \cdot (a + b) = a$
(B5)	Distributivity	$a \cdot (b + c) = (a \cdot b) + (a \cdot c)$
		$a + (b \cdot c) = (a + b) \cdot (a + c)$
(B6)	Universal bounds	$a + 0 = a, a + 1 = 1$
		$a \cdot 1 = a, a \cdot 0 = 0$
(B7)	Complementarity	$a + \bar{a} = 1$
		$a \cdot \bar{a} = 0$
		$\bar{1} = 0$
(B8)	Involution	$\bar{\bar{a}} = a$
(B9)	Dualization	$\overline{a + b} = \bar{a} \cdot \bar{b}$
		$\overline{a \cdot b} = \bar{a} + \bar{b}$

They are referred to as *inference rules*. Examples of some tautologies frequently used as inference rules are:

$$(a \wedge (a \Rightarrow b)) \Rightarrow b \qquad (modus\ ponens),$$

$$(\bar{b} \wedge (a \Rightarrow b)) \Rightarrow \bar{a} \qquad (modus\ tollens),$$

$$((a \Rightarrow b) \wedge (b \Rightarrow c)) \Rightarrow (a \Rightarrow c) \qquad (hypothetical\ syllogism).$$

Modus ponens, for instance, states that given two true propositions a and $a \Rightarrow b$ (the premises), the truth of the proposition b (the conclusion) may be inferred.

Every tautology remains a tautology when any of its variables is replaced with any arbitrary logic formula. This property is another example of a powerful rule of inference, referred to as a *rule of substitution*.

It is well established that propositional logic is isomorphic to set theory under the appropriate correspondence between components of these two mathematical systems. Furthermore, both of these systems are isomorphic to a Boolean algebra, which is a mathematical system defined by abstract (interpretation-free) entities and their axiomatic properties.

A *Boolean algebra* on a set B is defined as the quadruple

$$\mathcal{B} = (B, +, \cdot, \overline{}),$$

where the set B has at least two elements (bounds) 0 and 1; $+$ and $\cdot$ are binary operations on B, and $\overline{}$ is a unary operation on B for which the properties listed in Table 1.4 are satisfied.* Properties (B1)–(B4) are common to all lattices. Boo-

---

* Not all these properties are necessary for an axiomatic characterization of Boolean algebras; we present this larger set of properties in order to emphasize the relationship between Boolean algebras, set theory, and propositional logic.

lean algebras are therefore lattices that are distributive (B5), bounded (B6), and complemented (B7)–(B9). This means that each Boolean algebra can also be characterized in terms of a partial ordering on a set that is defined as follows: $a \leq b$ if and only if $a \cdot b = a$ or, alternatively, if and only if $a + b = b$.

The isomorphisms between Boolean algebra, set theory, and propositional logic guarantee that every theorem in any one of these theories has a counterpart in each of the other two theories. These counterparts can be obtained from one another by applying the substitutional correspondences in Table 1.5. All symbols used in this table have previously been defined in the text except for the symbol $\mathscr{F}(V)$; V denotes here the set of all combinations of truth values of given logic variables, and $\mathscr{F}(V)$ stands for the set of all logic functions defined in terms of these combinations. It is obviously required that the cardinalities of sets V and X be equal. These isomorphisms allow us, in effect, to cover all these theories by developing only one of them. We take advantage of this possibility by focusing the discussion in this book primarily on the theory of fuzzy sets rather than on fuzzy logic. For example, our study in Chap. 2 of the general operations on fuzzy sets is not repeated for operations of fuzzy logic, since the isomorphism between the two areas allows the properties of the latter to be obtained directly from the corresponding properties of fuzzy set operations.

Propositional logic is concerned only with those logic relationships that depend on the way in which propositions are composed from other propositions by logic operations. These latter propositions are treated as unanalyzed wholes. This is not adequate for many instances of deductive reasoning, for which the internal structure of propositions cannot be ignored.

Propositions are sentences expressed in some language. Each sentence representing a proposition can fundamentally be broken down into a *subject* and a *predicate*. In other words, a simple proposition can be expressed, in general, in the canonical form

$$x \text{ is } P,$$

where x is a symbol of a subject and P designates a predicate, which characterizes a property. For example, "Austria is a German-speaking country" is a proposition in which "Austria" stands for a subject (a particular country) and "a German

TABLE 1.5. CORRESPONDENCES DEFINING ISOMORPHISMS BETWEEN SET THEORY, BOOLEAN ALGEBRA, AND PROPOSITIONAL LOGIC.

Set theory	Boolean algebra	Propositional logic
$\mathscr{P}(X)$	B	$\mathscr{F}(V)$
$\cup$	$+$	$\vee$
$\cap$	$\cdot$	$\wedge$
$\bar{}$	$\bar{}$	$\bar{}$
X	1	1
$\varnothing$	0	0
$\subseteq$	$\leq$	$\Rightarrow$

speaking-country'' is a predicate that characterizes a specific property, namely, the property of being a country whose inhabitants speak German. This proposition is true.

Instead of dealing with particular propositions, we may use the general form "x is P," where x now stands for any subject from a designated universe of discourse X. The predicate P then plays the role of a function defined on X, which for each value of x forms a proposition. This function is usually called a *predicate* and is denoted by $P(x)$. Clearly, a predicate becomes a proposition that is either true or false when a particular subject from X is substituted for x.

It is useful to extend the concept of a predicate in two ways. First, it is natural to extend it to more than one variable. This leads to the notion of an n-ary predicate $P(x_1, x_2, \ldots, x_n)$, which for $n = 1$ represents a property and for $n \geq 2$ an n-ary relation among subjects from designated universal sets X_i ($i \in \mathbb{N}_n$). For example,

$$x_1 \text{ is a citizen of } x_2,$$

where x_1 stands for individual persons from a designated population X_1 and x_2 stands for individual countries from a designated set X_2 of countries, is a binary predicate. Here, elements of X_2 are usually called *objects* rather than subjects. For convenience, n-ary predicates for $n = 0$ are defined as propositions in the same sense as in propositional logic.

Another way of extending the scope of a predicate is to quantify its applicability with respect to the domain of its variables. Two kinds of quantification have been predominantly used for predicates; they are referred to as existential quantification and universal quantification.

Existential quantification of a predicate $P(x)$ is expressed by the form

$$(\exists x)P(x),$$

which represents the sentence "There exists an individual x (in the universal set X of the variable x) such that x is P" (or the equivalent sentence "Some $x \in X$ are P"). The symbol $\exists$ is called an *existential quantifier*. We have the following equality:

$$(\exists x)P(x) = \bigvee_{x \in X} P(x). \tag{1.1}$$

Universal quantification of a predicate P(x) is expressed by the form

$$(\forall x)P(x),$$

which represents the sentence "For every individual x (in the designated universal set), x is P" (or the equivalent sentence "All $x \in X$ are P"). The symbol $\forall$ is called a *universal quantifier*. Clearly, the following equality holds:

$$(\forall x)P(x) = \bigwedge_{x \in X} P(x). \tag{1.2}$$

For n-ary predicates, we may use up to n quantifiers of either kind, each applying to one variable. For instance,

$$(\exists x_1)(\forall x_2)(\exists x_3)P(x_1, x_2, x_3)$$

stands for the sentence "there exists an $x_1 \in X_1$ such that for all $x_2 \in X_2$ there exists x_3 such that $P(x_1, x_2, x_3)$." For example, if $X_1 = X_2 = X_3 = [0, 1]$ and $P(x_1, x_2, x_3)$ means $x_1 \leq x_2 \leq x_3$, then the sentence is true (assume $x_1 = 0$ and $x_3 = 1$).

The standard existential and universal quantification of predicates can be conveniently generalized by conceiving a quantifier Q applied to a predicate $P(x)$, $x \in X$, as a binary relation

$$Q \subset \{(\alpha, \beta) | \; \alpha, \beta \in \mathbb{N}, \alpha + \beta = |X|\},$$

where α and β specify the number of elements of X for which $P(x)$ is true or false, respectively. Formally,

$$\alpha = |\{x \in X \mid P(x) \text{ is true}\}| \;,$$

$$\beta = |\{x \in X \mid P(x) \text{ is false}\}| \;.$$

For example, when Q is defined by the condition $\alpha \neq 0$, we obtain the standard existential quantifier; when $\beta = 0$, Q becomes the standard universal quantifier; when $\alpha > \beta$, we obtain the so-called plurality quantifier, expressed by the word *most*.

New predicates (quantified or not) can be produced from given predicates by logic formulas in the same way as new logic variables are produced by logic formulas in propositional logic. These formulas, which are called *predicate formulas*, are the essence of *predicate logic*.

1.6 FUZZY LOGIC

The basic assumption upon which classical logic (or two-valued logic) is based— that every proposition is either true or false—has been questioned since Aristotle. In his treatise *On Interpretation*, Aristotle discusses the problematic truth status of matters that are future-contingent. Propositions about future events, he maintains, are neither actually true nor actually false but are potentially either; hence, their truth value is undetermined, at least prior to the event.

It is now well understood that propositions whose truth status is problematic are not restricted to future events. As a consequence of the Heisenberg principle of uncertainty, for example, it is known that truth values of certain propositions in quantum mechanics are inherently indeterminate due to fundamental limitations of measurement. In order to deal with such propositions, we must relax the true-false dichotomy of classical two-valued logic by allowing a third truth value, which may be called *indeterminate*.

The classical two-valued logic can be extended into *three-valued logic* in various ways. Several three-valued logics, each with its own rationale, are now well established. It is common in these logics to denote the truth, falsity, and indeterminacy by $1, 0$, and $\frac{1}{2}$, respectively. It is also common to define the negation $\bar{a}$ of a proposition a as $1 - a$; that is, $\bar{1} = 0$, $\bar{0} = 1$ and $\bar{\frac{1}{2}} = \frac{1}{2}$. Other primitives, such as $\wedge$, $\vee$, $\Rightarrow$, and $\Leftrightarrow$ differ from one logic to another. Five of the best known three-valued logics, labeled with the names of their originators, are defined in terms of these four primitives in Table 1.6

We can see from Table 1.6 that all the logic primitives listed for the five three-valued logics fully conform to the usual definitions of these primitives in the classical logic for $a, b \in \{0, 1\}$ and that they differ from each other only in their treatment of the new truth value $\frac{1}{2}$. We can also easily verify that none of these three-valued logics satisfies the law of contradiction ($a \wedge \bar{a} = 0$), the law of excluded middle ($a \vee \bar{a} = 1$), and some other tautologies of two-valued logic. The Bochvar three-valued logic, for example, clearly does not satisfy any of the tautologies of two-valued logic, since each of its primitives produces the truth value $\frac{1}{2}$ whenever at least one of the propositions a and b assumes this value. It is, therefore, common to extend the usual concept of a tautology to the broader concept of a *quasi-tautology*. We say that a logic formula in a three-valued logic that does not assume the truth value 0 (falsity) regardless of the truth values assigned to its proposition variables is a quasi-tautology. Similarly, we say that a logic formula that does not assume the truth value 1 (truth) is a *quasi-contradiction*.

Once the various three-valued logics were accepted as meaningful and useful, it became desirable to explore generalizations into *n-valued logics* for an arbitrary number of truth values ($n \geq 2$). Several *n*-valued logics were, in fact, developed in the 1930s. For any given *n*, the truth values in these generalized logics are usually labeled by rational numbers in the unit interval [0, 1]. These values are obtained by evenly dividing the interval between 0 and 1, exclusive. The set T_n of truth values of an *n*-valued logic is thus defined as

$$T_n = \left\{ 0 = \frac{0}{n-1}, \frac{1}{n-1}, \frac{2}{n-1}, \ldots, \frac{n-2}{n-1}, \frac{n-1}{n-1} = 1 \right\}.$$

These values can be interpreted as *degrees of truth*.

The first series of *n*-valued logics for which $n \geq 2$ was proposed by Łukasiewicz in the early 1930s as a generalization of his three-valued logic. It uses truth values in T_n and defines the primitives by the following equations:

$$\bar{a} = 1 - a,$$
$$a \wedge b = \min(a, b),$$
$$a \vee b = \max(a, b), \qquad\qquad (1.3)$$
$$a \Rightarrow b = \min(1, 1 + b - a),$$
$$a \Leftrightarrow b = 1 - |a - b|.$$

Łukasiewicz, in fact, used only negation and implication as primitives and defined the other logic operations in terms of these two primitives, as follows:

$$a \vee b = (a \Rightarrow b) \Rightarrow b,$$
$$a \wedge b = \overline{\bar{a} \vee \bar{b}},$$
$$a \Leftrightarrow b = (a \Rightarrow b) \wedge (b \Rightarrow a).$$

TABLE 1.6. PRIMITIVES OF SOME THREE-VALUED LOGICS.

a b	Łukasiewicz $\wedge$ $\vee$ $\Rightarrow$ $\Leftrightarrow$				Bochvar $\wedge$ $\vee$ $\Rightarrow$ $\Leftrightarrow$				Kleene $\wedge$ $\vee$ $\Rightarrow$ $\Leftrightarrow$				Heyting $\wedge$ $\vee$ $\Rightarrow$ $\Leftrightarrow$				Reichenbach $\wedge$ $\vee$ $\Rightarrow$ $\Leftrightarrow$			
0 0	0	0	1	1	0	0	1	1	0	0	1	1	0	0	1	1	0	0	1	1
0 $\frac{1}{2}$	0	$\frac{1}{2}$	1	$\frac{1}{2}$	$\frac{1}{2}$	$\frac{1}{2}$	$\frac{1}{2}$	$\frac{1}{2}$	0	$\frac{1}{2}$	1	$\frac{1}{2}$	0	$\frac{1}{2}$	1	0	0	$\frac{1}{2}$	1	$\frac{1}{2}$
0 1	0	1	1	0	0	1	1	0	0	1	1	0	0	1	1	0	0	1	1	0
$\frac{1}{2}$ 0	0	$\frac{1}{2}$	$\frac{1}{2}$	$\frac{1}{2}$	$\frac{1}{2}$	$\frac{1}{2}$	$\frac{1}{2}$	$\frac{1}{2}$	0	$\frac{1}{2}$	$\frac{1}{2}$	$\frac{1}{2}$	0	$\frac{1}{2}$	0	0	0	$\frac{1}{2}$	$\frac{1}{2}$	$\frac{1}{2}$
$\frac{1}{2}$ $\frac{1}{2}$	$\frac{1}{2}$	$\frac{1}{2}$	1	1	$\frac{1}{2}$	$\frac{1}{2}$	$\frac{1}{2}$	$\frac{1}{2}$	$\frac{1}{2}$	$\frac{1}{2}$	$\frac{1}{2}$	$\frac{1}{2}$	$\frac{1}{2}$	$\frac{1}{2}$	1	1	$\frac{1}{2}$	$\frac{1}{2}$	1	1
$\frac{1}{2}$ 1	$\frac{1}{2}$	1	1	$\frac{1}{2}$	$\frac{1}{2}$	$\frac{1}{2}$	$\frac{1}{2}$	$\frac{1}{2}$	$\frac{1}{2}$	1	1	$\frac{1}{2}$	$\frac{1}{2}$	1	1	$\frac{1}{2}$	$\frac{1}{2}$	1	1	$\frac{1}{2}$
1 0	0	1	0	0	0	1	0	0	0	1	0	0	0	1	0	0	0	1	0	0
1 $\frac{1}{2}$	$\frac{1}{2}$	1	$\frac{1}{2}$	$\frac{1}{2}$	$\frac{1}{2}$	$\frac{1}{2}$	$\frac{1}{2}$	$\frac{1}{2}$	$\frac{1}{2}$	1	$\frac{1}{2}$	$\frac{1}{2}$	$\frac{1}{2}$	1	$\frac{1}{2}$	$\frac{1}{2}$	$\frac{1}{2}$	1	$\frac{1}{2}$	$\frac{1}{2}$
1 1	1	1	1	1	1	1	1	1	1	1	1	1	1	1	1	1	1	1	1	1

It can be easily verified that Eqs. (1.3) become the definitions of the usual primitives of two-valued logic when $n = 2$ and that they define the primitives of Łukasiewicz's three-valued logic as given in Table 1.6.

For each $n \geq 2$, the n-valued logic of Łukasiewicz is usually denoted in the literature by L_n. The truth values of L_n are taken from T_n and its primitives are defined by Eqs. (1.3). The sequence $(L_2, L_3, \ldots, L_\infty)$ of these logics contains two extreme cases—logics L_2 and L_∞. Logic L_2 is clearly the classical two-valued logic discussed in Sec. 1.5. Logic L_∞ is an *infinite-valued logic* whose truth values are taken from the set T_∞ of all rational numbers in the unit interval $[0, 1]$.

When we do not insist on taking truth values only from the set T_∞ but rather accept as truth values any real numbers in the interval $[0, 1]$, we obtain an alternative infinite-valued logic. Primitives of both of these infinite-valued logics are defined by Eqs. (1.3); they differ in their sets of truth values. Whereas one of these logics uses the set T_∞ as truth values, the other employs the set of all real numbers in the interval $[0, 1]$. In spite of this difference, these two infinite-valued logics are established as essentially equivalent in the sense that they represent exactly the same tautologies. This equivalence holds, however, only for logic formulas involving propositions; for predicate formulas with quantifiers, some fundamental differences between the two logics emerge.

Unless otherwise stated, the term *infinite-valued logic* is usually used in the literature to indicate the logic whose truth values are represented by all the real numbers in the interval $[0, 1]$. This is also quite often called the *standard Łukasiewicz logic* L_1, where the subscript 1 is an abbreviation for $\aleph_1$ (read "aleph 1"), which is the symbol commonly used to denote the cardinality of the continuum.

Given the isomorphism that exists between logic and set theory as defined in Table 1.5, we can see that the standard Łukasiewicz logic L_1 is isomorphic to the original fuzzy set theory based on the min, max, and $1 - a$ operators for fuzzy set intersection, union, and complement, respectively, in the same way as the two-valued logic is isomorphic to the crisp set theory. In fact, the membership grades $\mu_A(x)$ for $x \in X$ by which a fuzzy set A on the universal set X is defined can be interpreted as the truth values of the proposition "x is a member of set A" in L_1. Conversely, the truth values for all $x \in X$ of any proposition "x is P"

in L_1, where P is a vague (fuzzy) predicate (such as tall, young, expensive, dangerous, and so on), can be interpreted as the membership degrees $\mu_P(x)$ by which the fuzzy set characterized by the property P is defined on X. The isomorphism then follows from the fact that the logic operations of L_1, defined by Eqs. (1.3), have exactly the same mathematical form as the corresponding standard operations on fuzzy sets.

The standard Łukasiewicz logic L_1 is only one of a variety of infinite-valued logics in the same sense as the standard fuzzy set theory is only one of a variety of fuzzy set theories, which differ from one another by the set operations they employ. For each particular infinite-valued logic, we can derive the isomorphic fuzzy set theory by the correspondence in Table 1.5; a similar derivation can be made of the infinite-valued logic that is isomorphic to a given particular fuzzy set theory. A thorough study of only one of these areas, therefore, reveals the full scope of both. We are free to examine either the classes of acceptable set operations or the classes of acceptable logic operations and their various combinations. We choose in this text to focus on set operations, which are fully discussed in Chap. 2. The isomorphic logic operations and their combinations, which we do not cover explicitly, are nevertheless utilized in some of the applications discussed in Chap. 6.

The insufficiency of any single infinite-valued logic (and therefore the desirability of a variety of these logics) is connected with the notion of a complete set of logic primitives. It is known that there exists no finite complete set of logic primitives for any infinite-valued logic. Hence, using a finite set of primitives that defines an infinite-valued logic, we can obtain only a subset of all the logic functions of the given primary logic variables. Because some applications require functions outside this subset, it may become necessary to resort to alternative logics.

Since, as argued in this section, the various many-valued logics have their counterparts in fuzzy set theory, they form the kernel of *fuzzy logic*, that is, a logic based on fuzzy set theory. In its full scale, however, fuzzy logic is actually an extension of many-valued logics. Its ultimate goal is to provide foundations for *approximate reasoning* with imprecise propositions using fuzzy set theory as the principle tool. This is analogous to the role of quantified predicate logic for reasoning with precise propositions.

The primary focus of fuzzy logic is on natural language, where approximate reasoning with imprecise propositions is rather typical. The following syllogism is an example of approximate reasoning in linguistic terms that cannot be dealt with by the classical predicate logic:

> Old coins are usually rare collectibles.
>
> Rare collectibles are expensive.
>
> Old coins are usually expensive.

This is a meaningful deductive inference. In order to deal with inferences such as this, fuzzy logic allows the use of *fuzzy predicates* (expensive, old, rare, dangerous, and so on), *fuzzy quantifiers* (many, few, almost all, usually, and the like), *fuzzy truth values* (quite true, very true, more or less true, mostly false, and so forth), and various other kinds of *fuzzy modifiers* (such as likely, almost impossible, or extremely unlikely).

Each simple fuzzy predicate, such as

$$x \text{ is } P$$

is represented in fuzzy logic by a fuzzy set, as described previously. Assume, for example, that *x* stands for the *age* of a person and that *P* has the meaning of *young*. Then, assuming that the universal set is the set of integers from 0 to 60 representing different ages, the predicate may be represented by a fuzzy set whose membership function is shown in Fig. 1.8(a). Consider now the truth value of a proposition obtained by a particular substitution for *x* into the predicate, such as

$$\text{Tina is young.}$$

The truth value of this proposition depends not only on the membership grade of Tina's age in the fuzzy set chosen to characterize the concept of a young person (Fig. 1.8(a)) but also depends upon the strength of truth (or falsity) claimed. Examples of some possible truth claims are:

$$\text{Tina is young is true.}$$

$$\text{Tina is young is false.}$$

$$\text{Tina is young is fairly true.}$$

$$\text{Tina is young is very false.}$$

Each of the possible truth claims is represented by an appropriate fuzzy set. All

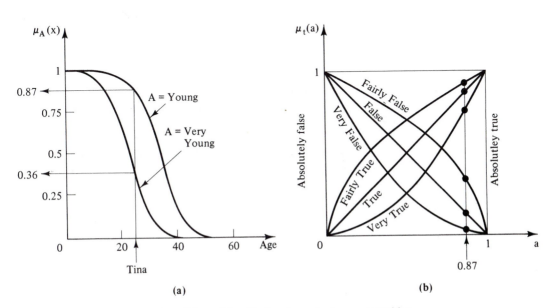

Figure 1.8. Truth values of a fuzzy proposition.

these sets are defined on the unit interval [0, 1]. Some examples are shown in Fig. 1.8(b), where a stands for the membership grade in the fuzzy set that represents the predicate involved and t is a common label representing each of the fuzzy sets in the figure that expresses truth values. Thus, in our case $a = \mu_{young}(x)$ for each $x \in X$. Returning now to Tina, who is 25 years old, we obtain $\mu_{young}(25) = .87$ (Fig. 1.8(a)), and the truth values of the propositions

> Tina is young is fairly true (true, very true, fairly false,
> false, very false)

are .9 (.87, .81, .18, .13, .1), respectively.

We may operate on fuzzy sets representing predicates with any of the basic fuzzy set operations of complementation, union, and intersection. Furthermore, these sets can be modified by special operations corresponding to linguistic terms such as very, extremely, more or less, quite, and so on. These terms are often called *linguistic hedges*. For example, applying the linguistic hedge *very* to the fuzzy set labeled as *young* in Fig. 1.8(a), we obtain a new fuzzy set representing the concept of a *very young* person, which is specified in the same figure.

In general, fuzzy quantifiers are represented in fuzzy logic by fuzzy numbers. These are manipulated in terms of the operations of fuzzy arithmetic, which is now well established.

From this brief outline of fuzzy logic we can see that it is operationally based on a great variety of manipulations with fuzzy sets, through which reasoning in natural language is approximated. The principles underlying these manipulations are predominantly *semantic* in nature. While full coverage of these principles is beyond the scope of this book, Chap. 6 contains illustrations of some aspects of fuzzy reasoning in the context of a few specific applications.

NOTES

1.1. The *theory of fuzzy sets* was founded by Lotfi Zadeh [1965a], primarily in the context of his interest in the analysis of complex systems [Zadeh, 1962, 1965b, 1973]. However, some of the key ideas of the theory were envisioned by Max Black, a philosopher, almost 30 years prior to Zadeh's seminal paper [Black, 1937].

1.2. The development of fuzzy set theory since its introduction in 1965 has been dramatic. Thousands of publications are now available in this new area. A survey of the status of the theory and its applications in the late 1970s is well covered in a book by Dubois and Prade [1980a]. Current contributions to the theory are scattered in many journals and books of collected papers, but the most important source is the specialized journal *Fuzzy Sets and Systems* (North-Holland). A very comprehensive bibliography of fuzzy set theory appears in a book by Kandel [1982]. An excellent annotated bibliography covering the first decade of fuzzy set theory was prepared by Gaines and Kohout [1977]. Books by Kaufmann [1975], Zimmermann [1985], and Kandel [1986] are useful supplementary readings on fuzzy set theory.

1.3. The concept of *L-fuzzy sets* was introduced by Goguen [1967]. A thorough investigation of properties of *fuzzy sets of type* 2 and higher types was done by Mizumoto and Tanaka [1976, 1981]. The concept of *fuzzy sets of level k,* which is due to Zadeh

[1971b], was investigated by Gottwald [1979]. Convex fuzzy sets were studied in greater detail by Lowen [1980] and Liu [1985].

1.4. One concept that is only mentioned in this book but not sufficiently developed is the concept of a *fuzzy number.* It is a basis for *fuzzy arithmetic,* which can be viewed as an extension of interval arithmetic [Moore, 1966, 1979]. Among other applications, fuzzy numbers are essential for expressing *fuzzy cardinalities* and, consequently, *fuzzy quantifiers* [Dubois and Prade, 1985c]. Fuzzy arithmetic is thus a basic tool for dealing with fuzzy quantifiers in approximate reasoning; it is also a basis for developing a *fuzzy calculus* [Dubois and Prade, 1982b]. We do not cover fuzzy arithmetic, since there now exists an excellent book devoted solely to this subject [Kaufmann and Gupta, 1985].

1.5. The *extension principle* was introduced by Zadeh [1975b]. A further elaboration of the principle was presented by Yager [1986].

1.6. Fuzzy extensions of some mathematical subject areas are beyond the scope of this introductory text and are thus not covered here. They include, for example, *fuzzy topological spaces* [Chang, 1968; Wong, 1975; Lowen, 1976], *fuzzy metric spaces* [Kaleva and Seikkala, 1984], and *fuzzy games* [Butnariu, 1978].

1.7. An excellent and comprehensive survey of many-valued logics was prepared by Rescher [1969]; it also contains an extensive bibliography on the subject. Various aspects of the relationship between many-valued logics and fuzzy logic are examined by numerous authors, including Baldwin [1979a, b, c], Baldwin and Guild [1980a, b], Baldwin and Pilsworth [1980], Dubois and Prade [1979a, 1984a], Gaines [1976, 1978, 1983], Giles [1977], Gottwald [1980], Lee and Chang [1971], Mizumoto [1981], Skala [1978], Turksen and Yao [1984], and White [1979]. Approximate reasoning based on fuzzy predicate logic is also investigated in some of these papers. Particularly good overview papers were prepared by Zadeh [1975c, 1984, 1985] and Gaines [1976]. Most aspects of approximate reasoning were developed by Zadeh [1971b, 1972, 1975b, c, 1976, 1978b, 1983a, b, 1984, 1985], but we should also mention an early and important paper by Goguen [1968–69].

1.8. An alternative set theory, which is referred to as the *theory of semisets,* was proposed and developed by Vopěnka and Hájek [1972] to represent sets with imprecise boundaries. Unlike fuzzy sets, however, semisets may be defined in terms of vague properties and not necessarily by explicit membership grade functions. While semisets are more general than fuzzy sets, they are required to be approximated by fuzzy sets in practical situations. The relationship between semisets and fuzzy sets is well characterized by Novák [1984]. The concept of semisets leads into a formulation of an *alternative (nonstandard) set theory* [Vopěnka, 1979].

1.9. For a general background on crisp sets and classical two-valued logic, we recommend the book *Set Theory and Related Topics* by S. Lipschutz (Shaum, New York, 1964). The book covers all topics that are needed for this text and contains many solved examples. For a more advanced treatment of the topics, we recommend the book *Set Theory and Logic* by R. R. Stoll (W.H. Freeman, San Francisco, 1961).

EXERCISES

1.1. For each of the properties of crisp set operations listed in Table 1.1, determine whether the property holds for the complement, union, and intersection operations originally proposed for fuzzy sets.

1.2. Compute the scalar cardinality and the fuzzy cardinality for each of the following fuzzy sets:

(a) $A = .4/v + .2/w + .5/x + .4/y + 1/z$;

(b) $B = 1/x + 1/y + 1/z$;

(c) $\mu_C(x) = \dfrac{x}{x + 1}$, $x \in \{0, 1, 2, \ldots, 10\}$.

1.3. Consider the fuzzy sets A, B, and C defined on the interval $X = [0, 10]$ of real numbers by the membership grade functions

$$\mu_A(x) = \frac{x}{x + 2}, \qquad \mu_B(x) = 2^{-x}, \qquad \mu_C(x) = \frac{1}{1 + 10(x - 2)^2}.$$

Determine mathematical formulas and graphs of the membership grade functions of each of the following:

(a) $\overline{A}, \overline{B}, \overline{C}$;

(b) $A \cup B, A \cup C, B \cup C$;

(c) $A \cap B, A \cap C, B \cap C$;

(d) $A \cup B \cup C, A \cap B \cap C$;

(e) $A \cap \overline{C}, \overline{B} \cap C, \overline{A \cup C}$.

1.4. Show that DeMorgan's laws are satisfied for the three pairs of fuzzy sets obtained from fuzzy sets A, B, and C in Exercise 1.6.

1.5. Propose an extension of the standard fuzzy set operations (min, max, $1 - a$) to interval-valued fuzzy sets.

1.6. Order the fuzzy sets defined by the following membership grade functions (assuming $x \geq 0$) by the inclusion (subset) relation:

$$\mu_A(x) = \frac{1}{1 + 20x}, \qquad \mu_B(x) = \left(\frac{1}{1 + 10x}\right)^{1/2}, \qquad \mu_C(x) = \left(\frac{1}{1 + 10x}\right)^{2}.$$

1.7. Let the membership grade functions of sets A, B, and C in Exercise 1.3 be defined on the set $X = \{0, 1, \ldots, 10\}$ and let $f(x) = x^2$ for all $x \in X$. Use the extension principle to derive $f(A)$, $f(B)$, and $f(C)$.

1.8. Define α-cuts of each of the fuzzy sets defined in Exercises 1.2 and 1.3 for $\alpha = .2$, .5, .9, 1.

1.9. Show that all α-cuts of any fuzzy set A defined on $\mathbb{R}^n$ ($n \geq 1$) are convex if and only if

$$\mu_A[\lambda \mathbf{r} + (1 - \lambda)\mathbf{s}] \geq \min[\mu_A(\mathbf{r}), \mu_A(\mathbf{s})]$$

for all $\mathbf{r}, \mathbf{s} \in \mathbb{R}^n$ and all $\lambda \in [0, 1]$.

1.10. For each of the three-valued logics defined in Table 1.6, determine the truth values of each of the following logic expressions for all combinations of truth values of logic variables a, b, c (assume that negation $\overline{a}$ is defined by $1 - a$):

(a) $(\overline{a} \wedge b) \Rightarrow c$;

(b) $(\overline{a} \vee \overline{b}) \Leftrightarrow \overline{(a \wedge b)}$;

(c) $(a \Rightarrow b) \Rightarrow (\overline{c} \Rightarrow a)$.

1.11. Define in the form of a table (analogous to Table 1.6) primitives, $\wedge$, $\vee$, $\Rightarrow$, and $\Leftrightarrow$, of the Łukasiewicz logics L_4 and L_5.

1.12. Repeat the example illustrated by Fig. 1.8, which is discussed in Sec. 1.6 (Tina is young, and so on) for yourself.

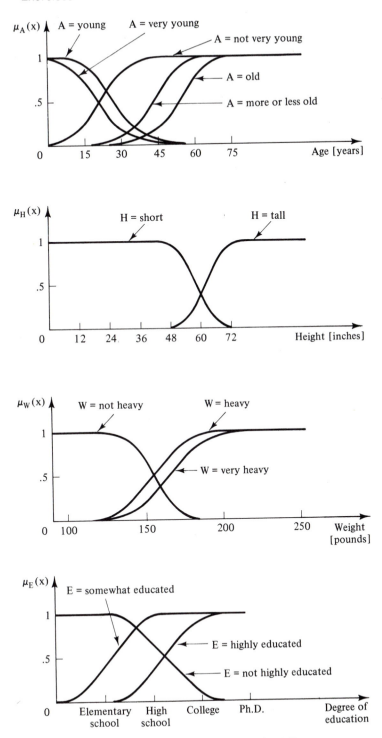

Figure 1.9.   Fuzzy sets for Exercise 1.13.

1.13. Assume four types of fuzzy predicates applicable to persons (age, height, weight, and level of education). Several specific fuzzy predicates for each of these types are represented by fuzzy sets whose membership functions are specified in Fig. 1.9. Apply these membership functions and the fuzzy truth values defined in Fig. 1.8(b) to some person x (perhaps yourself) to determine the truth values of various propositions such as the following:

x is highly educated and not very young is very true;
x is very young, tall, not heavy, and somewhat educated is true;
x is more or less old or highly educated is fairly true;
x is very heavy or old or not highly educated is fairly true;
x is short, not very young and highly educated is very true.

In your calculations, use standard fuzzy set operators (min, max, $1 - a$).

2

OPERATIONS ON FUZZY SETS

2.1 GENERAL DISCUSSION

As mentioned in Chap. 1, the original theory of fuzzy sets was formulated in terms of the following specific operators of set complement, union, and intersection:

$$\mu_{\bar{A}}(x) = 1 - \mu_A(x), \tag{2.1}$$

$$\mu_{A \cup B}(x) = \max[\mu_A(x), \mu_B(x)], \tag{2.2}$$

$$\mu_{A \cap B}(x) = \min[\mu_A(x), \mu_B(x)]. \tag{2.3}$$

Note that when the range of membership grades is restricted to the set $\{0, 1\}$, these functions perform precisely as the corresponding operators for crisp sets, thus establishing them as clear generalizations of the latter. It is now understood, however, that these functions are not the only possible generalizations of the crisp set operators. For each of the three set operations, several different classes of functions, which possess appropriate axiomatic properties, have subsequently been proposed. This chapter contains discussions of these desirable properties and defines some of the different classes of functions satisfying them.

Despite this variety of fuzzy set operators, however, the original complement, union, and intersection still possess particular significance. Each defines a special case within all the various classes of satisfactory functions. For instance, if the functions within a class are interpreted as performing union or intersection operations of various strengths, then the classical max union is found to be the strongest of these and the classical min intersection, the weakest. Furthermore, a particularly desirable feature of these original operators is their inherent prevention of the compounding of errors of the operands. If any error e is associated with the membership degrees $\mu_A(x)$ and $\mu_B(x)$, then the maximum error associated with the membership grade of x in $\overline{A}$, $A \cup B$, or $A \cap B$ remains e. Many of the alternative fuzzy set operator functions later proposed lack this characteristic.

Fuzzy set theory that is based on the operators given by Eqs. (2.1) through (2.3) is now usually referred to as *possibility theory*. This theory emerges, quite naturally, as a special case of fuzzy measures. It is covered in this latter context in Chap. 4. For convenience, let the operations defined by Eqs. (2.1) through (2.3) be called the *standard operations* of fuzzy set theory.

2.2 FUZZY COMPLEMENT

A complement of a fuzzy set A is specified by a function

$$c : [0, 1] \rightarrow [0, 1],$$

which assigns a value $c(\mu_A(x))$ to each membership grade $\mu_A(x)$. This assigned value is interpreted as the membership grade of the element x in the fuzzy set representing the negation of the concept represented by A. Thus, if A is the fuzzy set of tall men, its complement is the fuzzy set of men who are not tall. Obviously, there are many elements that can have some nonzero degree of membership in both a fuzzy set and in its complement.

In order for any function to be considered a fuzzy complement, it must satisfy at least the following two axiomatic requirements:

Axiom c1. $c(0) = 1$ and $c(1) = 0$, that is, c behaves as the ordinary complement for crisp sets (*boundary conditions*).

Axiom c2.   For all $a, b \in [0, 1]$, if $a < b$, then $c(a) \geq c(b)$, that is, c is *monotonic nonincreasing*.

Symbols a and b, which are used in Axiom c2 and the rest of this section as arguments of the function c, represent degrees of membership of some arbitrary elements of the universal set in a given fuzzy set. For example, $a = \mu_A(x)$ and $b = \mu_A(y)$ for some $x, y \in X$ and some fuzzy set A.

There are many functions satisfying Axioms c1 and c2. For any particular fuzzy set A, different fuzzy sets can be said to constitute its complement, each being produced by a different fuzzy complement function. In order to distinguish the complement resulting from the application of the classical fuzzy complement of Eq. (2.1) and these numerous others, the former is denoted in this text by $\overline{A}$; the latter, expressed by function c, is denoted by $C(A)$, where

$$C : \tilde{\mathscr{P}}(X) \rightarrow \tilde{\mathscr{P}}(X)$$

is a function such that $c(\mu_A(x)) = \mu_{C(A)}(x)$ for all $x \in X$.

Given a particular fuzzy complement c, function C may conveniently be used as a global operator representing c. Each function C transforms a fuzzy set A into its complement $C(A)$ as determined by the corresponding function c, which assigns to elements of X membership grades in the complement $C(A)$. Thus each fuzzy complement c implies a corresponding function C.

All functions that satisfy Axioms c1 and c2 form the most general class of fuzzy complements. It is rather obvious that the exclusion or weakening of either of these axioms would add to this class some functions totally unacceptable as complements. Indeed, a violation of Axiom c1 would include functions that do not conform to the ordinary complement for crisp sets. Axion c2 is essential since we intuitively expect that an increase in the degree of membership in a fuzzy set must result either in a decrease or, in the extreme case, in no change in the degree of membership in its complement. Let Axioms c1 and c2 be called the *axiomatic skeleton for fuzzy complements*.

In most cases of practical significance, it is desirable to consider various additional requirements for fuzzy complements. Each of them reduces the general class of fuzzy complements to a special subclass. Two of the most desirable requirements, which are usually listed in the literature among axioms of fuzzy complements, are the following:

Axiom c3. c is a *continuous* function.

Axiom c4. c is *involutive*, which means that $c(c(a)) = a$ for all $a \in [0, 1]$.

Functions that satisfy Axiom c3 form a special subclass of the general class of fuzzy complements; those satisfying Axiom c4 are necessarily continuous as well and, therefore, form a further nested subclass, as illustrated in Fig. 2.1. The classical fuzzy complement given by Eq. (2.1) is contained within the class of involutive complements.

Examples of general fuzzy complements that satisfy only the axiomatic skeleton are the threshold-type complements defined by

$$c(a) = \begin{cases} 1 & \text{for } a \leq t, \\ 0 & \text{for } a > t, \end{cases}$$

where $a \in [0, 1]$ and $t \in [0, 1)$; t is called the threshold of c. This function is illustrated in Fig. 2.2(a).

An example of a fuzzy complement that is continuous (Axiom c3) but not involutive (Axiom c4) is the function

$$c(a) = \tfrac{1}{2}(1 + \cos \pi a),$$

which is illustrated in Fig. 2.2(b). The failure of this function to satisfy the property of involution can be seen by noting that, for example, $c(.33) = .75$ but $c(.75) = .15 \neq .33$.

One class of involutive fuzzy complements is the *Sugeno class* defined by

$$c_\lambda(a) = \frac{1 - a}{1 + \lambda a},$$

where $\lambda \in (-1, \infty)$. For each value of the parameter λ, we obtain one particular

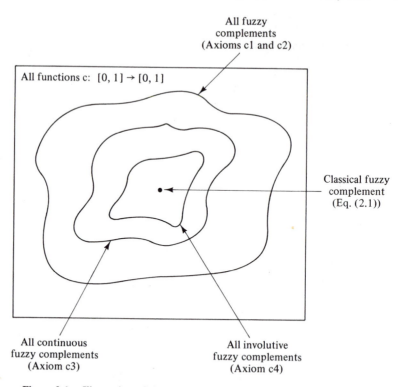

Figure 2.1. Illustration of the nested subset relationship of the basic classes of fuzzy complements.

involutive fuzzy complement. This class is illustrated in Fig. 2.3(a) for several different values of λ. Note how the shape of the function is affected as the value of λ is changed. For $\lambda = 0$, the function becomes the classical fuzzy complement defined by Eq. (2.1).

Another example of a class of involutive fuzzy complements is defined by

$$c_w(a) = (1 - a^w)^{1/w},$$

where $w \in (0, \infty)$; let us refer to it as the *Yager class* of fuzzy complements. Figure 2.3(b) illustrates this class of functions for various values of w. Here again, changing the value of the parameter w results in a deformation of the shape of the function. When $w = 1$, this function becomes the classical fuzzy complement of $c(a) = 1 - a$.

Several important properties are shared by all fuzzy complements. These concern the *equilibrium* of a fuzzy complement c, which is defined as any value a for which $c(a) = a$. In other words, the equilibrium of a complement c is that degree of membership in a fuzzy set A equaling the degree of membership in the complement $C(A)$. For instance, the equilibrium value for the classical fuzzy complement given by Eq. (2.1) is .5, which is the solution of the equation $1 - a = a$.

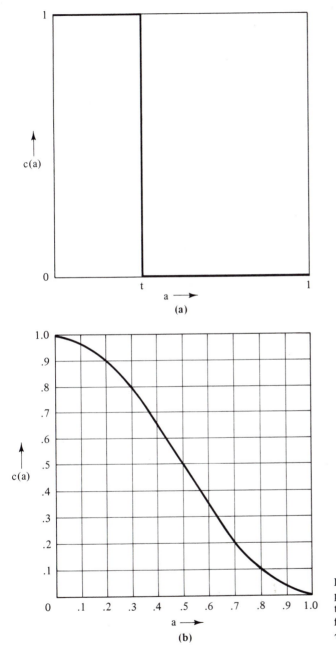

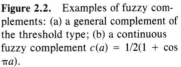

Figure 2.2. Examples of fuzzy complements: (a) a general complement of the threshold type; (b) a continuous fuzzy complement $c(a) = 1/2(1 + \cos \pi a)$.

Theorem 2.1. Every fuzzy complement has at most one equilibrium.

Proof: Let c be an arbitrary fuzzy complement. An equilibrium of c is a solution of the equation

$$c(a) - a = 0,$$

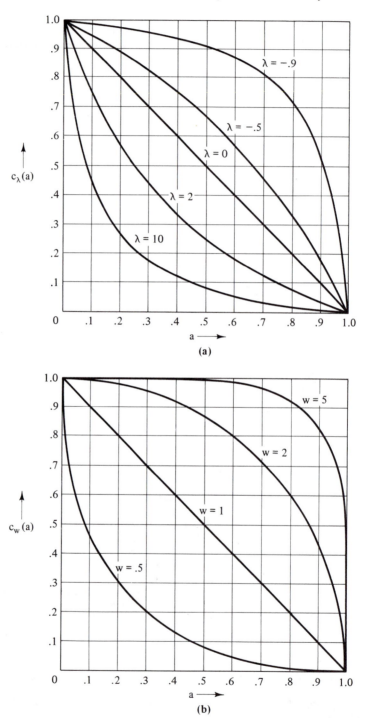

Figure 2.3. Examples from two classes of involutive fuzzy complements: (a) Sugeno class; (b) Yager class.

where $a \in [0, 1]$. We can demonstrate that any equation $c(a) - a = b$, where b is a real constant, must have at most one solution, thus proving the theorem. In order to do so, we assume that a_1 and a_2 are two different solutions of the equation $c(a) - a = b$ such that $a_1 < a_2$. Then, since $c(a_1) - a_1 = b$ and $c(a_2) - a_2 = b$, we get

$$c(a_1) - a_1 = c(a_2) - a_2. \tag{2.4}$$

However, because c is monotonic nonincreasing (by Axiom c2), $c(a_1) \geq c(a_2)$ and, since $a_1 < a_2$,

$$c(a_1) - a_1 > c(a_2) - a_2.$$

This inequality contradicts Eq. (2.4), thus demonstrating that the equation must have at most one solution. ■

Theorem 2.2.   Assume that a given fuzzy complement c has an equilibrium e_c which by Theorem 2.1 is unique. Then

$$a \leq c(a) \quad \text{if and only if} \quad a \leq e_c$$

and

$$a \geq c(a) \quad \text{if and only if} \quad a \geq e_c.$$

Proof: Let us assume that $a < e_c$, $a = e_c$, and $a > e_c$, in turn. Then, since c is monotonic nonincreasing by Axiom c2, $c(a) \geq c(e_c)$ for $a < e_c$, $c(a) = c(e_c)$ for $a = e_c$, and $c(a) \leq c(e_c)$ for $a > e_c$. Because $c(e_c) = e_c$, we can rewrite these expressions as $c(a) \geq e_c$, $c(a) = e_c$, and $c(a) \leq e_c$, respectively. In fact, due to our initial assumption we can further rewrite these as $c(a) > a$, $c(a) = a$, and $c(a) < a$, respectively. Thus, $a \leq e_c$ implies $c(a) \geq a$ and $a \geq e_c$ implies $c(a) \leq a$. The inverse implications can be shown in a similar manner. ■

Theorem 2.3.   If c is a continuous fuzzy complement, then c has a unique equilibrium.

Proof: The equilibrium e_c of a fuzzy complement c is the solution of the equation $c(a) - a = 0$. This is a special case of the more general equation $c(a) - a = b$, where $b \in [-1, 1]$ is a constant. By Axiom c1, $c(0) - 0 = 1$ and $c(1) - 1 = -1$. Since c is a continuous complement, it follows from the intermediate value theorem for continuous functions* that for each $b \in [-1, 1]$, there exists at least one a such that $c(a) - a = b$. This demonstrates the necessary existence of an equilibrium value for a continuous function, and Theorem 2.1 guarantees its uniqueness. ■

* See, for example, *Mathematical Analysis* (second ed.), by T. M. Apostol, Addison-Wesley, Reading, Mass., 1974, p. 85.

The equilibrium for each individual fuzzy complement c_λ of the Sugeno class is given by

$$e_{c_\lambda} = \begin{cases} \dfrac{\sqrt{1 + \lambda} - 1}{\lambda} & \text{for } \lambda \neq 0, \\\\ \dfrac{1}{2} & \text{for } \lambda = 0 \left(= \lim_{\lambda \to 0} \dfrac{\sqrt{1 + \lambda} - 1}{\lambda} \right). \end{cases}$$

This is clearly obtained by selecting the positive solution of the equation

$$\frac{1 - e_{c_\lambda}}{1 + \lambda e_{c_\lambda}} = e_{c_\lambda}.$$

The dependence of the equilibrium e_{c_λ} on the parameter λ is shown in Fig. 2.4.

If we are given a fuzzy complement c and a membership grade whose value is represented by a real number $a \in [0, 1]$, then any membership grade represented by the real number $^d a \in [0, 1]$ such that

$$c(^d a) - {}^d a = a - c(a), \tag{2.5}$$

is called a *dual point* of a with respect to c.

It follows directly from the proof of Theorem 2.1 that Eq. (2.5) has at most one solution for $^d a$ given c and a. There is, therefore, at most one dual point for each particular fuzzy complement c and membership grade of value a. Moreover, it follows from the proof of Theorem 2.3 that a dual point exists for each $a \in [0, 1]$ when c is a continuous complement.

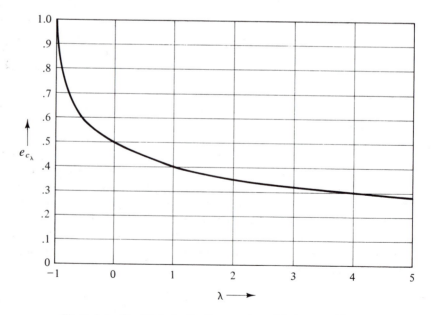

Figure 2.4. Equilibria for the Sugeno class of fuzzy complements.

Theorem 2.4.   If a complement c has an equilibrium e_c, then

$$^d e_c = e_c.$$

Proof: If $a = e_c$, then by our definition of equilibrium, $c(a) = a$ and thus $a - c(a) = 0$. Additionally, if $^d a = e_c$, then $c(^d a) = {}^d a$ and $c(^d a) - {}^d a = 0$. Therefore,

$$c(^d a) - {}^d a = a - c(a).$$

This satisfies Eq. (2.5) when $a = {}^d a = e_c$. Hence, the equilibrium of any complement is its own dual point. ∎

Theorem 2.5.   For each $a \in [0, 1]$, $^d a = c(a)$ if and only if $c(c(a)) = a$, that is, when the complement is involutive.

Proof: Let $^d a = c(a)$. Then, substitution of $c(a)$ for $^d a$ in Eq. (2.5) produces

$$c(c(a)) - c(a) = a - c(a).$$

Therefore, $c(c(a)) = a$. For the reverse implication, let $c(c(a)) = a$. Then, substitution of $c(c(a))$ for a in Eq. (2.5) yields

$$c(^d a) - {}^d a = c(c(a)) - c(a).$$

Because $^d a$ can be substituted for $c(a)$ everywhere in this equation to yield a tautology, $^d a = c(a)$. ∎

Thus, the dual point of any membership grade is equal to its complemented value whenever the complement is involutive. If the complement is not involutive, then either the dual point does not exist or it does not coincide with the complement point.

These results associated with the concepts of the equilibrium and the dual point of a fuzzy complement are referenced in the discussion of measures of fuzziness contained in Chap. 5.

2.3 FUZZY UNION

The union of two fuzzy sets A and B is specified in general by a function of the form

$$u : [0, 1] \times [0, 1] \to [0, 1].$$

For each element x in the universal set, this function takes as its argument the pair consisting of the element's membership grades in set A and in set B and yields the membership grade of the element in the set constituting the union of A and B. Thus,

$$\mu_{A \cup B}(x) = u[\mu_A(x), \mu_B(x)].$$

In order for any function of this form to qualify as a fuzzy union, it must satisfy at least the following axioms:

Axiom u1. $u(0, 0) = 0$; $u(0, 1) = u(1, 0) = u(1, 1) = 1$; that is, u behaves as the classical union with crisp sets (*boundary conditions*).

Axiom u2. $u(a, b) = u(b, a)$; that is, u is *commutative*.

Axiom u3. If $a \le a'$ and $b \le b'$, then $u(a, b) \le u(a', b')$; that is, u is *monotonic*.

Axiom u4. $u(u(a, b), c) = u(a, u(b, c))$; that is, u is *associative*.

Let us call this set of axioms the *axiomatic skeleton for fuzzy set unions*.

The first axiom insures that the function will define an operation that generalizes the classical crisp set union. The second axiom of commutativity (or symmetry) indicates indifference to the order in which the sets to be combined are considered. The third axiom is the natural requirement that a decrease in the degree of membership in set A or set B cannot produce an increase in the degree of membership in $A \cup B$. Finally, the fourth axiom of associativity ensures that we can take the union of any number of sets in any order of pairwise grouping desired; this axiom allows us to extend the operation of fuzzy set union to more than two sets.

It is often desirable to restrict the class of fuzzy unions by considering various additional requirements. Two of the most important requirements are expressed by the following axioms:

Axiom u5. u is a *continuous* function.

Axiom u6. $u(a, a) = a$; that is, u is *idempotent*.

The axiom of continuity prevents a situation in which a very small increase in the membership grade in either set A or set B produces a large change in the membership grade in $A \cup B$. Axiom u6 insures that the union of any set with itself yields precisely the same set.

Several classes of functions have been proposed whose individual members satisfy all the axiomatic requirements for the fuzzy union and neither, one, or both of the optional axioms. One of these classes of fuzzy unions is known as the *Yager class* and is defined by the function

$$u_w(a, b) = \min[1, (a^w + b^w)^{1/w}], \tag{2.6}$$

where values of the parameter w lie within the open interval $(0, \infty)$. This class of functions satisfies Axioms u1 through u5, but these functions are not, in general, idempotent. Special functions within this class are formed when certain values are chosen for the parameter w. For instance, for $w = 1$, the function becomes

$$u_1(a, b) = \min[1, a + b];$$

for $w = 2$, we obtain

$$u_2(a, b) = \min[1, \sqrt{a^2 + b^2}].$$

Since it is not obvious what form the function u_w given by Eq. (2.6) takes for $w \to \infty$, we use the following theorem.

Theorem 2.6. $\lim\limits_{w \to \infty} \min[1, (a^w + b^w)^{1/w}] = \max(a, b)$.

Proof: The theorem is obvious whenever (1) a or b equal 0, or (2) $a = b$, because the limit of $2^{1/w}$ as $w \to \infty$ equals 1. If $a \neq b$ and the min equals $(a^w + b^w)^{1/w}$, the proof reduces to the demonstration that

$$\lim_{w \to \infty} (a^w + b^w)^{1/w} = \max(a, b).$$

Let us assume, with no loss of generality, that $a < b$, and let $Q = (a^w + b^w)^{1/w}$. Then

$$\lim_{w \to \infty} \ln Q = \lim_{w \to \infty} \frac{\ln(a^w + b^w)}{w}.$$

Using l'Hospital's rule, we obtain

$$\lim_{w \to \infty} \ln Q = \lim_{w \to \infty} \frac{a^w \ln a + b^w \ln b}{a^w + b^w}$$

$$= \lim_{w \to \infty} \frac{(a/b)^w \ln a + \ln b}{(a/b)^w + 1} = \ln b.$$

Hence,

$$\lim_{w \to \infty} Q = \lim_{w \to \infty} (a^w + b^w)^{1/w} = b \ (= \max(a, b)).$$

It remains to show that the theorem is still valid when the min equals 1. In this case,

$$(a^w + b^w)^{1/w} \geq 1$$

or

$$a^w + b^w \geq 1$$

for all $w \in (0, \infty)$. When $w \to \infty$, the last inequality holds if $a = 1$ or $b = 1$ (since $a, b \in [0, 1]$). Hence, the theorem is again satisfied. ■

The various functions of the Yager class, which are defined by different choices of the parameter w, can be interpreted as performing union operations of various strengths. Table 2.1(a) illustrates how the values produced by the Yager functions for fuzzy unions decrease as the value of w increases. In fact, we may interpret the value $1/w$ as indicating the degree of interchangeability present in the union operation u_w. The notion of the set union operation corresponds to the logical OR (disjunction), in which some interchangeability between the two arguments of the statement "A or B" is assumed. Thus, the union of two fuzzy sets *young* and *tall* would represent the concept *young or tall*. With the Yager

TABLE 2.1. EXAMPLES OF FUZZY SET OPERATIONS FROM THE YAGER CLASS.

(a) Fuzzy unions

$b =$	0	.25	.5	.75	1
$a = 1$	1	1	1	1	1
.75	.75	1	1	1	1
.5	.5	.75	1	1	1
.25	.25	.5	.75	1	1
0	0	.25	.5	.75	1

$w = 1$ (soft)

$b =$	0	.25	.5	.75	1
$a = 1$	1	1	1	1	1
.75	.75	.79	.9	1	1
.5	.5	.56	.71	.9	1
.25	.25	.35	.56	.79	1
0	0	.25	.5	.75	1

$w = 2$

$b =$	0	.25	.5	.75	1
$a = 1$	1	1	1	1	1
.75	.75	.75	.75	.8	1
.5	.5	.5	.54	.75	1
.25	.25	.27	.5	.75	1
0	0	.25	.5	.75	1

$w = 10$

$b =$	0	.25	.5	.75	1
$a = 1$	1	1	1	1	1
.75	.75	.75	.75	.75	1
.5	.5	.5	.5	.75	1
.25	.25	.25	.5	.75	1
0	0	.25	.5	.75	1

$w \rightarrow \infty$ (hard)

(b) Fuzzy intersections

$b =$	0	.25	.5	.75	1
$a = 1$	0	.25	.5	.75	1
.75	0	0	.25	.5	.75
.5	0	0	0	.25	.5
.25	0	0	0	0	.25
0	0	0	0	0	0

$w = 1$ (strong)

$b =$	0	.25	.5	.75	1
$a = 1$	0	.25	.5	.75	1
.75	0	.21	.44	.65	.75
.5	0	.1	.29	.44	.5
.25	0	0	.1	.21	.25
0	0	0	0	0	0

$w = 2$

$b =$	0	.25	.5	.75	1
$a = 1$	0	.25	.5	.75	1
.75	0	.25	.5	.73	.75
.5	0	.25	.46	.5	.5
.25	0	.20	.25	.25	.25
0	0	0	0	0	0

$w = 10$

$b =$	0	.25	.5	.75	1
$a = 1$	0	.25	.5	.75	1
.75	0	.25	.5	.75	.75
.5	0	.25	.5	.5	.5
.25	0	.25	.25	.25	.25
0	0	0	0	0	0

$w \rightarrow \infty$ (weak)

fuzzy union for which $w = 1$, the membership grades in the two sets are summed to produce the membership grade in their union. Therefore, this union is very soft and indicates perfect interchangeability between the two arguments. On the other hand, the Yager function for which $w \rightarrow \infty$ (the classical fuzzy union) performs a very hard OR by selecting the largest degree of membership in either set. In this sense, then, the functions of the Yager class perform a union operation, which increases in strength as the value of the parameter w increases.

Some other proposed classes of fuzzy set unions along with the corresponding class of fuzzy set intersections are given in Table 2.2. They are identified by the names of their originators and the date of the publication in which they were introduced. While we do not deem it essential to examine all these various classes in this text, the information provided in Note 2.7 is sufficient to allow the reader to pursue such an examination.

2.4 FUZZY INTERSECTION

The discussion of fuzzy intersection closely parallels that of fuzzy union. Like fuzzy union, the general fuzzy intersection of two fuzzy sets A and B is specified by a function

$$i : [0, 1] \times [0, 1] \rightarrow [0, 1].$$

The argument to this function is the pair consisting of the membership grade of some element x in fuzzy set A and the membership grade of that same element in fuzzy set B. The function returns the membership grade of the element in the set $A \cap B$. Thus,

$$\mu_{A \cap B}(x) = i[\mu_A(x), \mu_B(x)].$$

A function of this form must satisfy the following axioms in order to be considered a fuzzy intersection:

Axiom i1. $i(1, 1) = 1; i(0, 1) = i(1, 0) = i(0, 0) = 0$; that is, i behaves as the classical intersection with crisp sets (*boundary conditions*).

Axiom i2. $i(a, b) = i(b, a)$; that is, i is *commutative*.

Axiom i3.   If $a \leq a'$ and $b \leq b'$, then $i(a, b) \leq i(a', b')$; that is, i is *monotonic*.

Axiom i4. $i(i(a, b), c) = i(a, i(b, c))$, that is, i is *associative*.

The justification for these essential axioms (the *axiomatic skeleton for fuzzy set intersections*) is similar to that given in the previous section for the required axioms of fuzzy union.

The most important additional requirements for fuzzy set intersections, which are desirable in certain applications, are expressed by the following two axioms:

TABLE 2.2. SOME CLASSES OF FUZZY SET UNIONS AND INTERSECTIONS.

Reference	Fuzzy Unions	Fuzzy Intersections	Range of Parameter
Schweizer & Sklar [1961]	$1 - \max[0, (1-a)^{-p} + (1-b)^{-p} - 1]^{1/p}$	$\max[0, a^{-p} + b^{-p} - 1]^{-1/p}$	$p \in (-\infty, \infty)$
Hamacher [1978]	$\dfrac{a + b - (2-\gamma)ab}{1 - (1-\gamma)ab}$	$\dfrac{ab}{\gamma + (1-\gamma)(a + b - ab)}$	$\gamma \in (0, \infty)$
Frank [1979]	$1 - \log_s\left[1 + \dfrac{(s^{1-a} - 1)(s^{1-b} - 1)}{s - 1}\right]$	$\log_s\left[1 + \dfrac{(s^a - 1)(s^b - 1)}{s - 1}\right]$	$s \in (0, \infty)$
Yager [1980]	$\min[1, (a^w + b^w)^{1/w}]$	$1 - \min[1, ((1-a)^w + (1-b)^w)^{1/w}]$	$w \in (0, \infty)$
Dubois & Prade [1980]	$\dfrac{a + b - ab - \min(a, b, 1 - \alpha)}{\max(1 - a, 1 - b, \alpha)}$	$\dfrac{ab}{\max(a,b,\alpha)}$	$\alpha \in (0, 1)$
Dombi [1982]	$\dfrac{1}{1 + \left[\left(\frac{1}{a} - 1\right)^{-\lambda} + \left(\frac{1}{b} - 1\right)^{-\lambda}\right]^{-1/\lambda}}$	$\dfrac{1}{1 + \left[\left(\frac{1}{a} - 1\right)^{\lambda} + \left(\frac{1}{b} - 1\right)^{\lambda}\right]^{1/\lambda}}$	$\lambda \in (0, \infty)$

Axiom i5. i is a *continuous* function.

Axiom i6. $i(a, a) = a$; that is, i is *idempotent*.

The implications of these two properties for fuzzy intersection operations are basically the same as those given in the previous section for fuzzy unions.

Some of the classes of functions that satisfy Axioms i1 through i4 are shown in Table 2.2. Let us examine one of these—the *Yager class*, which is defined by the function

$$i_w(a, b) = 1 - \min[1, ((1 - a)^w + (1 - b)^w)^{1/w}], \tag{2.7}$$

where values of the parameter w lie in the open interval $(0, \infty)$.

For each value of the parameter w, we obtain one particular fuzzy set intersection. Like the Yager class of fuzzy unions, all the functions of this class are continuous but most are not idempotent. For $w = 1$, the function of Eq. (2.7) is defined by

$$i_1(a, b) = 1 - \min[1, 2 - a - b];$$

for $w = 2$, we obtain

$$i_2(a, b) = 1 - \min[1, \sqrt{(1 - a)^2 + (1 - b)^2}];$$

similar values are obtained for other finite values of w. For $w \to \infty$, the form of the function given by Eq. (2.7) is not obvious and, therefore, we employ the following theorem.

Theorem 2.7. $\operatorname*{Lim}_{w \to \infty} i_w = \lim_{w \to \infty} (1 - \min[1, ((1 - a)^w + (1 - b)^w)^{1/w}])$
$= \min(a, b)$.

Proof: From the proof of Theorem 2.6, we know that

$$\lim_{w \to \infty} \min[1, [(1 - a)^w + (1 - b)^w]^{1/w}] = \max[1 - a, 1 - b].$$

Thus, $i_\infty(a, b) = 1 - \max[1 - a, 1 - b]$. Let us assume, with no loss of generality, that $a \le b$. Then, $1 - a \ge 1 - b$ and

$$i_\infty(a, b) = 1 - (1 - a) = a.$$

Hence, $i_\infty(a, b) = \min(a, b)$, which concludes the proof. ∎

As is the case with the functions in the Yager class of fuzzy unions, the choice of the parameter w determines the strength of the intersection operations performed by the Yager functions of Eq. (2.7). Table 2.1(b) illustrates the increasing values returned by the Yager intersections as the value of the parameter w increases. Thus, the value $1/w$ can be interpreted as the degree of strength of the intersection performed. Since the intersection is analogous to the logical AND (conjuction), it generally demands simultaneous satisfaction of the operands of A and B. The Yager intersection for which $w = 1$ returns a positive value only when the summation of the membership grades in the two sets exceeds 1. Thus, it performs a hard intersection with the strongest demand for simultaneous set membership. In contrast to this, the Yager function for which $w \to \infty$, which is the classical fuzzy set intersection, performs a soft intersection that allows the lowest degree of membership in either set to dictate the degree of membership in their intersection. In effect then, this operation shows the least demand for simultaneous set membership.

2.5 COMBINATIONS OF OPERATIONS

It is known that fuzzy set unions that satisfy the axiomatic skeleton (Axioms u1 through u4 given in Sec. 2.3) are bounded by the inequalities

$$\max(a, b) \leq u(a, b) \leq u_{\max}(a, b), \tag{2.8}$$

where

$$u_{\max}(a, b) = \begin{cases} a & \text{when } b = 0, \\ b & \text{when } a = 0, \\ 1 & \text{otherwise.} \end{cases}$$

Similarly, fuzzy set intersections that satisfy Axioms i1 through i4 (given in Sec. 2.4) are bounded by the inequalities

$$i_{\min}(a, b) \leq i(a, b) \leq \min(a, b), \tag{2.9}$$

where

$$i_{\min}(a, b) = \begin{cases} a & \text{when } b = 1, \\ b & \text{when } a = 1, \\ 0 & \text{otherwise.} \end{cases}$$

The inequalities $u(a, b) \geq \max(a, b)$ and $i(a, b) \leq \min(a, b)$ are often used as axioms for fuzzy unions and intersections, respectively, instead of the axioms of associativity. However, these inqualities as well as those for $u_{\max}$ and $i_{\min}$ can be derived from our axioms as shown in the following four theorems.

Theorem 2.8. For all $a, b \in [0, 1]$, $u(a, b) \geq \max(a, b)$.

Proof: Using associativity (Axiom u4), the equation

$$u(a, u(0, 0)) = u(u(a, 0), 0)$$

is valid. By applying the boundary condition $u(0, 0) = 0$ (Axiom u1), we can rewrite this equation as

$$u(a, 0) = u(u(a, 0), 0).$$

Assume now that the solution of this equation is $u(a, 0) = \alpha \neq a$. Substitution of α for $u(a, 0)$ in the equation yields $\alpha = u(\alpha, 0)$, which contradicts our assumption. Hence, the only solution of the equation is $u(a, 0) = a$. Now, by monotonicity of u (Axiom u3), we have

$$u(a, b) \geq u(a, 0) = a,$$

and, by employing commutativity (Axiom u2), we also have

$$u(a, b) = u(b, a) \geq u(b, 0) = b.$$

Hence, $u(a, b) \geq \max(a, b)$. ∎

Theorem 2.9.   For all $a, b \in [0, 1]$, $u(a, b) \leq u_{max}(a, b)$.

Proof: When $b = 0$, then $u(a, b) = a$ (see the proof of Theorem 2.8) and the theorem holds. Similarly, by commutativity, when $a = 0$, then $u(a, b) = b$, and the theorem again holds. Since $u(a, b) \in [0, 1]$, it follows from Theorem 2.8 that $u(a, 1) = u(1, b) = 1$. Now, by monotonicity we have

$$u(a, b) \leq u(a, 1) = u(1, b) = 1.$$

This concludes the proof. ∎

Theorem 2.10.   For all $a, b \in [0, 1]$, $i(a, b) \leq \min(a, b)$.

Proof: The proof of this theorem is similar to that of Theorem 2.8. First, we form the equation

$$i(a, i(1, 1)) = i(i(a, 1), 1)$$

based on the associativity of i. Then, using the boundary condition $i(1, 1) = 1$, we rewrite the equation as

$$i(a, 1) = i(i(a, 1), 1).$$

The only solution of this equation is $i(a, 1) = a$. Then, by monotonicity we have

$$i(a, b) \leq i(a, 1) = a,$$

and by commutativity

$$i(a, b) = i(b, a) \leq i(b, 1) = b,$$

which completes the proof. ∎

Theorem 2.11.   For all $a, b \in [0, 1]$, $i(a, b) \geq i_{min}(a, b)$.

Proof: The proof is analogous to the proof of Theorem 2.9. When $b = 1$, then $i(a, b) = a$ (see the proof of Theorem 2.10) and the theorem is satisfied. Similarly, commutativity ensures that when $a = 1$, $i(a, b) = b$ and the theorem holds. Since $i(a, b) \in [0, 1]$, it follows from Theorem 2.10 that $i(a, 0) = i(0, b) = 0$. By monotonicity,

$$i(a, b) \geq i(0, b) = i(a, 0) = 0$$

and the proof is complete. ∎

We can see that the standard max and min operations have a special significance: they represent, respectively, the lower bound of functions u (the strongest union) and the upper bound of functions i (the weakest intersection). Of all the possible pairs of fuzzy set unions and intersections, the max and min functions are closest to each other. That is, for all $a, b \in [0, 1]$, the inequality

$$\max(a, b) - \min(a, b) = |a - b| \leq u(a, b) - i(a, b)$$

is satisfied for any arbitrary pair of functions u and i that qualify as a fuzzy set union and intersection, respectively. The standard max and min operations therefore represent an extreme pair of all the possible pairs of fuzzy unions and intersections. Moreover, the functions of max and min are related to each other by DeMorgan's laws based on the standard complement $c(a) = 1 - a$, that is,

$$\max(a, b) = 1 - \min(1 - a, 1 - b),$$

$$\min(a, b) = 1 - \max(1 - a, 1 - b).$$

The operations $u_{\max}$ and $i_{\min}$ represent another pair of a fuzzy union and a fuzzy intersection which is extreme in the sense that for all $a, b \in [0, 1]$, the inequality

$$u_{\max}(a, b) - i_{\min}(a, b) \geq u(a, b) - i(a, b)$$

is satisfied for any arbitrary pair of fuzzy unions u and fuzzy intersections i. As is the case with the standard max and min operations, the operations $u_{\max}$ and $i_{\min}$ are related to each other by DeMorgan's laws under the standard complement, that is,

$$u_{\max}(a, b) = 1 - i_{\min}(1 - a, 1 - b),$$

$$i_{\min}(a, b) = 1 - u_{\max}(1 - a, 1 - b).$$

The Yager class of fuzzy unions and intersections, discussed in Secs. 2.3 and 2.4, covers the entire range of these operations as given by inequalities (2.8) and (2.9). The standard max and min operations are represented by $w \to \infty$, and the $u_{\max}$ and $i_{\min}$ operations at the other extreme are represented by $w \to 0$. Of the other classes of operations listed in Table 2.2, the full range of these operations is covered only by the Schweizer and Sklar class (with $p \to \infty$ for the standard operations and $p \to -\infty$ for the other extreme) and by the Dombi class (with $\lambda \to \infty$ representing the standard operations and $\lambda \to 0$ the other extreme).

The standard max and min operations are additionally significant in that they constitute the only fuzzy union and intersection operators that are continuous and idempotent. We express this fact by the following two theorems.

Theorem 2.12. $u(a, b) = \max(a, b)$ is the only continuous and idempotent fuzzy set union (i.e., the only function that satisfies Axioms u1 through u6).

Proof: By associativity, we can form the equation

$$u(a, u(a, b)) = u(u(a, a), b).$$

The application of idempotency (Axiom u6) allows us to replace $u(a, a)$ in this equation with a and thus to obtain

$$u(a, (u(a, b)) = u(a, b).$$

Similarly,

$$u(u(a, b), b) = u(a, u(b, b)) = u(a, b).$$

Hence,

$$u(a, u(a, b)) = u(u(a, b), b)$$

or, by commutativity,

$$u(a, u(a, b)) = u(b, u(a, b)). \qquad (2.10)$$

When $a = b$, idempotency is applicable and Eq. (2.10) is satisfied. Let $a < b$ and assume that $u(a, b) = \alpha$, where $\alpha \neq a$ and $\alpha \neq b$. Then, Eq. (2.10) becomes

$$u(a, \alpha) = u(b, \alpha).$$

Since u is continuous (Axiom u5) and monotonic nondecreasing (Axiom u3) with $u(0, \alpha) = \alpha$ and $u(1, \alpha) = 1$ (as determined in proofs of Theorems 2.8 and 2.9, respectively), there exists a pair $a, b \in [0, 1]$ such that

$$u(a, \alpha) < u(b, \alpha)$$

and, consequently, the assumption is not warranted.* Assume now that $u(a, b) = a = \min(a, b)$. This assumption is also unacceptable, since it violates the boundary conditions (Axiom u1) when $a = 0$ and $b = 1$. The final possibility is to consider $u(a, b) = b = \max(a, b)$. In this case, the boundary conditions are satisfied and Eq. (2.10) becomes

$$u(a, b) = u(b, b);$$

that is, it is satisfied for all $a < b$. Because of commutativity, the same argument can be repeated for $a > b$. Hence, max is the only function that satisfies Axioms u1 through u6. ∎

Theorem 2.13. $i(a, b) = \min(a, b)$ is the only continuous and idempotent fuzzy set intersection (i.e., the only function that satisfies Axioms i1 through i6).

Proof: This theorem can be proven in exactly the same way as Theorem 2.12 by replacing function u with function i and by applying Axioms i1 through i6 instead of Axioms u1 through u6. The counterpart of Eq. (2.10) is

$$i(a, i(a, b)) = i(b, i(a, b)).$$

We use the same reasoning as in the proof of Theorem 2.8, albeit with different boundary conditions (Axiom i1 instead of Axiom u1) to conclude that $u(a, b) = \min(a, b)$ is the only solution of this equation. ∎

---

* This is a consequence of the intermediate value theorem for continuous functions.

The operations of complement, union, and intersection defined on crisp subsets of X form a Boolean lattice on the power set $\mathcal{P}(X)$, as explained in Sec. 1.5; they possess the properties listed in Table 1.1 (or, in the abstracted form, in Table 1.4). The various fuzzy counterparts of these operations are defined on the power set $\tilde{\mathcal{P}}(X)$—the set of all *fuzzy* subsets of X. It is known that every possible selection of these three fuzzy operations violates some properties of the Boolean lattice on $\mathcal{P}(X)$. Different selected operations, however, may violate different properties of the Boolean lattice. Let us examine some possibilities.

It can be easily verified that the standard fuzzy operations satisfy all the properties of the Boolean lattice except the law of excluded middle $A \cup \overline{A} = X$ and the law of contradiction $A \cap \overline{A} = \emptyset$. These operations are said to form a *pseudo-complemented distributive lattice* on $\tilde{\mathcal{P}}(X)$. We know from Theorems 2.12 and 2.13 that the max and min operations are the only operations of fuzzy union and intersection that are idempotent. This means, in turn, that none of the other possible operations of fuzzy unions and intersections form a lattice on $\tilde{\mathcal{P}}(X)$. Some of them, however, satisfy the law of excluded middle and the law of contradiction, which for fuzzy sets have the form

$$u(a, c(a)) = 1 \quad \text{and} \quad i(a, c(a)) = 0$$

for all $a \in [0, 1]$. These latter operations are characterized by the following theorem.

Theorem 2.14. Fuzzy set operations of union, intersection, and continuous complement that satisfy the law of excluded middle and the law of contradiction are not idempotent or distributive.

Proof: Since the standard operations do not satisfy the two laws of excluded middle and of contradiction and, by Theorems 2.12 and 2.13, they are the only operations that are idempotent, operations that do satisfy these laws cannot be idempotent. Next, we must prove that these operations do not satisfy the distributive laws,

$$u(a, i(b, d)) = i(u(a, b), u(a, d)) \tag{2.11}$$

and

$$i(a, u(b, d)) = u(i(a, b), i(a, d)). \tag{2.12}$$

Let e denote the equilibrium of the complement c involved, that is, $c(e) = e$. Then, from the law of excluded middle, we obtain

$$u(e, c(e)) = u(e, e) = 1;$$

similarly, from the law of contradiction,

$$i(e, c(e)) = i(e, e) = 0.$$

Then, by applying e to the left hand side of Eq. (2.11), we obtain

$$u(e, i(e, e)) = u(e, 0).$$

We observe that e is neither 0 nor 1 because of the requirement that $c(0) = 1$ and $c(1) = 0$ (Axiom c1). By Theorem 2.8 and Theorem 2.9, we have $u(e, 0) = e$ and, consequently,

$$u(e, i(e, e)) = e \quad (\neq 1).$$

Now we apply e to the right hand side of Eq. (2.11) to obtain

$$i(u(e, e), u(e, e)) = i(1, 1) = 1.$$

This demonstrates that the distributive law (2.11) is violated.

Let us now apply e to the second distributive law (2.12). By Theorems 2.10 and 2.11, we obtain

$$i(e, u(e, e)) = i(e, 1) = e \quad (\neq 0),$$

and

$$u(i(e, e), i(e, e)) = u(0, 0) = 0,$$

which demonstrates that Eq. (2.12) is not satisfied. This completes the proof. ∎

It follows from Theorem 2.14 that we may, if it is desired, preserve the law of excluded middle and the law of contradiction in our choice of fuzzy union and intersection operations by sacrificing idempotency and distributivity. The reverse is also true. The context of each particular application determines which of these options is preferable.

It is trivial to verify that u_{max}, i_{min}, and the standard complement satisfy the law of excluded middle and the law of contradiction. Another combination of operations of this type is the following:

$$u(a, b) = \min(1, a + b),$$

$$i(a, b) = \max(0, a + b - 1),$$

$$c(a) = 1 - a.$$

As previously mentioned, these operations do not form a lattice on $\tilde{\mathcal{P}}(X)$.

Given two of the three operations u, i, and c, it is sometimes desirable to determine the third operation in such a way that DeMorgan's laws are satisfied. This amounts to solving the functional equation

$$c(u(a, b)) = i(c(a), c(b)) \tag{2.13}$$

with respect to the unknown operation. When c is continuous and involutive, we have

$$u(a, b) = c[i(c(a), c(b))] \tag{2.14}$$

and

$$i(a, b) = c[u(c(a), c(b))]. \tag{2.15}$$

Example 2.1.

Given $u(a, b) = \max(a, b)$ and $c(a) = (1 - a^2)^{1/2}$, determine i such that DeMorgan's laws are satisfied. Employing Eq. (2.15), we obtain

$$i(a, b) = (1 - u^2[(1 - a^2)^{1/2}, (1 - b^2)^{1/2}])^{1/2}$$

$$= (1 - \max^2[(1 - a^2)^{1/2}, (1 - b^2)^{1/2}])^{1/2}.$$

Solving Eq. (2.13) for c is more difficult and may result in more than one solution. For example, if the standard max and min operations are employed for u and i, respectively, then every involutive complement satisfies the equation. Hence, max, min, and any of the Sugeno complements (or Yager complements) defined in Sec. 2.2 satisfy DeMorgan's laws.

For the sake of simplicity, we have omitted an examination of the properties of one operation that is important in fuzzy logic—fuzzy implication, $\Rightarrow$. This operation can be expressed in terms of fuzzy disjunction, $\vee$, fuzzy conjunction, $\wedge$, and negation, $^-$, by using the equivalences

$$a \Rightarrow b = \bar{a} \vee b \quad \text{or} \quad a \Rightarrow b = \overline{a \wedge \bar{b}}.$$

By employing the correspondences between logic operations and set operations defined in Table 1.5, the equivalences just given can be fully studied in terms of the functions

$$u(c(a), b) \quad \text{or} \quad c(i(a, c(b))).$$

Different fuzzy implications are obtained when different fuzzy complements c and either different fuzzy unions u or different fuzzy intersections i are used.

2.6 GENERAL AGGREGATION OPERATIONS

Aggregation operations on fuzzy sets are operations by which several fuzzy sets are combined to produce a single set. In general, any *aggregation operation* is defined by a function

$$h : [0, 1]^n \rightarrow [0, 1]$$

for some $n \geq 2$. When applied to n fuzzy sets $A_1, A_2, \ldots, A_n$ defined on X, h produces an aggregate fuzzy set A by operating on the membership grades of each $x \in X$ in the aggregated sets. Thus,

$$\mu_A(x) = h(\mu_{A_1}(x), \mu_{A_2}(x), \ldots, \mu_{A_n}(x))$$

for each $x \in X$.

In order to qualify as an aggregation function, h must satisfy at least the following two axiomatic requirements, which express the essence of the notion of aggregation:

Axiom h1. $h(0, 0, \ldots, 0) = 0$ and $h(1, 1, \ldots, 1) = 1$ (*boundary conditions*).

Axiom h2.   For any pair $(a_i \mid i \in \mathbb{N}_n)$ and $(b_i \mid i \in \mathbb{N}_n)$, where $a_i \in [0, 1]$ and $b_i \in [0, 1]$, if $a_i \geq b_i$ for all $i \in \mathbb{N}_n$, then $h(a_i \mid i \in \mathbb{N}_n) \geq h(b_i \mid i \in \mathbb{N}_n)$, that is, h is *monotonic nondecreasing* in all its arguments.

Two additional axioms are usually employed to characterize aggregation operations despite the fact that they are not essential:

Axiom h3. h is a *continuous* function.

Axiom h4. h is a *symmetric* function in all its arguments, that is,

$$h(a_i \mid i \in \mathbb{N}_n) = h(a_{p(i)} \mid i \in \mathbb{N}_n)$$

for any permutation p on $\mathbb{N}_n$.

Axiom h3 guarantees that an infinitesimal variation in any argument of h does not produce a noticeable change in the aggregate. Axiom h4 reflects the usual assumption that the aggregated sets are equally important. If this assumption is not warranted in some application context, the symmetry axiom must be dropped.

We can easily see that fuzzy unions and intersections qualify as aggregation operations on fuzzy sets. Although they are defined for only two arguments, their property of associativity guaranteed by Axioms u4 and i4 provides a mechanism for extending their definition to any number of arguments. Hence, fuzzy unions and intersections can be viewed as special aggregation operations that are symmetric, usually continuous, and required to satisfy some additional boundary conditions. As a result of these additional requirements, fuzzy unions and intersections can produce only aggregates that are subject to restrictions (2.8) and (2.9). In particular, they do not produce any aggregates of $a_1, a_2, \ldots, a_n$ that produce values between $\min(a_1, a_2, \ldots, a_n)$ and $\max(a_1, a_2, \ldots, a_n)$. Aggregates that are not restricted in this way are, however, allowed by Axioms h1 through h4; operations that produce them are usually called *averaging operations*.

Averaging operations are therefore aggregation operations for which

$$\min(a_1, a_2, \ldots, a_n) \leq h(a_1, a_2, \ldots, a_n) \leq \max(a_1, a_2, \ldots, a_n). \quad (2.16)$$

In other words, the standard max and min operations represent boundaries between the averaging operations and the fuzzy unions and intersections, respectively.

One class of averaging operations that covers the entire interval between the min and max operations consists of *generalized means*. These are defined by the formula

$$h_\alpha(a_1, a_2, \ldots, a_n) = \left(\frac{a_1^\alpha + a_2^\alpha + \cdots + a_n^\alpha}{n} \right)^{1/\alpha}, \quad (2.17)$$

where $\alpha \in \mathbb{R}$ ($\alpha \neq 0$) is a parameter by which different means are distinguished.

Function h_α clearly satisfies Axioms h1 through h4 and, consequently, it represents a parameterized class of continuous and symmetric aggregation operations. It also satisfies the inequalities (2.16) for all $\alpha \in \mathbb{R}$, with its lower bound

$$h_{-\infty}(a_1, a_2, \ldots, a_n) = \min(a_1, a_2, \ldots, a_n)$$

and its upper bound

$$h_{\infty}(a_1, a_2, \ldots, a_n) = \max(a_1, a_2, \ldots, a_n).$$

For fixed arguments, function h_α is monotonic increasing with α. For $\alpha \to 0$, the function h_α becomes the *geometric mean*

$$h_0(a_1, a_2, \ldots, a_n) = (a_1 \cdot a_2 \cdots a_n)^{1/n};$$

furthermore,

$$h_{-1}(a_1, a_2, \ldots, a_n) = \frac{n}{\dfrac{1}{a_1} + \dfrac{1}{a_2} + \cdots + \dfrac{1}{a_n}}$$

is the *harmonic mean* and

$$h_1(a_1, a_2, \ldots, a_n) = \frac{1}{n}(a_1 + a_2 + \cdots + a_n)$$

is the *arithmetic mean*.

Since it is not obvious that h_α represents the geometric mean for $\alpha \to 0$, we use the following theorem.

Theorem 2.15.   Let h_α be given by Eq. (2.17). Then,

$$\lim_{\alpha \to 0} h_\alpha = (a_1 \cdot a_2 \cdots a_n)^{1/n}.$$

Proof: First, we determine

$$\lim_{\alpha \to 0} \ln h_\alpha = \lim_{\alpha \to 0} \frac{\ln(a_1^\alpha + a_2^\alpha + \cdots + a_n^\alpha) - \ln n}{\alpha}.$$

Using l'Hospital's rule, we now have

$$\lim_{\alpha \to 0} \ln h_\alpha = \lim_{\alpha \to 0} \frac{a_1^\alpha \ln a_1 + a_2^\alpha \ln a_2 + \cdots + a_n^\alpha \ln a_n}{a_1^\alpha + a_2^\alpha + \cdots + a_n^\alpha}$$

$$= \frac{\ln a_1 + \ln a_2 + \cdots + \ln a_n}{n} = \ln(a_1 \cdot a_2 \cdots a_n)^{1/n}.$$

Hence,

$$\lim_{\alpha \to 0} h_\alpha = (a_1 \cdot a_2 \cdots a_n)^{1/n}. \quad \blacksquare$$

When it is desirable to accommodate variations in the importance of individual aggregated sets, the function h_α can be generalized into *weighted generalized means*, as defined by the formula

$$h_\alpha(a_1, a_2, \ldots, a_n; w_1, w_2, \ldots, w_n) = \left(\sum_{i=1}^{n} w_i a_i^\alpha \right)^{1/\alpha}, \qquad (2.18)$$

where $w_i \geq 0$ ($i \in \mathbb{N}_n$) are weights that express the relative importance of the aggregated sets; it is required that

$$\sum_{i=1}^{n} w_i = 1.$$

The weighted means are obviously not symmetric. For fixed arguments and weights, the function h_α given by Eq. (2.18) is monotonic increasing with α.

The full scope of fuzzy aggregation operations is summarized in Fig. 2.5. Included in this diagram are only the generalized means, which cover the entire range of averaging operators, and those parameterized classes of fuzzy unions and intersections given in Table 2.2 that cover the full ranges specified by the inequalities (2.8) and (2.9). For each class of operators, the range of the respective parameter is indicated. Given one of these families of operations, the identification of a suitable operation for a specific application is equivalent to the estimation of the parameter involved.

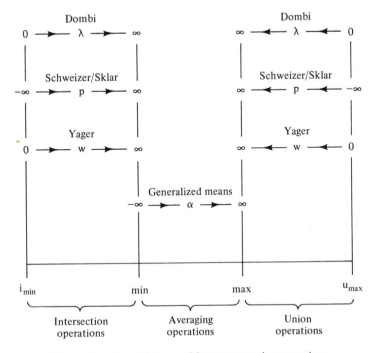

Figure 2.5.   The full scope of fuzzy aggregation operations.

NOTES

2.1. In the seminal paper by Zadeh [1965a], fuzzy set theory is formulated in terms of the standard operations of complement, union, and intersection, but other possibilities of combining fuzzy sets are also hinted at.

2.2. The first axiomatic treatment of fuzzy set operations was presented by Bellman and Giertz [1973]. They demonstrated the uniqueness of the max and min operators in terms of axioms that consist of our axiomatic skeletons for u and i, the axioms of continuity, distributivity, strict increase of $u(a, a)$ and $i(a, a)$ in a, and lower and upper bounds $u(a, b) \geq \max(a, b)$ and $i(a, b) \leq \min(a, b)$. They concluded, however, that the operation of fuzzy complement is not unique even when all reasonable requirements (boundary conditions, monotonicity, continuity, and involution) are employed as axioms. A thorough investigation of properties of the max and min operators was done by Voxman and Goetschel [1983].

2.3. The Sugeno class of fuzzy complements results from special measures (called λ-measures) introduced by Sugeno [1977]. The Yager class of fuzzy complements is derived from his class of fuzzy unions defined by Eq. (2.6) by requiring that $A \cup C(A) = X$, where A is a fuzzy set defined on X. This requirement can be expressed more specifically by requiring that $u_w(a, c_w(a)) = 1$ for all $a \in [0, 1]$ and all $w > 0$.

2.4. Different approaches to the study of fuzzy complements were used by Lowen [1978], Esteva, Trillas, and Domingo [1981], and Ovchinnikov [1981, 1983]. Yager [1979b, 1980a] investigated fuzzy complements for the purpose of developing useful measures of fuzziness (Sec. 5.2). Our presentation of fuzzy complements in Sec. 2.2 is based upon a paper by Higashi and Klir [1982], which is also motivated by the aim of developing measures of fuzziness.

2.5. The Yager class of fuzzy unions and intersections was introduced in a paper by Yager [1980b], which contains some additional characteristics of these classes. Yager [1982b] also addressed the question of the meaning of the parameter w in his class and the problem of selecting appropriate operations for various purposes.

2.6. The axiomatic skeletons that we use for characterizing fuzzy intersections and unions are known in the literature as *triangular norms* (or *t-norms*) and *triangular conorms* (or *t-conorms*), respectively [Schweizer and Sklar, 1960, 1961, 1983]. These concepts were originally introduced by Menger [1942] in his study of statistical metric spaces. In current literature on fuzzy set theory, the terms *t*-norms and *t*-conorms are used routinely.

2.7. Classes of functions given in Table 2.2 that can be employed for fuzzy unions and intersections were proposed by Schweizer and Sklar [1961, 1963, 1983], Hamacher [1978], Frank [1979], Yager [1980b], Dubois and Prade [1980b], and Dombi [1982]. Additional theoretical studies of fuzzy set operations were done by Trillas, Alsina, and Valverde [1982], Czogała and Drewniak [1984], Klement [1984], and Silvert [1979]. A good overview of the various classes of fuzzy set operations was prepared by Dubois and Prade [1982a]; they also overviewed properties of various combinations of fuzzy set operations [Dubois and Prade, 1980b].

2.8. The issue of which operations on fuzzy sets are suitable in various situations was studied by Zimmermann [1978a], Thole, Zimmermann, and Zysno [1979], Zimmermann and Zysno [1980], Yager [1979a, 1982b], and Dubois and Prade [1980b].

2.9. One class of operators not covered in this book are fuzzy implication operators. They were extensively studied by Bandler and Kohout [1980a, b] and Yager [1983b].

2.10. An excellent overview of the whole spectrum of aggregation operations on fuzzy sets was prepared by Dubois and Prade [1985a]; it covers fuzzy unions and intersections as well as averaging operations. In another paper, Dubois and Prade [1984b] presented a similar overview in the context of decision-making applications. The class of generalized means defined by Eq. (2.17) is covered in a paper by Dyckhoff and Pedrycz [1984].

EXERCISES

2.1. Using Sugeno complements for $\lambda = 1, 2, 10$ and Yager complements for $w = 1, 2, 3$, determine complements of the following fuzzy sets:
   (a) the fuzzy number defined in Fig. 1.2;
   (b) the fuzzy sets defined in Exercise 1.3;
   (c) some of the fuzzy sets defined in Fig. 1.9.

2.2. Does the function $c(a) = (1 - a)^w$ qualify for each $w > 0$ as a fuzzy complement? Plot the function for some values $w > 1$ and some values $w < 1$.

2.3. Prove that the Sugeno complements are monotonic nonincreasing (Axiom c2) for all $\lambda \in (-1, \infty)$.

2.4. Show that the Sugeno complements are involutive for all $\lambda \in (-1, \infty)$. Show that the Yager complements are involutive for $w \in (1, \infty)$.

2.5. Show that the equilibria e_{c_w} for the Yager fuzzy complements are given by the formula

$$e_{c_w} = (1/2)^{1/w}.$$

Plot this function for $w \in (0, 10]$.

2.6. Prove that Axioms u1 through u5 (or i1 through i5) are satisfied by all fuzzy unions (or intersections) in the Yager class.

2.7. Prove that the following properties are satisfied by all fuzzy unions in the Yager class:
   (a) $u_w(a, 0) = a$;   **(b)** $u_w(a, 1) = 1$;
   (c) $u_w(a, a) \geq a$;   **(d)** if $w \leq w'$, then $u_w(a, b) \geq u_{w'}(a, b)$;
   (e) $\lim\limits_{w \to 0} u_w(a, b) = u_{max}(a, b)$.

2.8. Prove that the following properties are satisfied by all fuzzy intersections in the Yager class:
   (a) $i_w(a, 0) = 0$;   **(b)** $i_w(a, 1) = a$;
   (c) $i_w(a, a) \leq a$;   **(d)** if $w \leq w'$, then $i_w(a, b) \leq i_{w'}(a, b)$;
   (e) $\lim\limits_{w \to 0} i_w(a, b) = i_{min}(a, b)$.

2.9. Show that $u_w(a, c_w(a)) = 1$ for all $a \in [0, 1]$ and all $w > 0$, where u_w and c_w denote the Yager union and complement, respectively (Note 2.3).

2.10. For each class of fuzzy set unions and intersections defined in Table 2.2 and several values of the parameter involved (values 1 and 2, for instance), determine membership functions of the respective unions and intersections in a form similar to Table 2.1.

2.11. For each of the classes of fuzzy unions defined by the parameterized functions in Table 2.2, show that the function decreases with an increase in the parameter.

2.12. For each of the classes of fuzzy intersections defined by the parameterized functions in Table 2.2, show that the function increases with any increase in the parameter.

2.13. The proof of Theorem 2.13 is outlined in Sec. 2.5. Describe the proof in full detail.

2.14. Show that the following operations satisfy the law of excluded middle and the law of contradiction:

(a) u_{max}, i_{min}, $c(a) = 1 - a$;

(b) $u(a, b) = \min(1, a + b)$, $i(a, b) = \max(0, a + b - 1)$, $c(a) = 1 - a$.

2.15. Show that the following operations on fuzzy sets satisfy DeMorgan's laws:

(a) u_{max}, i_{min}, $c(a) = 1 - a$;

(b) max, min, c_λ is a Sugeno complement for some $\lambda \in (-1, \infty)$;

(c) max, min, c_w is a Yager complement for some $w \in (0, \infty)$;

2.16. Determine the membership function based on the generalized means (in a form similar to Table 2.1) for $\alpha = -2, -1, 0, 1, 2$; assume only two arguments and, then, repeat one of the cases for three arguments.

2.17. Show that the generalized means defined by Eq. (2.17) become min and max operations for $\alpha \rightarrow -\infty$ and $\alpha \rightarrow \infty$, respectively.

2.18. Demonstrate that the generalized means h_α defined by Eq. (2.17) are monotonic increasing with α for fixed arguments.

3

FUZZY RELATIONS

3.1 CRISP AND FUZZY RELATIONS

A *crisp relation* represents the presence or absence of association, interaction, or interconnectedness between the elements of two or more sets. This concept can be generalized to allow for various degrees or strengths of relation or interaction between elements. Degrees of association can be represented by membership grades in a *fuzzy relation* in the same way as degrees of set membership are represented in the fuzzy set. In fact, just as the crisp set can be viewed as a restricted case of the more general fuzzy set concept, the crisp relation can be considered to be a restricted case of the fuzzy relation.

Throughout this chapter the concepts and properties of crisp relations are briefly discussed as a refresher and in order to demonstrate their generalized application to fuzzy relations.

The *Cartesian product* of two crisp sets X and Y, denoted by $X \times Y$, is the crisp set of all ordered pairs such that the first element in each pair is a member of X and the second element is a member of Y. Formally,

$$X \times Y = \{(x, y) \mid x \in X \text{ and } y \in Y\}.$$

Note that if $X \neq Y$, then $X \times Y \neq Y \times X$.

The Cartesian product can be generalized for a family of crisp sets $\{X_i \mid i \in \mathbb{N}_n\}$ and denoted either by $X_1 \times X_2 \times \cdots \times X_n$ or by $\times_{i \in \mathbb{N}_n} X_i$. Elements of the Cartesian product of n crisp sets are n-tuples $(x_1, x_2, \ldots, x_n)$ such that $x_i \in X_i$ for all $i \in \mathbb{N}_n$. Thus,

$$\underset{i \in \mathbb{N}_n}{\times} X_i = \{(x_1, x_2, \ldots, x_n) \mid x_i \in X_i \text{ for all } i \in \mathbb{N}_n\}.$$

It is possible for all the sets X_i to be equal, that is, to be a single set X. In this case, the Cartesian product of a set X with itself n times is usually denoted by X^n.

A *relation* among crisp sets $X_1, X_2, \ldots, X_n$ is a subset of the Cartesian product $\times_{i\in\mathbb{N}_n} X_i$. It is denoted either by $R(X_1, X_2, \ldots, X_n)$ or by the abbreviated form $R(X_i \mid i \in \mathbb{N}_n)$. Thus,

$$R(X_1, X_2, \ldots, X_n) \subset X_1 \times X_2 \times \cdots \times X_n,$$

so that for relations among sets $X_1, X_2, \ldots, X_n$, the Cartesian product $X_1 \times X_2 \times \cdots \times X_n$ represents the universal set. Because a relation is itself a set, the basic set concepts such as containment or subset, union, intersection, and complement can be applied without modification to relations.

Each crisp relation R can be defined by a characteristic function that assigns a value of 1 to every tuple of the universal set belonging in the relation and a 0 to every tuple that does not belong. Thus,

$$\mu_R (x_1, x_2, \ldots, x_n) = \begin{cases} 1 & \text{if and only if} \quad (x_1, x_2, \ldots, x_n) \in R, \\ 0 & \text{otherwise.} \end{cases}$$

The membership of a tuple in a relation signifies that the elements of the

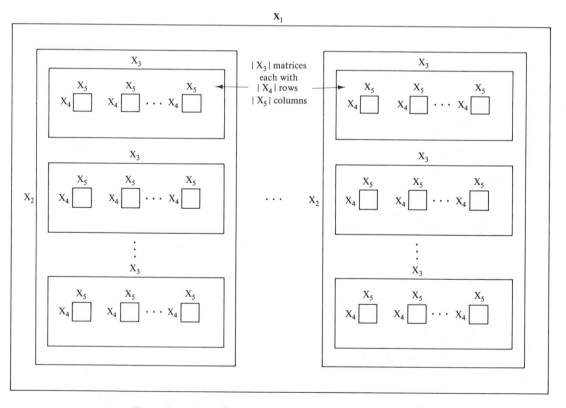

Figure 3.1.   A possible representation of quinary relation by 5-dimensional array.

tuple are related or associated with one another. For instance, let R represent the relation of marriage between the set of all men and the set of all women. Of all the possible pairings of men and women, then, only those pairs whose members are married to each other will be assigned a value of 1, indicating that they belong in this relation. A relation between two sets is called *binary*; if three, four, or five sets are involved, the relation is called *ternary*, *quaternary*, or *quinary*, respectively. In general, a relation defined on n sets is called *n-ary*, or *n-dimensional*.

A relation can be written as a set of ordered tuples. Another convenient way of representing a relation $R(X_1, X_2, \ldots, X_n)$ involves an n-dimensional membership array: $\mathbf{M}_R = [\mu_{i_1,i_2,\ldots,i_n}]$. Each element of the first dimension i_1 of this array corresponds to exactly one member of X_1, each element of dimension i_2 to exactly one member of X_2, and so on. If the n-tuple $(x_1, x_2, \ldots, x_n) \in X_1 \times X_2 \times \cdots \times X_n$ corresponds to the element $\mu_{i_1,i_2,\ldots,i_n}$ of $\mathbf{M}_R$, then

$$\mu_{i_1,i_2,\ldots,i_n} = \begin{cases} 1 & \text{if and only if } (x_1, x_2, \ldots, x_n) \in R, \\ 0 & \text{otherwise.} \end{cases}$$

Example 3.1.

Let R be a relation among the three sets $X = \{$English, French$\}$, $Y = \{$dollar, pound, franc, mark$\}$ and $Z = \{$United States, France, Canada, Britain, Germany$\}$, which associates a country with a currency and language as follows:

$R(X, Y, Z) = \{$(English, dollar, United States), (French, franc, France), (English, dollar, Canada), (French, dollar, Canada), (English, pound, Britain)$\}$.

This relation can also be represented by the following three-dimensional membership array:

	U.S.	Fra	Can	Brit	Ger
Dollar	1	0	1	0	0
Pound	0	0	0	1	0
Franc	0	0	0	0	0
Mark	0	0	0	0	0

English

	U.S.	Fra	Can	Brit	Ger
Dollar	0	0	1	0	0
Pound	0	0	0	0	0
Franc	0	1	0	0	0
Mark	0	0	0	0	0

French

To illustrate the convenience of representing n-ary relations by n-dimensional arrays (especially important for computer processing), a possible structure for a five-dimensional array based on sets X_i ($i \in \mathbb{N}_5$) is shown in Fig. 3.1. The full array (five-dimensional) can be viewed as a "library" that consists of $|X_1|$ "books" (four-dimensional arrays) distinguished from each other by elements of X_1. Each "book" consists of $|X_2|$ "pages" (three-dimensional arrays) distinguished from each other by elements of X_2. Each "page" consists of $|X_3|$ matrices (two-dimensional arrays) distinguished from each other by elements of X_3. Each matrix consists of $|X_4|$ rows (one-dimensional arrays) distinguished from each other by elements of X_4. Each row consists of $|X_5|$ individual entries distinguished from each other by elements of X_5.

Just as the characteristic function of a crisp set can be generalized to allow for degrees of set membership, the characteristic function of a crisp relation can be generalized to allow tuples to have degrees of membership within the relation.

Thus, a *fuzzy relation* is a fuzzy set defined on the Cartesian product of crisp sets $X_1, X_2, \ldots, X_n$, where tuples $(x_1, x_2, \ldots, x_n)$ may have varying degrees of membership within the relation. The membership grade is usually represented by a real number in the closed interval $[0, 1]$ and indicates the strength of the relation present between the elements of the tuple.

A fuzzy relation can also conveniently be represented by an n-dimensional membership array whose entries correspond to n-tuples in the universal set. These entries take values representing the membership grades of the corresponding n-tuples.

Example 3.2.

Let R be a fuzzy relation between the two sets $X = \{$New York City, Paris$\}$ and $Y = \{$Beijing, New York City, London$\}$ that represents the relational concept "very far." This relation can be written in list notation as

$R(X, Y) = 1/$NYC, Beijing $+ 0/$NYC, NYC $+ .6/$NYC, London $.9/$Paris, Beijing $+ .7/$Paris, NYC $+ .3/$Paris, London.

This relation can also be represented by the following two-dimensional membership array:

	NYC	Paris
Beijing	1	.9
NYC	0	.7
London	.6	.3

Ordinary fuzzy relations (with the valuation set $[0, 1]$) can obviously be extended to L-fuzzy relations (with an arbitrary ordered valuation set L) in the same way as fuzzy sets are extended to L-fuzzy sets. Similarly, type t, level k, and interval-valued fuzzy relations can be defined.

Consider the Cartesian product of all sets in the family $X = \{X_i \mid i \in \mathbb{N}_n\}$. For each sequence ($n$-tuple)

$$\mathbf{x} = (x_i \mid i \in \mathbb{N}_n) \in \underset{i \in \mathbb{N}_n}{\times} X_i$$

and each sequence (r-tuple, $r \subset n$)

$$\mathbf{y} = (y_j \mid j \in J) \in \underset{j \in J}{\times} X_j,$$

where $J \subset \mathbb{N}_n$, let $\mathbf{y}$ be called a *subsequence** of $\mathbf{x}$ if and only if $y_j = x_j$ for all $j \in J$. Let $\mathbf{y} \prec \mathbf{x}$ denote that $\mathbf{y}$ is a subsequence of $\mathbf{x}$.

Given a relation $R(X_1, X_2, \ldots, X_n)$, let $[R \downarrow Y]$ denote the *projection* of R that disregards all variables in X except those in the set

$$Y = \{X_j \mid j \in J \subset \mathbb{N}_n\}.$$

---

* More appropriate names are usually used in various specific contexts. For example, if n-tuples $\mathbf{x}$ represent states of a system, $\mathbf{y}$ is called a *substate* of $\mathbf{x}$; if they are strings of symbols in a formal language, $\mathbf{y}$ is called a *substring* of $\mathbf{x}$.

Then, $[R \downarrow Y]$ is a fuzzy set (relation) whose membership function is defined on the Cartesian product of sets in Y by the equation

$$\mu_{[R \downarrow Y]}(\mathbf{y}) = \max_{\mathbf{x} > \mathbf{y}} \mu_R(\mathbf{x}), \qquad (3.1)$$

where μ_R is the membership function of the given n-ary relation R.

Example 3.3.

Consider the sets $X_1 = \{x, y\}$, $X_2 = \{a, b\}$ and $X_3 = \{*, \$\}$ and the ternary fuzzy relation

$$R(X_1, X_2, X_3) = .9/x, a, * + .4/x, b, * + 1/y, a, * + .7/y, a, \$ + .8/y, b, \$$$

defined on $X_1 \times X_2 \times X_3$. Let $R_{i,j} = [R \downarrow \{X_i, X_j\}]$ and $R_i = [R \downarrow \{X_i\}]$ for all $i, j \in \mathbb{N}_3$. Then,

$$R_{1,2} = .9/x, a + .4/x, b + 1/y, a + .8/y, b,$$

$$R_{1,3} = .9/x, * + 1/y, * + .8/y, \$,$$

$$R_{2,3} = 1/a, * + .4/b, * + .7/a, \$ + .8/b, \$,$$

$$R_1 = .9/x + 1/y,$$

$$R_2 = 1/a + .8/b,$$

$$R_3 = 1/* + .8/\$.$$

A detailed calculation of $\mu_{R_{1,3}}(x_1, x_3)$ is shown in Table 3.1.

The projection defined by Eq. (3.1) can be generalized by replacing the max operator with any of the class of operators for fuzzy unions that are discussed in Sec. 2.3.

Another operation on relations, which is in some sense an inverse to the projection, is called a *cylindric extension*. Let X and Y denote the same families of sets as employed in the definition of projection. Let R be a relation defined on the Cartesian product of sets in the family Y, and let $[R \uparrow X - Y]$ denote the cylindric extension of R into sets X_i $(i \in \mathbb{N}_n)$ that are in X but are not in Y. Then,

$$\mu_{[R \uparrow X - Y]}(\mathbf{x}) = \mu_R(\mathbf{y}), \qquad (3.2)$$

for each $\mathbf{x}$ such that $\mathbf{x} > \mathbf{y}$.

TABLE 3.1. CALCULATION OF THE PROJECTION $R_{1,3}$ IN EXAMPLE 3.3.

$(x_1 \; x_2, \; x_3)$	$\mu_R(x_1, x_2, x_3)$	
$(x, a, *)$	.9	$\mu_{R_{1,3}}(x, *) = \max[\mu_R(x, a, *), \mu_R(x, b, *)]$
$(x, b, *)$	.4	$= \max[.9, .4] = .9$
$(y, a, *)$	1	$\mu_{R_{1,3}}(y, *) = \max[\mu_R(y, a, *), \mu_R(y, b, *)]$
$(y, a, \$)$	.7	$= \max[1, 0] = 1$
$(y, b, \$)$		$\mu_{R_{1,3}}(y, \$) = \max[\mu_R(y, a, \$), \mu_R(y, b, \$)]$
	.8	$= \max[.7, .8] = .8$

The cylindric extension clearly produces the *largest fuzzy relation* (in the sense of membership grades of elements of the extended Cartesian product) that is compatible with the given projection. Such a relation is the least specific of all relations compatible with the projection. The cylindric extension thus *maximizes the nonspecificity* in deriving the n-dimensional relation from one of its r-dimensional projections ($r < n$). That is, it guarantees that no information that is not included in the projection is employed in determining the extended relation. Hence, the cylindric extension is totally *unbiased*.

Example 3.4.

Membership functions of cylindric extensions of all the projections in Example 3.3 with respect to X are shown in Table 3.2. For instance,

$$\mu_{[R_{1,2} \uparrow \{X_3\}]}(x, a, *) = \mu_{[R_{1,2} \uparrow \{X_3\}]}(x, a, \$) = \mu_{R_{1,2}}(x, a) = .9.$$

We can see that none of the cylindric extensions specified in Table 3.2 are equal to the original fuzzy relation from which the projections involved in the cylindric extensions were determined. This means that some information was lost when the given relation was replaced by any one of its projections in this example.

Relations that can be reconstructed from one of their projections by the cylindric extension exist, but they are rather rare. It is more common that a relation can be exactly reconstructed from several of its projections by taking the set intersection of their cylindric extensions. The resulting relation is usually called a *cylindric closure*. When projections are determined by the max operator (Eq. (3.1)), the min operator is normally used for the set intersection. Hence, given a set of projections $\{R_i \mid i \in I\}$ of a relation,

$$\mu_{\text{cyl}\{R_i\}}(\mathbf{x}) = \min_{i \in I} \mu_{[R_i \uparrow X - Y_i]}(\mathbf{x}), \qquad (3.3)$$

where cyl$\{R_i\}$ denotes the cylindric closure based on these projections.

The cylindric closure is clearly the largest fuzzy relation defined on the extended Cartesian product that is compatible with all the projections. Hence, it is unbiased in the same sense as the cylindric extension. In general, the min operator in Eq. (3.3) can be replaced with any of the class of intersection operators, as discussed in Sec. 2.4.

TABLE 3.2. CYLINDRIC EXTENSIONS OF PROJECTIONS CALCULATED IN EXAMPLE 3.3.

| | | Membership functions of the cylindric extension of | | | | |
(x_1, x_2, x_3)	$R_{1,2}$	$R_{1,3}$	$R_{2,3}$	R_1	R_2	R_3
$(x, a, *)$	.9	.9	1	.9	1	1
$(x, a, \$)$	.9	0	.7	.9	1	.8
$(x, b, *)$	.4	.9	.4	.9	.8	1
$(x, b, \$)$	.4	0	.8	.9	.8	.8
$(y, a, *)$	1	1	1	1	1	1
$(y, a, \$)$	1	.8	.7	1	1	.8
$(y, b, *)$	.8	1	.4	1	.8	1
$(y, b, \$)$	.8	.8	.8	1	.8	.8

Example 3.5.

The cylindric closure of the first three projections in Table 3.2 is the following fuzzy ternary relation:

$$\text{cyl}\{R_{1,2}, R_{1,3}, R_{2,3}\} = .9/x, a, * + .4/x, b, * + 1/y, a, * + .7/y, a, \$ + .4/y, b, *$$
$$+ .8/y, b, \$.$$

We can see that the cylindric closure is almost the same as the original relation $R(X_1, X_2, X_3)$ given in Example 3.3. The only error in the cylindric closure involves the triple $(y, b, *)$, which has a membership grade of 0 in R and .4 in the cylindric closure. Hence, the three projections do not capture the original relation fully but approximate it quite well.

3.2 BINARY RELATIONS

Any relation between two sets X and Y is known as a *binary relation*. It is usually denoted by $R(X, Y)$. Since binary relations are important for many applications, we cover them in more detail. When $X \neq Y$, binary relations $R(X, Y)$ are often referred to as *bipartite graphs*; when $X = Y$, they are called *directed graphs*, or *digraphs*.

In addition to membership matrices, another useful representation of binary relations $R(X, Y)$ are *sagittal diagrams*. Each of the sets X, Y is represented by a set of nodes in the diagram; nodes corresponding to one set are clearly distinguished from nodes representing the other set. Elements of $X \times Y$ with nonzero membership grades in $R(X, Y)$ are represented in the diagram by lines connecting the respective nodes. These lines are labeled with the value of the membership grade $\mu_{R(X,Y)}(x, y)$. An example of the sagittal diagram together with the corresponding membership matrix is shown in Fig. 3.2.

The symbol R representing a binary relation is often used in the following alternative way: we write xRy when $(x, y) \in R(X, Y)$; for fuzzy relations, we write α/xRy when $\mu_R(x, y) = \alpha$.

The *domain* of a crisp binary relation is written as dom $R(X, Y)$ and is defined as the crisp subset of X whose members participate in the relation. Formally,

$$\text{dom } R(X, Y) = \{x \mid x \in X, (x, y) \in R \quad \text{for some} \quad y \in Y\}.$$

If $R(X, Y)$ is a fuzzy relation, its domain is the fuzzy set dom $R(X, Y)$ whose membership function is defined by

$$\mu_{\text{dom } R}(x) = \max_{y \in Y} \mu_R(x, y),$$

for each $x \in X$. Thus, each element of set X belongs in the domain of R to the degree equal to the strength of its strongest relation to any member of set Y.

The *range* of a crisp binary relation $R(X, Y)$ is denoted by ran $R(X, Y)$ and is defined as the subset of Y whose members participate in the relation. Thus,

$$\text{ran } R(X, Y) = \{y \mid y \in Y, (x, y) \in R \quad \text{for some} \quad x \in X\}.$$

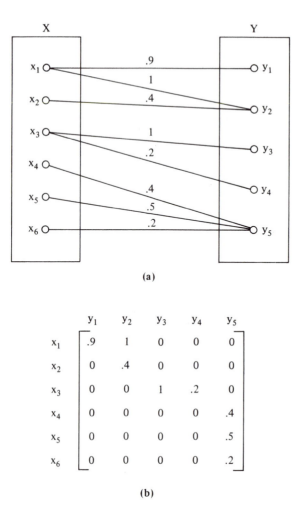

(a)

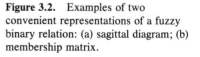

(b)

Figure 3.2. Examples of two convenient representations of a fuzzy binary relation: (a) sagittal diagram; (b) membership matrix.

When $R(X, Y)$ is a fuzzy relation, its range is a fuzzy set ran $R(X, Y)$ whose membership function is defined by

$$\mu_{\text{ran } R}(y) = \max_{x \in X} \mu_R(x, y),$$

for each $y \in Y$. Therefore, the strength of the strongest relation that each element of Y has to an element of X is equal to the degree of that element's membership in the range of R. In addition, the *height* of a fuzzy relation R is a number $h(R)$ defined by

$$h(R) = \max_{y \in Y} \max_{x \in X} \mu_R(x, y),$$

that is, the largest membership grade attained by any pair (x, y) in R. If $h(R) = 1$, then the relation is called *normal*; otherwise it is called *subnormal*.

The concepts of domain and range provide a basis for defining the following classifications of binary relations. These classifications apply to both crisp and fuzzy relations. In the latter case, however, references to the domain and range of the relation are assumed to refer to the support of the domain and range, respectively.

Let R be a binary relation on sets X and Y. If the domain of R is equal to the support of set X, then the relation is called *completely specified*; otherwise it is called *incompletely specified*. If the range of R is equal to the support of set Y, then R is called a relation from X *onto* Y; otherwise it is called a relation from X *into* Y.

If each member of the domain of a binary relation R appears exactly once in R, the relation is called a *mapping* or a *function*. When at least one member of the domain is related to more then one element of the range, the relation is not a mapping and is instead called *one-to-many*. If $R(X, Y)$ is a mapping, let this important property be denoted by $R(X \rightarrow Y)$. If $\mu_{R(X \rightarrow Y)}(x, y) > 0$, then y is called the *image* of x in R.

If R is a mapping and, in addition, at least one of the members of its range appears more than once in R, then the relation is called *many-to-one*. This name refers to the fact that many elements from the domain map to a single element of the range. If, instead, each element of the range appears exactly once in the mapping, it is called a *one-to-one* relation.

Example 3.6.

The domain and range of the fuzzy relation $R(X, Y)$ specified in Fig. 3.2 are the sets X and Y, respectively. Hence, the relation is completely specified and onto. It is also normal (since $h(R) = 1$), but it is not a mapping since elements x_1 and x_3 of the domain appear more than once in R.

Given a discrimination level α as a real number in the closed interval $[0, 1]$, an α-cut can be defined for a fuzzy relation exactly as for a fuzzy set and can be denoted by R_α. Using the max operator for set unions, every fuzzy relation $R(X, Y)$ can be represented by its *resolution form*

$$R = \bigcup_\alpha \alpha R_\alpha, \qquad \alpha \in \Lambda_R \text{ (level set of } R\text{)}, \tag{3.4}$$

where αR_α is a fuzzy relation defined by

$$\mu_{\alpha R_\alpha}(x, y) = \alpha \cdot \mu_{R_\alpha}(x, y),$$

for every $(x, y) \in X \times Y$. In this way, it is possible to represent a fuzzy relation as the series of crisp relations comprising its α-cuts, each scaled by the value α.

Example 3.7.

Let a binary fuzzy relation R be defined by the following membership matrix:

$$\mathbf{M}_R = \begin{bmatrix} .7 & .4 & 0 \\ .9 & 1 & .4 \\ 0 & .7 & 1 \\ .7 & .9 & 0 \end{bmatrix}$$

The level set of R consists of five numbers: $\Lambda_R = \{0, .4, .7, .9, 1\}$. When $\alpha = 0$, the term αR_α in Eq. (3.4) represents the empty set and can thus be ignored. Hence, we obtain the following resolution form of R, where $\cup$ stands for the max operator:

$$
\mathbf{M}_R = .4 \begin{bmatrix} 1 & 1 & 0 \\ 1 & 1 & 1 \\ 0 & 1 & 1 \\ 1 & 1 & 0 \end{bmatrix} \cup .7 \begin{bmatrix} 1 & 0 & 0 \\ 1 & 1 & 0 \\ 0 & 1 & 1 \\ 1 & 1 & 0 \end{bmatrix} \cup .9 \begin{bmatrix} 0 & 0 & 0 \\ 1 & 1 & 0 \\ 0 & 0 & 1 \\ 0 & 1 & 0 \end{bmatrix} \cup 1 \begin{bmatrix} 0 & 0 & 0 \\ 0 & 1 & 0 \\ 0 & 0 & 1 \\ 0 & 0 & 0 \end{bmatrix}
$$

$$
\qquad\qquad R_{.4} \qquad\qquad\qquad R_{.7} \qquad\qquad\qquad R_{.9} \qquad\qquad\qquad R_1
$$

$$
= \begin{bmatrix} .4 & .4 & 0 \\ .4 & .4 & .4 \\ 0 & .4 & .4 \\ .4 & .4 & 0 \end{bmatrix} \cup \begin{bmatrix} .7 & 0 & 0 \\ .7 & .7 & 0 \\ 0 & .7 & .7 \\ .7 & .7 & 0 \end{bmatrix} \cup \begin{bmatrix} 0 & 0 & 0 \\ .9 & .9 & 0 \\ 0 & 0 & .9 \\ 0 & .9 & 0 \end{bmatrix} \cup \begin{bmatrix} 0 & 0 & 0 \\ 0 & 1 & 0 \\ 0 & 0 & 1 \\ 0 & 0 & 0 \end{bmatrix}
$$

The *inverse* of a crisp relation $R(X, Y)$ is written as $R^{-1}(X, Y)$ and is a subset of $Y \times X$ such that

$$
R^{-1}(X, Y) = \{(y, x) \mid (x, y) \in R\},
$$

where $x \in X$ and $y \in Y$. Clearly, dom $R(X, Y) = $ ran $R^{-1}(X, Y)$ and dom $R^{-1}(X, Y) = $ ran $R(X, Y)$. For a fuzzy relation $R(X, Y)$, the inverse fuzzy relation $R^{-1}(X, Y)$ is defined by

$$
\mu_{R^{-1}}(y, x) = \mu_R(x, y),
$$

for all $(x, y) \in X \times Y$. A membership matrix $\mathbf{M}_{R^{-1}}$ representing $R^{-1}(X, Y)$ is the transpose of the matrix $\mathbf{M}_R$ for $R(X, Y)$, that is, the rows of $\mathbf{M}_{R^{-1}}$ equal the columns of $\mathbf{M}_R$ and the columns of $\mathbf{M}_{R^{-1}}$ equal the rows of $\mathbf{M}_R$. Clearly,

$$
(R^{-1})^{-1} = R
$$

for any binary fuzzy relation.

Example 3.8.

Let $R(X, Y)$ be a fuzzy relation on $X = \{x, y, z\}$ and $Y = \{a, b\}$ such that

$$
\mathbf{M}_R = \begin{matrix} x \\ y \\ z \end{matrix} \begin{matrix} a \quad\ b \\ \begin{bmatrix} .3 & .2 \\ 0 & 1 \\ .6 & .4 \end{bmatrix} \end{matrix}
$$

The inverse of $R(X, Y)$ is then specified by the membership matrix

$$
\mathbf{M}_{R^{-1}} = \mathbf{M}_R^T = \begin{matrix} a \\ b \end{matrix} \begin{matrix} x \quad y \quad z \\ \begin{bmatrix} .3 & 0 & .6 \\ .2 & 1 & .4 \end{bmatrix} \end{matrix}
$$

where $\mathbf{M}_R^T$ denotes the transpose of $\mathbf{M}_R$.

Consider two crisp binary relations $P(X, Y)$ and $Q(Y, Z)$ defined with a common set Y. The *composition* of these two relations is denoted by

$$R(X, Z) = P(X, Y) \circ Q(Y, Z),$$

and is defined as a subset $R(X, Z)$ of $X \times Z$ such that $(x, z) \in R$ if and only if there exists at least one $y \in Y$ such that $(x, y) \in P$ and $(y, z) \in Q$. It follows from this definition of composition that the following three properties are satisfied for binary relations P, Q, R:

$$P \circ Q \neq Q \circ R,$$

$$(P \circ Q)^{-1} = Q^{-1} \circ P^{-1},$$

$$(P \circ Q) \circ R = P \circ (Q \circ R).$$

Just as the classical set operations such as union and intersection have a variety of generalizations for fuzzy sets, the composition operation for fuzzy relations can take several forms. The most common of these is the *max-min composition*. Denoted again by $P(X, Y) \circ Q(Y, Z)$, this operation is defined by

$$\mu_{P \circ Q}(x, z) = \max_{y \in Y} \min[\mu_P(x, y), \mu_Q(y, z)] \qquad (3.5)$$

for all $x \in X$ and $z \in Z$. This operation satisfies the same three properties just listed for the composition of crisp relations.

Example 3.9.

Consider the two binary relations $P(X, Y)$ and $Q(Y, Z)$ specified by their sagittal diagrams in Fig. 3.3(a). Let $R(X, Z) = P(X, Y) \circ Q(Y, Z)$. Then, $R(X, Z)$ is represented by the membership function specified in Fig. 3.3(b). For example,

$$\mu_R(1, \beta) = \max[\min(.7, .8), \min(.5, 1)] = \max[.7, .5] = .7,$$

$$\mu_R(4, \beta) = \max[\min(.4, 1), \min(.3, .9)] = \max[.4, .3] = .4.$$

Note that the pairs of $X \times Z$ that have nonzero membership grades in the composition $R = P \circ Q$ are those joined by lines intersecting at a common element of Y in Fig. 3.3(a).

If the composition for crisp relations $P(X, Y)$ and $Q(Y, Z)$ is thought of as representing the existence of a relational chain between elements of X and Z, then the max-min composition for fuzzy relations can be interpreted as indicating the strength of such a relational chain. This strength is represented by the membership grade of the pair (x, z) in the composition. The strength of each chain equals the strength of its weakest link and the strength of the relation between elements x and z is then the strength of the strongest chain between them.

An alternative common form of this operation on fuzzy relations is the *max-product composition*. It is denoted by $P(X, Y) \odot Q(Y, Z)$ and defined by

$$\mu_{P \odot Q}(x, z) = \max_{y \in Y}[\mu_P(x, y) \cdot \mu_Q(y, z)]$$

for all $x \in X$ and $z \in Z$. This operation also satisfies the three properties listed earlier.

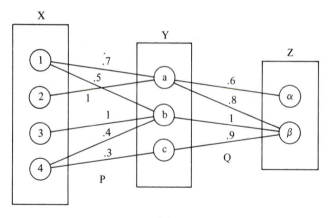

(a)

Composition: $R = P \circ Q$		
x	z	$\mu_R(x, z)$
1	α	.6
1	β	.7
2	α	.6
2	β	.8
3	β	1
4	β	.4

(b)

Join: $S = P * Q$			
x	y	z	$\mu_S(x, y, z)$
1	a	α	.6
1	a	β	.7
1	b	β	.5
2	a	α	.6
2	a	β	.8
3	b	β	1
4	b	β	.4
4	c	β	.3

(c)

Figure 3.3. Composition and join of binary relations (Examples 3.9 and 3.11).

Compositions of binary relations are conveniently performed in terms of membership matrices of the relations. Let $\mathbf{M}_P = [p_{i,k}]$, $\mathbf{M}_Q = [q_{k,j}]$ and $\mathbf{M}_R = [r_{i,j}]$ be membership matrices of binary relations such that $R = P \circ Q$. We can then write, using this matrix notation,

$$[r_{i,j}] = [p_{i,k}] \circ [q_{k,j}],$$

where

$$r_{i,j} = \max_k \min(p_{i,k}, q_{k,j}).$$

Observe that the same elements of $\mathbf{M}_P$ and $\mathbf{M}_Q$ are used in the calculation of $\mathbf{M}_R$ as would be used in the regular multiplication of matrices, but the product and sum operations are here replaced with the min and max operations, respectively; for the max-product composition, only the sum operation is replaced with the max operation. In general, the product and sum operations may be replaced with other appropriate operations to produce different forms of composition.

Example 3.10.

The following matrix equations illustrate the max-min composition and max-product composition for the same pair of binary relations:

$$\begin{bmatrix} .3 & .5 & .8 \\ 0 & .7 & 1 \\ .4 & .6 & .5 \end{bmatrix} \circ \begin{bmatrix} .9 & .5 & .7 & .7 \\ .3 & .2 & 0 & .9 \\ 1 & 0 & .5 & .5 \end{bmatrix} = \begin{bmatrix} .8 & .3 & .5 & .5 \\ 1 & .2 & .5 & .7 \\ .5 & .4 & .5 & .6 \end{bmatrix}$$

$$\begin{bmatrix} .3 & .5 & .8 \\ 0 & .7 & 1 \\ .4 & .6 & .5 \end{bmatrix} \odot \begin{bmatrix} .9 & .5 & .7 & .7 \\ .3 & .2 & 0 & .9 \\ 1 & 0 & .5 & .5 \end{bmatrix} = \begin{bmatrix} .8 & .15 & .4 & .45 \\ 1 & .14 & .5 & .63 \\ .5 & .2 & .28 & .54 \end{bmatrix}$$

A similar operation on two binary relations, which differs from the composition in that it yields triples instead of pairs, is known as the *relational join*. Let us denote it by $P * Q$. For crisp relations $P(X, Y)$ and $Q(Y, Z)$, it is defined as

$$P(X, Y) * Q(Y, Z) = \{(x, y, z) \mid (x, y) \in P \quad \text{and} \quad (y, z) \in Q\}$$

for each $x \in X$, $y \in Y$, and $z \in Z$. For fuzzy relations $P(X, Y)$ and $Q(Y, Z)$, the relational join corresponding to the max-min composition is defined by

$$\mu_{P*Q}(x, y, z) = \min[\mu_P(x, y), \mu_Q(y, z)] \tag{3.6}$$

for each $x \in X$, $y \in Y$, and $z \in Z$.

We can see that the join operation forms a ternary relation from two binary relations. This is a major difference from the operation of composition, which produces another binary relation. In fact, the max-min composition is obtained by aggregating appropriate elements of the corresponding join by the max operator. Formally,

$$\mu_{P \circ Q}(x, z) = \max \mu_{P*Q}(x, y, z), \tag{3.7}$$

for each $x \in X$ and $z \in Z$. Observe that although a composition can be determined from the corresponding join, the reverse determination cannot be made, that is, the join cannot be determined from the composition.

Example 3.11.

The join $S = P * Q$ of relations P and Q given in Fig. 3.3(a) is represented by the membership function specified in Fig. 3.3(c). To convert this join into the corresponding composition $R = P \circ Q$ (Example 3.9) by Eq. (3.7), the two indicated pairs of values of $\mu_S(x, y, z)$ in Fig. 3.3(c) are aggregated by the max operator. For instance,

$$\mu_R(1, \beta) = \max[\mu_S(1, a, \beta), \mu_S(1, b, \beta)]$$

$$= \max[.7, .5] = .7$$

The operation of relational join can be modified by replacing the min operator in Eq. (3.6) with other operators eligible for fuzzy set intersection, as discussed in Sec. 2.4. For example, when the min operator is replaced with the arithmetic product, we obtain a join operator corresponding to the max-product composition.

3.3 BINARY RELATIONS ON A SINGLE SET

In addition to defining a binary relation that exists between two different sets, it also is possible to define a crisp or fuzzy binary relation among the elements of a single set X. A binary relation of this type can be denoted by $R(X, X)$ or $R(X^2)$ and is a subset of $X \times X = X^2$. These relations are often referred to as *directed graphs* or *digraphs*.

Binary relations $R(X, X)$ can be expressed by the same forms as general binary relations (matrices, sagittal diagrams, tables). In addition, however, they can be conveniently expressed in terms of simple diagrams with the following properties: (1) each element of the set X is represented by a single node in the diagram; (2) directed connections between nodes indicate pairs of elements of X for which the grade of membership in R is nonzero; and (3) each connection in the diagram is labeled by the actual membership grade of the corresponding pair in R. An example of this diagram for a relation $R(X, X)$ defined on $X = \{1, 2, 3, 4\}$ is shown in Fig. 3.4, where it can be compared with the other forms of representation of binary relations.

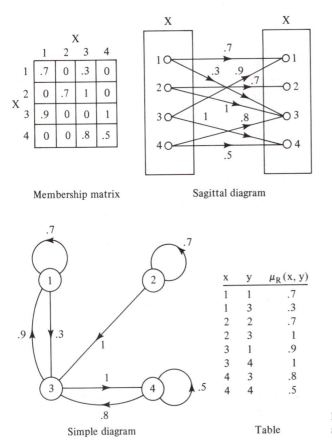

Figure 3.4. Forms of representation of a fuzzy relation $R(X, X)$.

Various significant types of relations $R(X, X)$ are distinguished on the basis of three different characteristic properties: reflexivity, symmetry and transitivity. First, let us consider crisp relations.

A crisp relation $R(X, X)$ is *reflexive* if and only if $(x, x) \in R$ for every $x \in X$, that is, if every element of X is related to itself. Otherwise, $R(X, X)$ is called *irreflexive*. If $(x, x) \not\in R$ for every $x \in X$, the relation is called *antireflexive*.

A crisp relation $R(X, X)$ is *symmetric* if and only if for every $(x, y) \in R$, it is also the case that $(y, x) \in R$, where $x, y, \in X$. Thus, whenever an element x is related to an element y through a symmetric relation, then y will also be related to x. If this is not the case for each $x, y \in X$, then the relation is called *asymmetric*. If $(x, y) \in R$ and $(y, x) \in R$ implies $x = y$, then the relation is called *antisymmetric*. If either $(y, x) \in R$ or $(x, y) \in R$, whenever $x \neq y$, then the relation is called *strictly antisymmetric*.

A crisp relation $R(X, X)$ is called *transitive* if and only if $(x, z) \in R$ whenever both $(x, y) \in R$ and $(y, z) \in R$ for at least one $y \in X$. In other words, the relation of x to y and of y to z implies the relation of x to z in a transitive relation. A relation that does not satisfy this property is called *nontransitive*. If $(x, z) \not\in R$ whenever both $(x, y) \in R$ and $(y, z) \in R$, then the relation is called *antitransitive*.

The properties of reflexivity, symmetry and transitivity are illustrated for crisp relations $R(X, X)$ in Fig. 3.5. We can readily see that these properties are preserved under inversion of the relation.

Example 3.12.

Let R be a crisp relation defined on $X \times X$, where X is the set of all university courses and R represents the relation "is a prerequisite of." R is antireflexive because a course is never a prerequisite of itself. Further, if one course is a prerequisite of another, the reverse will never be true. Therefore, R is antisymmetric. Finally, if a course is a prerequisite for a second course which is itself prerequisite for a third, then the first course is also a prerequisite for the third course. Thus, the relation R is transitive.

These three properties can be extended for fuzzy relations $R(X, X)$, by defining them in terms of the membership function of the relation. Thus, $R(X, X)$ is *reflexive* if and only if

$$\mu_R(x, x) = 1$$

for all $x \in X$. If this is not the case for some $x \in X$, then the relation is called *irreflexive*; if it is not satisfied for all $x \in X$, the relation is called *antireflexive*. A

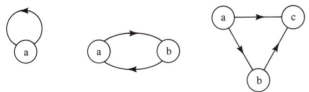

Reflexivity   Symmetry

Transitivity

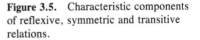

Figure 3.5. Characteristic components of reflexive, symmetric and transitive relations.

weaker form of reflexivity, referred to as ϵ-*reflexivity*, is sometimes defined by requiring that

$$\mu_R(x, x) \geq \epsilon,$$

where $0 < \epsilon < 1$.

A fuzzy relation is *symmetric* if and only if

$$\mu_R(x, y) = \mu_R(y, x)$$

for all $x, y \in X$. Whenever this equality is not satisfied for some $x, y \in X$, the relation is called *asymmetric*. If the equality fails to hold for all the members of the support of the relation, then the relation is called *antisymmetric*, and if it is not satisfied for all $x, y \in X$, then the relation is called *strictly antisymmetric*.

A fuzzy relation $R(X, X)$ is *transitive* (or, more specifically, *max-min transitive*) if and only if

$$\mu_R(x, z) \geq \max_{y \in Y} \min[\mu_R(x, y), \mu_R(y, z)]$$

is satisfied for each pair $(x, z) \in X^2$. A relation failing to satisfy this inequality for some members of X is called *nontransitive*, and if

$$\mu_R(x, z) < \max_{y \in Y} \min[\mu_R(x, y), \mu_R(y, z)],$$

for all $(x, z) \in X^2$, then the relation is called *antitransitive*.

Transitivity of fuzzy relations can be defined in various ways alternative to the form given above. A second common form is known as the *max-product transitivity* and is defined by

$$\mu_R(x, z) \geq \max_{y \in Y}[\mu_R(x, y) \cdot \mu_R(y, z)],$$

for all $(x, z) \in X^2$.

Example 3.13.

Let R be the fuzzy relation defined on the set of cities and representing the concept *very near*. We may assume that a city is certainly (i.e., to a degree of 1) very near to itself. The relation is therefore reflexive. Further, if city A is very near to city B, then B is certainly very near to A to the same degree. Therefore, the relation is also symmetric. Finally, if city A is very near to city B to some degree, say .7, and city B is very near to city C to some degree, say .8, it is possible (although not necessary) that city A is very near to city C to a smaller degree, say .5. Therefore, the relation is nontransitive.

By considering the three variants of reflexivity, four variants of symmetry, and three variants of transitivity defined for binary relations, we can distinguish $3 \times 4 \times 3 = 36$ different types of binary relations on a single set. Some of the most important of these types are discussed in Secs. 3.4 through 3.6; their names and properties are given in Fig. 3.6.

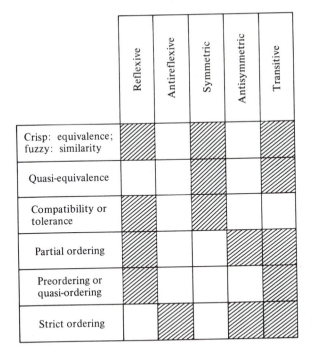

Figure 3.6.   Some important types of binary relations $R(X, X)$.

The *transitive closure* of a crisp relation $R(X, X)$ is defined as the relation that is transitive, contains $R(X, X)$, and has the fewest possible members. For fuzzy relations, this last requirement is generalized such that the elements of the transitive closure have the smallest possible membership grades that still allow the first two requirements to be met.

Given a relation $R(X, X)$, its transitive closure $R_T(X, X)$ can be determined by a simple algorithm that consists of the following three steps:

1. $R' = R \cup (R \circ R)$.
2. If $R' \neq R$, make $R = R'$ and go to Step 1.
3. Stop: $R' = R_T$.

This algorithm is applicable to both crisp and fuzzy relations. However, the type of composition and set union in Step 1 must be compatible with the definition of transitivity employed. When max-min composition and the max operator for set union are used, we call R_T the *transitive max-min closure*.

Example 3.14.

Using the algorithm just given, we can determine the transitive max-min closure $R_T(X, X)$ for a fuzzy relation $R(X, X)$ defined by the membership matrix

$$\mathbf{M}_R = \begin{bmatrix} .7 & .5 & 0 & 0 \\ 0 & 0 & 0 & 1 \\ 0 & .4 & 0 & 0 \\ 0 & 0 & .8 & 0 \end{bmatrix}$$

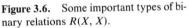

Applying Step 1 of the algorithm, we obtain

$$
\mathbf{M}_{R \circ R} =
\begin{bmatrix}
.7 & .5 & 0 & .5 \\
0 & 0 & .8 & 0 \\
0 & 0 & 0 & .4 \\
0 & .4 & 0 & 0
\end{bmatrix}
\quad
\mathbf{M}_{R \cup (R \circ R)} =
\begin{bmatrix}
.7 & .5 & 0 & .5 \\
0 & 0 & .8 & 1 \\
0 & .4 & 0 & .4 \\
0 & .4 & .8 & 0
\end{bmatrix}
= \mathbf{M}_{R'}
$$

Since $R' \neq R$, we take R' as a new relation R and, repeating the previous procedure, we obtain:

$$
\mathbf{M}_{R \circ R} =
\begin{bmatrix}
.7 & .5 & .5 & .5 \\
0 & .4 & .8 & .4 \\
0 & .4 & .4 & .4 \\
0 & .4 & .4 & .4
\end{bmatrix}
\quad
\mathbf{M}_{R \cup (R \circ R)} =
\begin{bmatrix}
.7 & .5 & .5 & .5 \\
0 & .4 & .8 & 1 \\
0 & .4 & .4 & .4 \\
0 & .4 & .8 & .4
\end{bmatrix}
= \mathbf{M}_{R'}
$$

Since $R' \neq R$ at this stage, we must again repeat the procedure with the new relation. If we do this, however, the last matrix does not change. Thus,

$$
\begin{bmatrix}
.7 & .5 & .5 & .5 \\
0 & .4 & .8 & 1 \\
0 & .4 & .4 & .4 \\
0 & .4 & .8 & .4
\end{bmatrix}
$$

is the membership matrix of the transitive closure R_T corresponding to the given relation $R(X, X)$.

3.4 EQUIVALENCE AND SIMILARITY RELATIONS

A crisp binary relation $R(X, X)$ that is reflexive, symmetric, and transitive is called an *equivalence relation*. For each element x in X, we can define a crisp set A_x, which contains all the elements of X that are related to x by the equivalence relation. Formally,

$$
A_x = \{ y \mid (x, y) \in R(X, X) \}.
$$

A_x is clearly a subset of X. The element x is itself contained in A_x due to the reflexivity of R; because R is transitive and symmetric, each member of A_x is related to all the other members of A_x. Furthermore, no member of A_x is related to any element of X not included in A_x. This set A_x is referred to as an *equivalence class* of $R(X, X)$ with respect to x. The members of each equivalence class can be considered to be equivalent to each other and only to each other under the relation R. The family of all such equivalence classes defined by the relation, which is usually denoted by X/R, forms a partition on X.

Example 3.15.

Let $X = \{1, 2, \ldots, 10\}$. The Cartesian product $X \times X$ contains 100 members: $(1, 1), (1, 2), (1, 3), \ldots, (10, 10)$. Let $R(X, X) = \{(x, y) \mid x$ and y have the same remainder when divided by 3$\}$. The relation is easily shown to be reflexive, symmetric, and transitive and is therefore an equivalence relation on X. The three equivalence classes defined by this relation are:

$$A_1 = A_4 = A_7 = A_{10} = \{1, 4, 7, 10\},$$

$$A_2 = A_5 = A_8 = \{2, 5, 8\},$$

$$A_3 = A_6 = A_9 = \{3, 6, 9\}.$$

Hence, in this example, $X/R = \{\{1, 4, 7, 10\}, \{2, 5, 8\}, \{3, 6, 9\}\}$.

A fuzzy binary relation that is reflexive, symmetric, and transitive is known as a *similarity relation*. While the max-min form of transitivity is assumed in the following discussion, the concepts can be generalized to the alternative definitions of fuzzy transitivity.

While an equivalence relation clearly groups elements that are equivalent under the relation into disjoint classes, the interpretation of a similarity relation can be approached in two different ways. First, the similarity relation can be considered to effectively group elements into crisp sets whose members are "similar" to each other to some specified degree. Obviously, when this degree is equal to 1, the grouping is then an equivalence class. Alternatively, however, we may wish to consider the degree of similarity that the elements of X have to some specified element $x \in X$. Thus, for each $x \in X$, a *similarity class* can be defined as a fuzzy set in which the membership grade of any particular element represents the similarity of that element to the element x. If all the elements in the class are similar to x to the degree of 1 and similar to all elements outside the set to the degree of 0, then the grouping again becomes an equivalence class. The following discussion briefly elaborates on each of these approaches in turn.

As mentioned in Sec. 3.2, any fuzzy relation can be represented by a resolution form given by Eq. (3.4). If the relation represented in this way is a similarity relation, then each α-cut R_α in Eq. (3.4) is an equivalence relation. Effectively then, we may take any similarity relation and, by taking an α-cut R_α for any value of α, create a crisp equivalence relation that represents the presence of similarity between the elements to the degree α. Each of these equivalence relations forms a partition. Let $\pi(R_\alpha)$ denote the partition corresponding to the equivalence relation R_α. Clearly, two elements x and y belong to the same block of the partition $\pi(R_\beta)$ if and only if $\mu_R(x, y) \geq \alpha$.

Each similarity relation is associated with the set

$$\pi(R) = \{\pi(R_\alpha) \mid \alpha \in \Lambda_\alpha\}$$

of partitions. These partitions are nested in the sense that $\pi(R_\alpha)$ is a refinement of $\pi(R_\beta)$ if and only if $\alpha \geq \beta$.

Example 3.16.

The fuzzy relation $R(X, X)$ represented by the membership matrix

$$
\begin{array}{c c c c c c c c}
 & a & b & c & d & e & f & g \\
\begin{array}{c} a \\ b \\ c \\ d \\ e \\ f \\ g \end{array} &
\left[\begin{array}{ccccccc}
1 & .8 & 0 & .4 & 0 & 0 & 0 \\
.8 & 1 & 0 & .4 & 0 & 0 & 0 \\
0 & 0 & 1 & 0 & 1 & .9 & .5 \\
.4 & .4 & 0 & 1 & 0 & 0 & 0 \\
0 & 0 & 1 & 0 & 1 & .9 & .5 \\
0 & 0 & .9 & 0 & .9 & 1 & .5 \\
0 & 0 & .5 & 0 & .5 & .5 & 1
\end{array}\right]
\end{array}
$$

is a similarity relation on $X = \{a, b, c, d, e, f, g\}$. The level set of R is $\Lambda_R = \{0, .4,$
$.5, .8, .9, 1\}$. Therefore, R is associated with a sequence of five nested partitions
$\pi(R_\alpha)$, for $\alpha \in \Lambda_R$ and $\alpha > 0$. Their refinement relationship can be conveniently
diagrammed by a *partition tree*, as shown in Fig. 3.7.

The equivalence classes formed by the levels of refinement of a similarity
relation can be interpreted as grouping elements that are similar to each other
and only to each other to a degree not less than α. Thus, in Example 3.16, c, e,
f, and g are all similar to each other to a degree of .5, but only c and e are similar
to each other to a degree of 1.

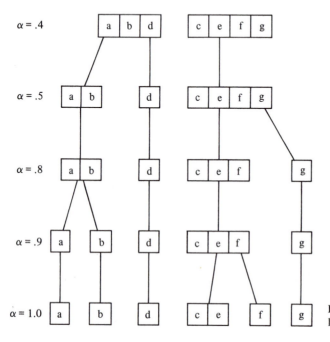

Figure 3.7. Partition tree for the similarity relation in Example 3.16.

Just as equivalence classes are defined by an equivalence relation, *similarity classes* are defined by a similarity relation. For a given similarity relation $R(X, X)$, the similarity class for each $x \in X$ is a fuzzy set in which the membership grade of each element $y \in X$ is simply the strength of that element's relation to x, or $\mu_R(x, y)$. Thus, the similarity class for an element x represents the degree to which all the other members of X are similar to x. Except in the restricted case of equivalence classes themselves, similarity classes are fuzzy and are therefore not generally disjoint.

Similarity classes are conveniently represented by membership matrices. Given a similarity relation R, the similarity class for each element is defined by the row of the membership matrix of R that corresponds to that element. For instance, the similarity classes for the element c and the element e of the similarity relation in Example 3.16 are equal. The relation, therefore, defines six different similarity classes.

Binary relations that are symmetric and transitive but not reflexive are usually referred to as *quasi-equivalence relations*. They are, however, of only marginal significance and we therefore omit a detailed discussion of them.

3.5 COMPATIBILITY OR TOLERANCE RELATIONS

A binary relation $R(X, X)$ that is reflexive and symmetric is usually called a *compatibility relation* or a *tolerance relation*. When $R(X, X)$ is a reflexive and symmetric fuzzy relation, it is sometimes called a *proximity relation*.

An important concept associated with compatibility relations are compatibility classes (also called tolerance classes). Given a crisp compatibility relation $R(X, X)$, a *compatibility class* is a subset A of X such that xRy for all $x, y \in A$. A *maximal compatibility class* or *maximal compatible* is a compatibility class that is not properly contained within any other compatibility class. The family consisting of all the maximal compatibles induced by R on X is called a *complete cover* of X with respect to R.

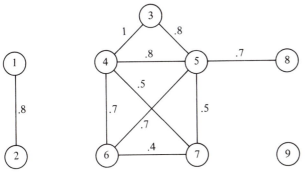

Figure 3.8. Graph of the compatibility relation in Example 3.17.

When R is a fuzzy compatibility relation, compatibility classes are defined in terms of a specified membership degree α. An α-*compatibility class* is a subset A_α of X such that $\mu_R(x, y) \geq \alpha$ for all $x, y \in A_\alpha$. *Maximal α-compatibles* and *complete α-cover* are obvious generalizations of the corresponding concepts for crisp ($\alpha = 1$) compatibility relations.

Compatibility relations are often conveniently viewed as *reflexive undirected graphs*. In this context, reflexivity implies that each node of the graph has a loop connecting the node to itself; the loops are usually omitted from the visual representations of the graphs, although they are assumed to be present. Connections between nodes, as defined by the relation, are not directed, since the property of symmetry guarantees that all existing connections appear in both directions. Each connection is labeled with the value of the corresponding membership grade $\mu_R(x, y) = \mu_R(y, x)$.

Example 3.17.

Consider a fuzzy relation $R(X, X)$ defined on $X = \mathbb{N}_9$ by the following membership matrix:

$$
\begin{array}{c|ccccccccc}
 & 1 & 2 & 3 & 4 & 5 & 6 & 7 & 8 & 9 \\
\hline
1 & 1 & .8 & 0 & 0 & 0 & 0 & 0 & 0 & 0 \\
2 & .8 & 1 & 0 & 0 & 0 & 0 & 0 & 0 & 0 \\
3 & 0 & 0 & 1 & 1 & .8 & 0 & 0 & 0 & 0 \\
4 & 0 & 0 & 1 & 1 & .8 & .7 & .5 & 0 & 0 \\
5 & 0 & 0 & .8 & .8 & 1 & .7 & .5 & .7 & 0 \\
6 & 0 & 0 & 0 & .7 & .7 & 1 & .4 & 0 & 0 \\
7 & 0 & 0 & 0 & .5 & .5 & .4 & 1 & 0 & 0 \\
8 & 0 & 0 & 0 & 0 & .7 & 0 & 0 & 1 & 0 \\
9 & 0 & 0 & 0 & 0 & 0 & 0 & 0 & 0 & 1 \\
\end{array}
$$

Since the matrix is symmetric and all entries on the main diagonal are equal to 1, the relation represented is reflexive and symmetric; it is, therefore, a compatibility relation. The graph of the relation is shown in Fig. 3.8; its complete α-covers for $\alpha > 0$ and $\alpha \in \Lambda_R = \{0, .4, .5, .7, .8, 1\}$ are depicted in Fig. 3.9.

The complete α-covers of compatibility relations $R(X, X)$ may, for some values of α, form partitions of X; in general, however, this is not the case due to the lack of transitivity. For example, the complete α-covers illustrated in Fig. 3.9 form partitions of $\mathbb{N}_9$ for $\alpha \geq .8$. It is obvious that similarity relations are special cases of compatibility relations for which all complete α-covers form partitions of X. Since the lack of transitivity distinguishes compatibility relations from similarity relations, the transitive closures of compatibility relations are similarity relations.

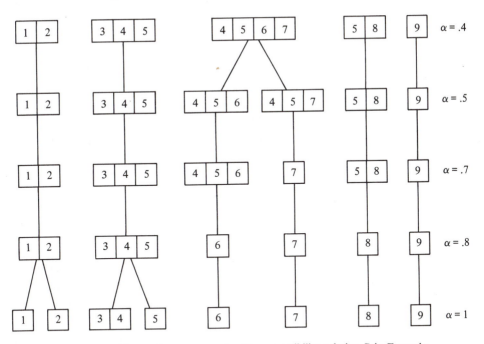

Figure 3.9. All complete α-covers for the compatibility relation R in Example 3.17.

3.6 ORDERINGS

While similarity and compatibility relations are characterized by symmetry, ordering relations require asymmetry (or antisymmetry) and transitivity. There are several types of ordering relations.

A crisp binary relation $R(X, X)$ that is reflexive, antisymmetric, and transitive is called a *partial ordering*. The common symbol $\leq$ is suggestive of the properties of this class of relations. Thus, $x \leq y$ denotes $(x, y) \in R$ and signifies that x *precedes* y. The inverse partial ordering $R^{-1}(X, X)$ is suggested by the symbol $\geq$. If $y \geq x$, indicating that $(y, x) \in R^{-1}$, then we say that y *succeeds* x. When $x \leq y$, x is also referred to as a *predecessor* of y, while y is called a *successor* of x. When $x \leq y$ and there is no z such that $x \leq z$ and $z \leq y$, then x is called an *immediate predecessor* of y, and y is called an *immediate successor* of x. If we need to distinguish several partial orderings, such as P, Q and R, we use the symbols $\overset{P}{\leq}$, $\overset{Q}{\leq}$, $\overset{R}{\leq}$, respectively.

Observe that a partial ordering $\leq$ on X does not guarantee that all pairs of elements x, y in X are comparable in the sense that either $x \leq y$ or $y \leq x$. Thus, for some x, $y \in X$, it is possible that x is neither a predecessor nor a successor of y. Such pairs are called *noncomparable* with respect to $\leq$.

The following are definitions of some fundamental concepts associated with partial orderings.

- If $x \in X$ and $x \leq y$ for every $y \in X$, then x is called the *first member* of X with respect to the relation denoted by $\leq$.
- If $x \in X$ and $y \leq x$ for every $y \in X$, then x is called the *last member* of X with respect to the partial ordering relation.
- If $x \in X$ and $y \leq x$ implies $x = y$, then x is called a *minimal member* of X with respect to the relation.
- If $x \in X$ and $x \leq y$ implies $x = y$, then x is called a *maximal member* of X with respect to the relation.

Using these concepts, every partial ordering satisfies the following properties:

1. There exists at most one first member and at most one last member.
2. There exists at least one maximal member and at least one minimal member.
3. If a first member exists, then only one minimal member exists and it is identical with the first member.
4. If a last member exists, then only one maximal member exists and it is identical with the last member.
5. The first and last members of a partial ordering relation correspond to the last and first members of the inverse partial ordering, respectively.

Let X again be a set on which a partial ordering is defined and let A be a subset of X ($A \subseteq X$). If $x \in X$ and $x \leq y$ for every $y \in A$, then x is called a *lower bound* of A on X with respect to the partial ordering. If $x \in X$ and $y \leq x$ for every $y \in A$, then x is called an *upper bound* of A on X with respect to the relation. If a particular lower bound succeeds every other lower bound of A, then it is called the *greatest lower bound*, or *infimum*, of A. If a particular upper bound precedes every other upper bound of A, then it is called the *least upper bound*, or *supremum*, of A.

A partial ordering on a set X that contains a greatest lower bound and a least upper bound for every subset of X is called a *lattice*.

A partial ordering $\leq$ on X is said to be *connected* if and only if for all x, $y \in X$, $x \neq y$ implies either $x \leq y$ or $y \leq x$. When a partial ordering is connected, then all pairs of elements of X are comparable by the ordering. Such an ordering is usually called a *linear ordering*; some alternative names used in the literature are *total ordering*, *simple ordering*, and *complete ordering*.

Every partial ordering on a set X can be conveniently represented by a diagram in which each element of X is expressed by a single node that is connected only to the nodes representing its immediate predecessors and immediate successors. The connections are directed in order to distinguish predecessors from successors; the arrow $\leftarrow$ indicates the inequality $\leq$. Diagrams of this sort are called *Hasse diagrams*.

Example 3.18.

Three crisp partial orderings P, Q, and R on the set $X = \{a, b, c, d, e\}$ are defined by their membership matrices (crisp) and their Hasse diagrams in Fig. 3.10. The underlined entries in each matrix indicate the relationship of immediate predecessor and successor that is employed in the corresponding Hasse diagram. P has no special properties, Q is a lattice, and R is an example of a lattice that represents a linear ordering.

A fuzzy binary relation R on a set X is a *fuzzy partial ordering* if and only if it is reflexive, antisymmetric, and transitive under some form of fuzzy transitivity. Any fuzzy partial ordering can be resolved into a series of crisp partial orderings in the same way in which this is done for similarity relations, that is, by taking a series of α-cuts that produce increasing levels of refinement.

When a fuzzy partial ordering is defined on a set X, then two fuzzy sets are associated with each element x in X. The first is called the *dominating class* of x. It is denoted by $R_{\geq[x]}$ and is defined by

$$\mu_{R_{\geq[x]}}(y) = \mu_R(x, y),$$

where $y \in X$. In other words, the dominating class of x contains the members of X to the degree to which they dominate x.

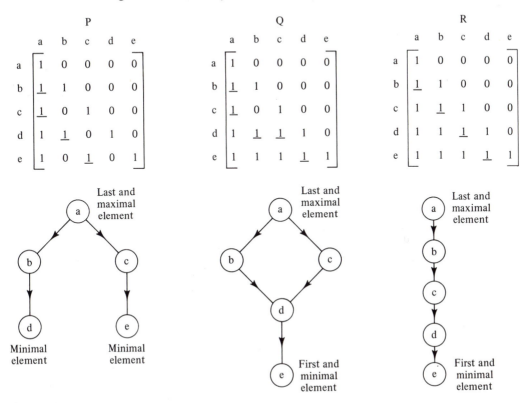

Figure 3.10.   Examples of partial orderings (Example 3.18).

The second fuzzy set of concern is the class *dominated* by x, which is denoted by $R_{\leq[x]}$ and defined by

$$\mu_{R_{\leq[x]}}(y) = \mu_R(y, x),$$

where $y \in X$. The class dominated by x contains the elements of X to the degree to which they are dominated by x.

An element $x \in X$ is *undominated* if and only if

$$\mu_R(x, y) = 0$$

for all $y \in X$ and $x \neq y$; an element x is *undominating* if and only if

$$\mu_R(y, x) = 0$$

for all $y \in X$ and $y \neq x$.

For a crisp subset A of a set X on which a fuzzy partial ordering R is defined, the *fuzzy upper bound* for A is the fuzzy set denoted by $U(R, A)$ and defined by

$$U(R, A) = \bigcap_{x \in A} R_{\geq[x]},$$

where $\cap$ denotes an appropriate fuzzy intersection. This definition reduces to that of the conventional upper bound when the partial ordering is crisp. If a *least upper bound* of the set A exists, it is the unique element x in $U(R, A)$ such that

$$\mu_{U(R,A)}(x) > 0 \quad \text{and} \quad \mu_R(x,y) > 0,$$

for all elements y in the support of $U(R, A)$.

Example 3.19.

The following membership matrix defines a fuzzy partial ordering R on the set $X = \{a, b, c, d, e\}$:

$$
\begin{array}{c}
 \\
a \\
b \\
c \\
d \\
e
\end{array}
\begin{array}{c}
\begin{array}{ccccc}
a & b & c & d & e
\end{array} \\
\left[
\begin{array}{ccccc}
1 & .7 & 0 & 1 & .7 \\
0 & 1 & 0 & .9 & 0 \\
.5 & .7 & 1 & 1 & .8 \\
0 & 0 & 0 & 1 & 0 \\
0 & .1 & 0 & .9 & 1
\end{array}
\right]
\end{array}
$$

The dominating class for each element is given by the row of the matrix corresponding to that element. The columns of the matrix give the dominated class for each element. Under this ordering, the element d is undominated and the element c is undominating. For the subset $A = \{a, b\}$, the upper bound is the fuzzy set produced by the intersection of the dominating classes for a and b. Employing the min operator for fuzzy intersection, we obtain

$$U(R, \{a, b\}) = .7/b + .9/d.$$

The unique least upper bound for the set A is the element b.

Several other concepts of crisp orderings generalize easily to the fuzzy case. A *fuzzy preordering* is a fuzzy relation that is reflexive and transitive. Unlike a partial ordering, the preordering is not necessarily antisymmetric.

A *fuzzy weak ordering* R is an ordering satisfying all the properties of a fuzzy linear ordering except antisymmetry. Alternatively, it can be thought of as a fuzzy preordering in which either $\mu_R(x, y) > 0$ or $\mu_R(y, x) > 0$ for all $x \neq y$. A *fuzzy strict ordering* is antireflexive, antisymmetric, and transitive; it can clearly be derived from any partial ordering R by replacing the values $\mu_R(x, x) = 1$ with zeros for all $x \in X$.

3.7 MORPHISMS

If two crisp binary relations $R(X, X)$ and $Q(Y, Y)$ are defined on sets X and Y respectively, then a function $h: X \rightarrow Y$ is said to be a *homomorphism* from (X, R) to (Y, Q) if

$$(x_1, x_2) \in R \quad \text{implies} \quad (h(x_1), h(x_2)) \in Q,$$

for all $x_1, x_2 \in X$. In other words, a homomorphism implies that for every two elements of set X that are related under the relation R, their homomorphic images $h(x_1)$, $h(x_2)$ in set Y are related under the relation Q.

When $R(X, X)$ and $Q(Y, Y)$ are fuzzy binary relations, this implication can be generalized to

$$\mu_R(x_1, x_2) \leq \mu_Q[h(x_1), h(x_2)],$$

for all $x_1, x_2 \in X$ and their images $h(x_1)$, $h(x_2) \in Y$. Thus, the strength of relation between two elements under R is equaled or exceeded by the strength of relation between their homomorphic images under Q.

Note that it is possible for a relation to exist under Q between the homomorphic images of two elements that are themselves unrelated under R. When this is never the case under a homomorphic function h, then h is called a *strong homomorphism*. It satisfies the two implications

$$(x_1, x_2) \in R \quad \text{implies} \quad (h(x_1), h(x_2)) \in Q$$

for all $x_1, x_2 \in X$, and

$$(y_1, y_2) \in Q \quad \text{implies} \quad (x_1, x_2) \in R$$

for all $y_1, y_2 \in Y$, where $x_1 \in h^{-1}(y_1)$ and $x_2 \in h^{-1}(y_2)$. Observe that when h is many-to-one, the inverse of h for each element of Y is a set of elements from X instead of a single element of X.

If relations $R(X, X)$ and $Q(Y, Y)$ are fuzzy, then the criteria that a many-to-one function h must satisfy in order to be a strong homomorphism are somewhat modified. The function h imposes a partition π_h on the set X such that any two elements $x_1, x_2 \in X$ belong to the same block of the partition if and only if h maps them to the same element of Y. Let $A = \{a_1, a_2, \ldots, a_n\}$ and $B = \{b_1, b_2, \ldots, b_n\}$ be two blocks of this partition π_h and let all elements of A be mapped to some element $y_1 \in Y$ and all elements of B be mapped to some element $y_2 \in Y$. Then the function h is said to be a strong homomorphism from (X, R) to (Y, Q) if and only if the degree of the strongest relation between any element of A and any element of B in the fuzzy relation R equals the strength of the relation between y_1 and y_2 in the fuzzy relation Q. Formally,

$$\max_{i,j} \mu_R(a_i, b_j) = \mu_Q(y_1, y_2).$$

This equality must be satisfied for each pair of blocks of the partition π_h.

Example 3.20.

The following membership matrices represent fuzzy relations $R(X, X)$ and $Q(Y, Y)$ defined on sets $X = \{a, b, c, d\}$ and $Y = \{\alpha, \beta, \gamma\}$, respectively:

	a	b	c	d
a	0	.5	0	0
b	0	0	.9	0
c	1	0	0	.5
d	0	.6	0	0

	α	β	γ
α	.5	.9	0
β	1	0	.9
γ	1	.9	0

The second relation, Q, is the homomorphic image of R under the homomorphic function h, which maps elements a and b to element α, element c to element β, and element d to element γ. Note that the pair (γ, β) in Q does not correspond to any pair in the relation R. Figure 3.11(a) illustrates this fuzzy homomorphism.

Consider now two relations $R(X, X)$ and $Q(Y, Y)$ defined on $X = \{a, b, c, d, e, f\}$ and $Y = \{\alpha, \beta, \gamma\}$, respectively, which are represented by the following membership matrices:

	a	b	c	d	e	f
a	.8	.4	0	0	0	0
b	0	.5	0	.7	0	0
c	0	0	.3	0	0	0
d	0	.5	0	0	.9	.5
e	0	0	0	1	0	0
f	0	0	0	0	1	.8

	α	β	γ
α	.7	0	.9
β	.4	.8	0
γ	1	0	1

In this case the second relation, Q, is a strong homomorphic image of R under the homomorphic function h, which maps element a to element β, elements b, c, and d to element α, and elements e and f to element γ. The set X is therefore partitioned into the three equivalence classes $\{a\}$, $\{b, c, d\}$, and $\{e, f\}$. Figure 3.11(b) depicts this strong homomorphism.

If a homomorphism exists from (X, R) to (Y, Q), then the relation $Q(Y, Y)$ preserves some of the properties of the relation $R(X, X)$—namely, that all the pairs $(x_1, x_2) \in X \times X$ that are members of R have corresponding homomorphic images $(h(x_1), h(x_2)) \in Y \times Y$ that are members of Q. Other members of Q may exist, however, that are not the counterparts of any member of R. This is not the case when the homomorphism is strong. Here more properties of the relation R are preserved in the relation Q. In fact, Q represents a simplification of R in which elements belonging to the same block of the partition π_h created by the function h on the set X are no longer distinguished. These functions are useful for performing various kinds of simplifications of systems that preserve desirable properties in sets, such as ordering or similarity.

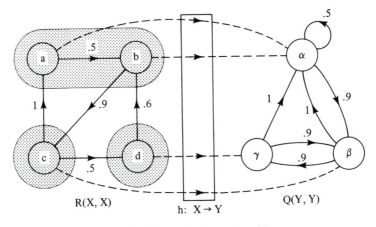

(a) Ordinary fuzzy homomorphism

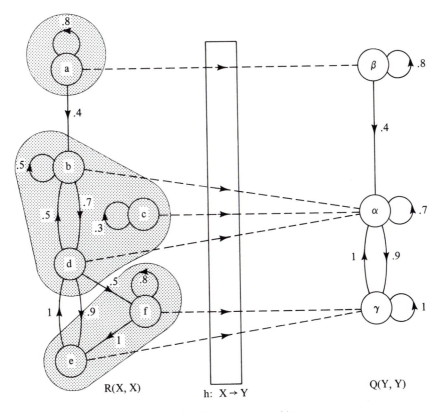

(b) Strong fuzzy homomorphism

Figure 3.11.   Fuzzy homomorphisms (Example 3.20).

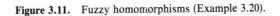

If $h : X \rightarrow Y$ is a homomorphism from (X, R) to (Y, Q) and if h is completely specified, one-to-one, and onto, then it is called an *isomorphism*. This is effectively a translation or direct relabeling of elements of the set X into elements of the set Y that preserves all the properties of R in Q. If $Y \subseteq X$, then h is called an *endomorphism*. A function that is both an isomorphism and an endomorphism is called an *automorphism*. In this case, the function maps the set X to itself and the relations R and Q are equal. Each of these terms applies without modification to fuzzy homomorphisms.

The notion of a crisp or fuzzy homomorphism can be generalized from binary relations on a single set to relations on two or more different sets. For instance, a homomorphism between two relations such as $R_1(X_1, Y_1)$ and $R_2(X_2, Y_2)$ would consist of two functions, $h_1 : X_1 \rightarrow X_2$ and $h_2 : Y_1 \rightarrow Y_2$. The properties described earlier must then be satisfied by each of these functions.

3.8 FUZZY RELATION EQUATIONS

The notion of fuzzy relation equations is associated with the concept of composition of binary relations. As explained in Sec. 3.2, the composition of two fuzzy binary relations $P(X, Y)$ and $Q(Y, Z)$ can be defined, in general, in terms of an operation on the membership matrices of P and Q that resembles matrix multiplication. This operation involves exactly the same combinations of matrix entries as in the regular matrix multiplication. However, the multiplications and additions that are applied to these combinations in the matrix multiplication are replaced with other operations; these alternative operations represent, in each given context, the appropriate operations of fuzzy set intersection and union, respectively. In the max-min composition, for example, the multiplications and additions are replaced with the min and max operations, respectively.

For the sake of simplicity and clarity, let our further discussion in this section be limited to the max-min form of composition. This form is not only viewed as the most fundamental composition of fuzzy relations, but it is also the form that has been studied the most extensively and that has been utilized in numerous applications.

Consider three fuzzy binary relations $P(X, Y)$, $Q(Y, Z)$, and $R(X, Z)$, which are defined on the sets

$$X = \{x_i \mid i \in I\}, \ Y = \{y_j \mid j \in J\}, \ Z = \{z_k \mid k \in K\},$$

where we assume that $I = \mathbb{N}_n$, $J = \mathbb{N}_m$, and $K = \mathbb{N}_s$. Let the membership matrices of P, Q, and R be denoted by

$$\mathbf{P} = [p_{i,j}], \quad \mathbf{Q} = [q_{j,k}], \quad \mathbf{R} = [r_{i,k}],$$

respectively, where

$$p_{i,j} = \mu_P(x_i, y_j), \quad q_{j,k} = \mu_Q(y_j, z_k), \quad r_{i,k} = \mu_R(x_i, z_k),$$

for all $i \in I \ (= \mathbb{N}_n)$, $j \in J \ (= \mathbb{N}_m)$ and $k \in K \ (= \mathbb{N}_s)$. This means that all entries in the matrices $\mathbf{P}$, $\mathbf{Q}$, and $\mathbf{R}$ are real numbers in the unit interval $[0, 1]$.

Assume now that the three relations constrain each other in such a way that

$$\mathbf{P} \circ \mathbf{Q} = \mathbf{R}, \tag{3.8}$$

where $\circ$ denotes the max-min composition. This means that

$$\max_{j \in J} \min (p_{i,j}, q_{j,k}) = r_{i,k}, \tag{3.9}$$

for all $i \in I$ and $k \in K$. That is, the matrix equation (3.8) encompasses $n \times s$ simultaneous equations of the form of Eq. (3.9). When two of the components in each of the equations are given and one is unknown, these equations are referred to as *fuzzy relation equations*.

When matrices $\mathbf{P}$ and $\mathbf{Q}$ are given and matrix $\mathbf{R}$ is to be determined from Eq. (3.8), the problem is trivial. It is solved simply by performing the max-min multiplication-like operation on $\mathbf{P}$ and $\mathbf{Q}$, as defined by Eq. (3.9). Clearly, the solution in this case exists and is unique. The problem becomes far from trivial when one of the two matrices on the left-hand side of Eq. (3.8) is unknown. In this case, the solution is neither guaranteed to exist nor to be unique.

Since $\mathbf{R}$ in Eq. (3.8) is obtained by composing $\mathbf{P}$ and $\mathbf{Q}$, it is suggestive to view the problem of determining $\mathbf{P}$ (or, alternatively, $\mathbf{Q}$) from $\mathbf{R}$ and $\mathbf{Q}$ (or, alternatively, $\mathbf{R}$ and $\mathbf{P}$) as a *decomposition* of $\mathbf{R}$ with respect to $\mathbf{Q}$ (or, alternatively, with respect to $\mathbf{P}$). Since many problems in various contexts can be formulated as problems of decomposition, the utility of any method for solving Eq. (3.8) is quite high. The use of fuzzy relation equations in some applications is illustrated in Chap. 6.

In the following discussion, let us assume that a pair of specific matrices $\mathbf{R}$ and $\mathbf{Q}$ from Eq. (3.8) is given and that we wish to determine the set of all particular matrices of the form $\mathbf{P}$ that satisfy Eq. (3.8). Let each particular matrix $\mathbf{P}$ that satisfies Eq. (3.8) be called its *solution* and let

$$S(\mathbf{Q}, \mathbf{R}) = \{\mathbf{P} \mid \mathbf{P} \circ \mathbf{Q} = \mathbf{R}\} \tag{3.10}$$

denote the set of all solutions (the *solution set*).

It is easy to see that this problem can be decomposed, without loss of generality, into a set of simpler problems expressed by matrix equations

$$\mathbf{p}_i \circ \mathbf{Q} = \mathbf{r}_i \tag{3.11}$$

for all $i \in I$, where

$$\mathbf{p}_i = [p_{i,j} \mid j \in J] \quad \text{and} \quad \mathbf{r}_i = [r_{i,k} \mid k \in K].$$

Indeed, each of the equations in (3.9) contains unknowns $p_{i,j}$ identified only by one particular value of the index i; that is, the unknowns $p_{i,j}$ distinguished by different values of i do not appear together in any of the individual equations. Observe that $\mathbf{p}_i$, $\mathbf{Q}$, and $\mathbf{r}_i$ in Eq. (3.11) represent, respectively, a fuzzy set on Y, a fuzzy relation on $Y \times Z$, and a fuzzy set on Z.

Let

$$S_i(\mathbf{Q}, \mathbf{r}_i) = \{\mathbf{p}_i \mid \mathbf{p}_i \circ \mathbf{Q} = \mathbf{r}_i\} \tag{3.12}$$

denote, for each $i \in I$, the solution set of one of the simpler problems expressed by Eq. (3.11). Then, the matrices **P** in Eq. (3.10) can be viewed as one-column matrices

$$\mathbf{P} = \begin{bmatrix} \mathbf{p}_1 \\ \mathbf{p}_2 \\ \vdots \\ \mathbf{p}_n \end{bmatrix}$$

where $\mathbf{p}_i \in S_i(\mathbf{Q}, \mathbf{r}_i)$ for all $i \in I\ (= \mathbb{N}_n)$.

It follows immediately from Eq. (3.9) that if

$$\max_{j \in J} q_{j,k} < r_{i,k} \tag{3.13}$$

for some $i \in I$ and some $k \in K$, then no values $p_{i,j} \in [0, 1]$ exist ($j \in J$) that satisfy Eq. (3.8) and, therefore, no matrix **P** exists that satisfies the matrix equation. This proposition can be stated more concisely as follows: if

$$\max_{j \in J} q_{j,k} < \max_{i \in I} r_{i,k} \tag{3.14}$$

for some $k \in K$, then $S(\mathbf{Q}, \mathbf{R}) = \varnothing$. This proposition allows us, in certain cases, to determine quickly that Eq. (3.8) has no solution; its negation, however, is only a necessary and not a sufficient condition for the existence of a solution of Eq. (3.8), that is, for $S(\mathbf{Q}, \mathbf{R}) \neq \varnothing$.

Example 3.21.

Consider the matrix equation

$$\begin{bmatrix} p_{11} & p_{12} & p_{13} \\ p_{21} & p_{22} & p_{23} \end{bmatrix} \circ \begin{bmatrix} .9 & .5 \\ .7 & .8 \\ 1 & .4 \end{bmatrix} = \begin{bmatrix} .6 & .3 \\ .2 & 1 \end{bmatrix}$$

whose general form is

$$[p_{i,j}] \circ [q_{j,k}] = [r_{i,k}],$$

where $i \in \mathbb{N}_2, j \in \mathbb{N}_3$, and $k \in \mathbb{N}_2$. The first matrix in this equation is unknown, and our problem is to determine all particular configurations of its entries for which the equation is satisfied.

The given matrix represents the following four equations of the form of Eq. (3.9):

$$\max[\min(p_{11}, .9), \min(p_{12}, .7), \min(p_{13}, 1)] = .6, \tag{a}$$

$$\max[\min(p_{11}, .5), \min(p_{12}, .8), \min(p_{13}, .4)] = .3, \tag{b}$$

$$\max[\min(p_{21}, .9), \min(p_{22}, .7), \min(p_{23}, 1)] = .2, \tag{c}$$

$$\max[\min(p_{21}, .5), \min(p_{22}, .8), \min(p_{23}, .4)] = 1. \tag{d}$$

Observe that equations (a) and (b) contain only unknowns p_{11}, p_{12}, and p_{13}, whereas equations (c) and (d) contain only unknowns p_{21}, p_{22}, and p_{23}. This means that each of these pairs of equations can be solved independently of the other. Hence, as previously argued, the given matrix equation can be decomposed into two simpler matrix equations,

$$[p_{11}\ p_{12}\ p_{13}] \circ \begin{bmatrix} .9 & .5 \\ .7 & .8 \\ 1 & .4 \end{bmatrix} = [.6\ .3]$$

and

$$[p_{21}\ p_{22}\ p_{23}] \circ \begin{bmatrix} .9 & .5 \\ .7 & .8 \\ 1 & .4 \end{bmatrix} = [.2\ 1]$$

We see, however, that the second equation, which is associated with $i = 2$, satisfies the inequality of Eq. (3.13) for $k = 2$ (and $i = 2$) and, therefore, has no solution. In fact,

$$\max(q_{12},\ q_{22},\ q_{32}) < r_{22},$$

or, specifically,

$$\max(.5,\ .8,\ .4) < 1.$$

Thus, the given matrix equation has no solution.

Since Eq. (3.8) can be decomposed without loss of generality into Eq. (3.11), we need only methods for solving equations of the latter form in order to arrive at a solution. We may, therefore, restrict our further discussion in this section to matrix equations of the form

$$\mathbf{p} \circ \mathbf{Q} = \mathbf{r}, \tag{3.15}$$

where

$$\mathbf{p} = [p_j \,|\, j \in J], \quad \mathbf{Q} = [q_{j,k} \,|\, j \in J, k \in K], \quad \mathbf{r} = [r_k \,|\, k \in K].$$

For the sake of consistency with our previous discussion, let us again assume that $\mathbf{p}$, $\mathbf{Q}$, and $\mathbf{r}$ represent, respectively, a fuzzy set on Y, a fuzzy relation on $Y \times Z$, and a fuzzy set on Z. Moreover, let $J = \mathbb{N}_n$ and $K = \mathbb{N}_s$ and let

$$S(\mathbf{Q}, \mathbf{r}) = \{\mathbf{p} \,|\, \mathbf{p} \circ \mathbf{Q} = \mathbf{r}\} \tag{3.16}$$

denote the solution set of Eq. (3.15).

In order to describe a method for solving Eq. (3.15), we need to introduce some additional concepts and convenient notation. First, let $\mathscr{P}$ denote the set of all possible vectors

$$\mathbf{p} = [p_j \,|\, j \in J]$$

such that $p_j \in [0, 1]$ for all $j \in J$ and let a partial ordering on $\mathscr{P}$ be defined as follows: for any pair $^1\mathbf{p}$, $^2\mathbf{p} \in \mathscr{P}$,

$$^1\mathbf{p} \leq {}^2\mathbf{p} \quad \text{if and only if} \quad {}^1p_j \leq {}^2p_j$$

for all $j \in J$. Given an arbitrary pair $^1\mathbf{p}$, $^2\mathbf{p} \in \mathscr{P}$ such that $^1\mathbf{p} \leq {}^2\mathbf{p}$, let

$$\langle {}^1\mathbf{p}, {}^2\mathbf{p} \rangle = \{\mathbf{p} \in \mathscr{P} \,|\, {}^1\mathbf{p} \leq \mathbf{p} \leq {}^2\mathbf{p}\}.$$

For any pair $^1\mathbf{p}$, $^2\mathbf{p} \in \mathscr{P}$, $(\langle {}^1\mathbf{p}, {}^2\mathbf{p} \rangle, \leq)$ is clearly a lattice.

Consider now some properties of the solution set $S(\mathbf{Q}, \mathbf{r})$. Employing the partial ordering on $\mathscr{P}$, let an element $\hat{\mathbf{p}}$ of $S(\mathbf{Q}, \mathbf{r})$ be called a *maximal solution* of Eq. (3.15) if, for all $\mathbf{p} \in S(\mathbf{Q}, \mathbf{r})$, $\mathbf{p} \geq \hat{\mathbf{p}}$ implies $\mathbf{p} = \hat{\mathbf{p}}$; similarly, let an element $\check{\mathbf{p}}$ of $S(\mathbf{Q}, \mathbf{r})$ be called a *minimal solution* of Eq. (3.15) if, for all $\mathbf{p} \in S(\mathbf{Q}, \mathbf{r})$, $\mathbf{p} \leq \check{\mathbf{p}}$ implies $\mathbf{p} = \check{\mathbf{p}}$.

It is well established that whenever the solution set $S(\mathbf{Q}, \mathbf{r})$ is not empty, it always contains a *unique maximum solution*, $\hat{p}$, and it may contain several minimal solutions. Let $\check{S}(\mathbf{Q}, \mathbf{r})$ denote the set of all minimal solutions. It is known that the solution set $S(\mathbf{Q}, \mathbf{r})$ is fully characterized by the maximum and minimal solutions in the following sense: it consists exactly of the maximum solution $\hat{p}$, all the minimal solutions, and all elements of $\mathscr{P}$ that are between $\hat{p}$ and each of the minimal solutions. Formally,

$$S(\mathbf{Q}, \mathbf{r}) = \bigcup_{\check{\mathbf{p}}} \langle \check{\mathbf{p}}, \hat{\mathbf{p}} \rangle, \tag{3.17}$$

where the union is taken for all $\check{\mathbf{p}} \in \check{S}(\mathbf{Q}, \mathbf{r})$. For a quick orientation, the meaning of Eq. (3.17) is illustrated visually in Fig. 3.12.

Equation (3.17) enables us to solve Eq. (3.15) solely by determining its unique maximum solution $\hat{\mathbf{p}}$ and the set $\check{S}(\mathbf{Q}, \mathbf{r})$ of its minimal solutions.

When $S(\mathbf{Q}, \mathbf{r}) \neq \varnothing$, the maximum solution

$$\hat{\mathbf{p}} = (\hat{p}_j \mid j \in J)$$

of Eq. (3.15) is determined as follows:

$$\hat{p}_j = \min_k \sigma(q_{j,k}, r_k), \tag{3.18}$$

where

$$\sigma(q_{j,k}, r_k) = \begin{cases} r_k & \text{if } q_{j,k} > r_k, \\ 1 & \text{otherwise.} \end{cases}$$

When $\hat{\mathbf{p}}$ determined in this way does not satisfy Eq. (3.15), then $S(\mathbf{Q}, \mathbf{r}) = \varnothing$. That is, the existence of the maximum solution $\hat{\mathbf{p}}$, as determined by Eq. (3.18), is a necessary and sufficient condition for $S(\mathbf{Q}, \mathbf{r}) \neq \varnothing$.

Once $\hat{\mathbf{p}}$ is determined by Eq. (3.18), we must check to see if it satisfies the given matrix equation (3.15). If it does not, then the equation has no solution ($S(\mathbf{Q}, \mathbf{r}) = \varnothing$). Otherwise, $\hat{\mathbf{p}}$ is the maximum solution of the equation, and we next determine the set $\check{S}(\mathbf{Q}, \mathbf{r})$ of its minimal solutions.

The method we describe here for determining all minimal solutions of Eq. (3.15) is based on the assumption that the components of the vector $\mathbf{r}$ in Eq. (3.15) are ordered such that $r_1 > r_2 > \cdots > r_s$. If the components are not initially ordered in this way, we permute them appropriately and perform the same permutation on the columns of the matrix $\mathbf{Q}$. This procedure clearly yields an equivalent matrix equation, which has exactly the same set of solutions as the original equation.

Assume now that $\mathbf{Q}$ and $\mathbf{r}$ of Eq. (3.15) are given and that we wish to determine the set $\check{S}(\mathbf{Q}, \mathbf{r})$ of all minimal solutions of the equation. Assume further that components of $\mathbf{r}$ are arranged in decreasing order and that $\hat{\mathbf{p}}$ has been determined by Eq. (3.18) and has been verified as the maximum solution of Eq.

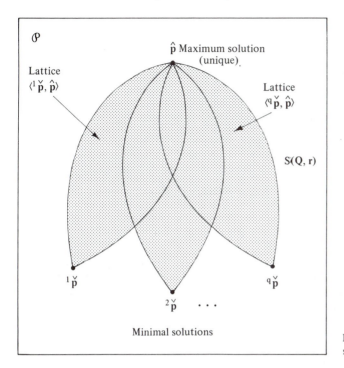

Figure 3.12.   Structure of the solution set (shaded area) of Eq. (3.17).

(3.15). At this point, Eq. (3.15) may be reduced in some cases. When $\hat{p}_j = 0$ for some $j \in J$, we may eliminate this component from $\hat{\mathbf{p}}$ as well as the jth row from matrix $\mathbf{Q}$, since, clearly, $\hat{p}_j = 0$ implies $p_j = 0$ for each $\mathbf{p} \in S(\mathbf{Q}, \mathbf{r})$. Furthermore, when $r_k = 0$ for some $k \in K$, we may eliminate this component from $\mathbf{r}$ and the kth column from matrix $\mathbf{Q}$, since each $\mathbf{p} \leq \hat{\mathbf{p}}$ ($\mathbf{p} \in \mathcal{P}$) must satisfy in this case the max-min equation represented by $\mathbf{p}$, the kth column of $\mathbf{Q}$, and $r_k = 0$. Although this reduction is not necessary, the reduced equation is easier to deal with. When we obtain solutions of the reduced equation, we simply extend them by inserting zeros at the locations that were eliminated in the reduction step.

For convenience, assume for our further discussion that Eq. (3.15) is a reduced equation and $\hat{\mathbf{p}}$ is its maximum solution (i.e., $\hat{p}_j \neq 0$ for all $j \in J = \mathbb{N}_m$ and $r_k \neq 0$ for all $k \in K = \mathbb{N}_s$). Given $\mathbf{Q}$, $\mathbf{r}$, and $\hat{\mathbf{p}}$, the set $\check{S}(\mathbf{Q}, \mathbf{r})$ of all minimal solutions of Eq. (3.15) can be determined by the following procedure:*

1. Determine the sets

$$J_k(\hat{\mathbf{p}}) = \{j \in J \mid \min(\hat{p}_j, q_{j,k}) = r_k\}$$

for all $k \in K$ and construct their Cartesian product

$$J(\hat{\mathbf{p}}) = \underset{k \in K}{\times} J_k(\hat{\mathbf{p}}).$$

* The procedure is based on several nontrivial theorems, which we do not consider essential to present here. Appropriate references are given in Note 3.3.

Denote elements (s-tuples) of $J(\hat{\mathbf{p}})$ by

$$\beta = (\beta_k \mid k \in K).$$

2. For each $\beta \in J(\hat{\mathbf{p}})$ and each $j \in J$, determine the set

$$K(\beta, j) = \{k \in K \mid \beta_k = j\}.$$

3. For each $\beta \in J(\hat{\mathbf{p}})$, generate the m-tuple

$$g(\beta) = (g_j(\beta) \mid j \in J)$$

by taking

$$g_j(\beta) = \begin{cases} \max_{k \in K(\beta, j)} r_k & \text{if } K(\beta, j) \neq \varnothing, \\ \\ 0 & \text{otherwise.} \end{cases}$$

4. From all the m-tuples $g(\beta)$ generated in Step 3, select only the minimal ones by pairwise comparison. The resulting set of m-tuples is the set $\check{S}(\mathbf{Q}, \mathbf{r})$ of all minimal solutions of Eq. (3.15).

Example 3.22.

Given

$$\mathbf{Q} = \begin{bmatrix} .1 & .4 & .5 & .1 \\ .9 & .7 & .2 & 0 \\ .8 & 1 & .5 & 0 \\ .1 & .3 & .6 & 0 \end{bmatrix} \quad \text{and} \quad \mathbf{r} = [.8, .7, .5, 0],$$

determine all solutions of Eq. (3.15).

First, we determine the maximum solution $\hat{\mathbf{p}}$ by Eq. (3.18):

$$\hat{p}_1 = \min(1, 1, 1, 0) = 0,$$

$$\hat{p}_2 = \min(.8, 1, 1, 1) = .8,$$

$$\hat{p}_3 = \min(1, .7, 1, 1) = .7,$$

$$\hat{p}_4 = \min(1, 1, .5, 1) = .5.$$

Thus, $\hat{\mathbf{p}} = [0, .8, .7, .5]$. We can easily check that $\hat{\mathbf{p}} \in S(\mathbf{Q}, \mathbf{r})$; hence, $S(\mathbf{Q}, \mathbf{r}) \neq \varnothing$.

Since $\hat{p}_1 = 0$, we may reduce the matrix equation by excluding p_1 and the first row of matrix $\mathbf{Q}$; since $r_4 = 0$, we may make a further reduction by excluding r_4 and the fourth column of $\mathbf{Q}$. The reduced equation has the form

$$[p_1, p_2, p_3] \circ \begin{bmatrix} .9 & .7 & .2 \\ .8 & 1 & .5 \\ .1 & .3 & .6 \end{bmatrix} = [.8, .7, .5].$$

We must remember that p_1, p_2, and p_3 in this reduced equation represent p_2, p_3, and p_4 of the original equation, respectively.

TABLE 3.3.   ILLUSTRATION TO EXAMPLE 3.22.

$K(\beta, j)$	j 1	2	3	$g(\beta)$
$\beta = 1\,1\,2$	$\{1, 2\}$	$\{3\}$	$\varnothing$	$(.8, .5, 0)$
$1\,1\,3$	$\{1, 2\}$	$\varnothing$	$\{3\}$	$(.8, 0, .5)$
$1\,2\,2$	$\{1\}$	$\{2, 3\}$	$\varnothing$	$(.8, .7, 0)$
$1\,2\,3$	$\{1\}$	$\{2\}$	$\{3\}$	$(.8, .7, .5)$

Next, we apply the four steps of the procedure for determining the set $\check{S}(\mathbf{Q}, \mathbf{r})$ of all minimal solutions of this reduced matrix equation:

1. Employing the maximum solution $\hat{\mathbf{p}} = [.8, .7, .5]$ of the reduced equation, we obtain $J_1(\hat{\mathbf{p}}) = \{1\}$, $J_2(\hat{\mathbf{p}}) = \{1, 2\}$, $J_3(\hat{\mathbf{p}}) = \{2, 3\}$; hence, $J(\hat{\mathbf{p}}) = \{1\} \times \{1, 2\} \times \{2, 3\}$.
2. The sets $K(\beta, j)$ that we must determine for all $\beta \in J(\hat{\mathbf{p}})$ and all $j \in J$ are listed in Table 3.3.
3. For each $\beta \in J(\hat{\mathbf{p}})$, we generate the triples $g(\beta)$, which are also listed in Table 3.3.
4. Two of the triples $g(\beta)$ in Table 3.3 are minimal: $(.8, .5, 0)$ and $(.8, 0, .5)$. These therefore comprise all the minimal solutions of the reduced matrix equation. By adding 0 as the first component to each of these triples, we obtain the minimal solutions of the original matrix equation. Hence,

$$\check{S}(\mathbf{Q}, \mathbf{r}) = \{{}^1\check{\mathbf{p}} = (0, .8, .5, 0), {}^2\check{\mathbf{p}} = (0, .8, 0, .5)\}.$$

The set $S(\mathbf{Q}, \mathbf{r})$ of all solutions of the given matrix equation is now fully captured by the maximum solution

$$\hat{\mathbf{p}} = (0, .8, .7, .5),$$

and the two minimal solutions

$${}^1\check{\mathbf{p}} = (0, .8, .5, 0),$$

$${}^2\check{\mathbf{p}} = (0, .8, 0, .5).$$

According to Eq. (3.17), we have

$$S(\mathbf{Q}, \mathbf{r}) = \{\mathbf{p} \in \mathcal{P} \mid {}^1\check{\mathbf{p}} \le \mathbf{p} \le \hat{\mathbf{p}}\} \cup \{\mathbf{p} \in \mathcal{P} \mid {}^2\check{\mathbf{p}} \le \mathbf{p} \le \hat{\mathbf{p}}\}.$$

Let us now summarize concisely the described procedure for solving finite max-min fuzzy relation equations.

Basic Procedure

1. Decompose Eq. (3.8) into equations of the form

$$\mathbf{p} \circ \mathbf{Q} = \mathbf{r}, \tag{a}$$

one for each row in **P** and **R** (**p** is associated with index j, **Q** with indices j and k, and **r** with index k).

2. For each equation (a), if

$$\max_{j \in J} q_{j,k} < \max r_k$$

for some k, then the equation has no solution: $S(\mathbf{Q}, \mathbf{r}) = \varnothing$ and the procedure terminates; otherwise proceed to Step 3.

3. Determine $\hat{\mathbf{p}}$ by *Procedure 1*.

4. If $\hat{\mathbf{p}}$ is not a solution of Eq. (a), then the equation has no solution: $S(\mathbf{Q}, \mathbf{r})$ $= \varnothing$ and the procedure terminates; otherwise, proceed to Step 5.

5. For each $\hat{p}_j = 0$ and $r_k = 0$, exclude these components as well as the corresponding rows j and columns k from matrix **Q** in (a): This results in the reduced equation

$$\mathbf{p}' \circ \mathbf{Q}' = \mathbf{r}' \tag{b}$$

where we assume $j \in J'$, $k \in K'$.

6. Determine all minimal solutions of the reduced equation (b) by *Procedure 2*: this results in $\check{S}(\mathbf{Q}', \mathbf{r}')$.

7. Determine the solution set of the reduced equation (b):

$$S(\mathbf{Q}', \mathbf{r}') = \bigcup_{\check{\mathbf{p}}'} \langle \check{\mathbf{p}}', \hat{\mathbf{p}}' \rangle,$$

where the union is taken over all $\check{\mathbf{p}}' \in \check{S}(\mathbf{Q}', \mathbf{r}')$.

8. Extend all solutions in $S(\mathbf{Q}', \mathbf{r}')$ by zeros that were excluded in Step 5: this results in the solution set $S(\mathbf{Q}, \mathbf{r})$ of equation (a).

9. Repeat Steps 2–8 for all equations of type (a) that are embedded in Eq. (3.8): this results in all matrices **P** that satisfy Eq. (3.8).

Procedure 1

Form the vector $\hat{\mathbf{p}} = [\hat{p}_j \mid j \in J]$ in which

$$\hat{p}_j = \min_k \sigma(q_{j,k}, r_k),$$

where

$$\sigma(q_{j,k}, r_k) = \begin{cases} r_k & \text{if } q_{j,k} > r_k, \\ 1 & \text{otherwise.} \end{cases}$$

Procedure 2

1. Permute elements of $\mathbf{r}'$ and the corresponding columns of $\mathbf{Q}'$ appropriately to arrange them in decreasing order.

2. Determine sets

$$J_k(\hat{\mathbf{p}}') = \{j \in J' \mid \min(\hat{p}_j', q_{j,k}') = r_k'\}$$

for all $k \in K'$ and form

$$J(\hat{\mathbf{p}}') = \underset{k \in K'}{\times} J_k(\hat{\mathbf{p}}').$$

3. For each $\beta \in J(\hat{\mathbf{p}}')$ and each $j \in J'$, determine the set

$$K(\beta, j) = \{k \in K' \mid \beta_k = j\}.$$

4. For each $\beta \in J(\hat{\mathbf{p}}')$, generate the tuple

$$g(\beta) = (g_j(\beta) \mid j \in J')$$

by taking

$$g_j(\beta) = \begin{cases} \max r'_k & \text{if } K(\beta, j) \neq \varnothing, \, k \in K(\beta, j), \\ 0 & \text{otherwise.} \end{cases}$$

5. From all tuples $g(\beta)$ generated in Step 4, select only the minimal ones: this results in $\check{S}(\mathbf{Q}', \mathbf{r}')$.

NOTES

3.1. The basic ideas of *fuzzy relations* and the concepts of *similarity* and *fuzzy orderings* were introduced by Zadeh [1971a]. Binary fuzzy relations were further investigated by Rosenfeld [1971], Yeh and Bang [1975], Yager [1981], and Ovchinnikov [1984]; they are also extensively covered in one of the early books on fuzzy set theory [Kaufmann, 1975]. *Relational morphisms* were studied in great detail for both crisp and fuzzy relations by Bandler and Kohout [1986a, b].

3.2. The concepts of *projection*, *cylindric extension*, and *cylindric closure* were introduced for *n*-dimensional crisp relations by Ashby [1964]. Their fuzzy set generalizations are due to Zadeh [1975a, b]; these concepts are essential for procedures of approximate reasoning.

3.3. The notion of *fuzzy relation equations* based upon the max-min composition was first proposed and investigated by Sanchez [1976]; it was further studied by Prevot [1981] and Czogała, Drewniak, and Pedrycz [1982] for finite sets. The method for solving fuzzy relation equations described in Sec. 3.8 is based on the paper by Higashi and Klir [1984a]; all theorems by which the method is justified can be found in this paper. Fuzzy relation equations for other types of composition were studied by Pedrycz [1983a], Sanchez [1984b], and Miyakoshi and Shimbo [1985]. Generalizations to *L*-fuzzy relations were explored by numerous authors, for example, DiNola and Sessa [1983], Sessa [1984], and Drewniak [1984].

EXERCISES

3.1. The fuzzy relation R is defined on sets $X_1 = \{a, b, c\}$, $X_2 = \{s, t\}$, $X_3 = \{x, y\}$, $X_4 = \{i, j\}$ as follows:

$$R(X_1, X_2, X_3, X_4) = .4/b, t, y, i + .6/a, s, x, i + .9/b, s, y, i + 1/b, s, y, j$$
$$+ .6/a, t, y, j + .2/c, s, y, i.$$

(a) Compute the projections $R_{1,2,4}$, $R_{1,3}$, and R_4.

(b) Compute the cylindric extensions $[R_{1,2,4} \uparrow \{X_3\}]$, $[R_{1,3} \uparrow \{X_2, X_4\}]$, $[R_4 \uparrow \{X_1, X_2, X_3\}]$.

(c) Compute the cylindric closure from the three cylindric extensions in (b).

(d) Is the cylindric closure from (c) equal to the original relation R?

3.2. Given any n-ary relation, how many different projections of the relation can be taken?

3.3. Express the relation defined in Exercise 3.1 in terms of a four-dimensional array.

3.4. Consider matrices $\mathbf{M}_1$, $\mathbf{M}_2$, $\mathbf{M}_3$ in Table 3.4 as pages in a three-dimensional array that represents a fuzzy ternary relation. Determine:

(a) all two-dimensional projections;

(b) cylindric extensions and cylindric closure of the two-dimensional projections;

(c) all one-dimensional projections;

(d) cylindric extensions and cylindric closure of the one-dimensional projections;

(e) two three-dimensional arrays expressing the difference between each of the cylindric closures and the original ternary relation.

TABLE 3.4. MATRIX REPRESENTATIONS OF FUZZY BINARY RELATIONS ON $X \times Y$ EMPLOYED IN EXERCISES (ASSUME EITHER $X = \{x_i \mid i \in \mathbb{N}_n\}$, $Y = \{y_j \mid j \in \mathbb{N}_n\}$, or $X = Y = \{x_i \mid i \in \mathbb{N}_n\}$, $n = 3, 4, 5, 7$).

$$\mathbf{M}_1 = \begin{bmatrix} 1 & 0 & .7 \\ .3 & .2 & 0 \\ 0 & .5 & 1 \end{bmatrix} \quad \mathbf{M}_2 = \begin{bmatrix} .6 & .6 & 0 \\ 0 & .6 & .1 \\ 0 & .1 & 0 \end{bmatrix} \quad \mathbf{M}_3 = \begin{bmatrix} 1 & 0 & .7 \\ 0 & 1 & 0 \\ .7 & 0 & 1 \end{bmatrix}$$

$$\mathbf{M}_4 = \begin{bmatrix} 1 & .2 & 0 & 0 \\ 0 & 0 & .4 & .3 \\ 1 & .2 & 0 & 0 \\ 0 & 0 & .4 & .3 \end{bmatrix} \quad \mathbf{M}_5 = \begin{bmatrix} .3 & .6 & 0 & 1 \\ .7 & 0 & 1 & .5 \\ .5 & 0 & 0 & .2 \\ 0 & 0 & 1 & 0 \end{bmatrix}$$

$$\mathbf{M}_6 = \begin{bmatrix} .7 & .4 & 0 & 1 \\ .7 & 0 & .6 & .2 \\ .5 & .2 & 0 & .2 \\ 0 & 0 & .6 & .3 \end{bmatrix} \quad \mathbf{M}_7 = \begin{bmatrix} 0 & .5 & .5 & .4 \\ .3 & 0 & .8 & 0 \\ 1 & 0 & .5 & 0 \\ 0 & .3 & 0 & .1 \end{bmatrix}$$

$$\mathbf{M}_8 = \begin{bmatrix} 0 & 0 & .5 & 0 & 0 \\ .6 & .2 & 0 & .3 & .7 \\ 0 & .7 & 0 & 0 & 0 \\ 0 & 0 & 1 & 0 & 1 \\ .1 & .9 & 0 & .9 & 0 \end{bmatrix} \quad \mathbf{M}_9 = \begin{bmatrix} .3 & .4 & .5 & 1 & .7 \\ .7 & .3 & .2 & .3 & .6 \\ 0 & .6 & .7 & .3 & .5 \\ .2 & .6 & 1 & .1 & .1 \\ .8 & .1 & .2 & .9 & .4 \end{bmatrix}$$

$$\mathbf{M}_{10} = \begin{bmatrix} 0 & .2 & 0 & .4 & .5 \\ .3 & 0 & 0 & .4 & 0 \\ 0 & 0 & .7 & 0 & 0 \\ 0 & .3 & 0 & 0 & .2 \\ .1 & 0 & 0 & .6 & 0 \end{bmatrix} \quad \mathbf{M}_{11} = \begin{bmatrix} 1 & 1 & .8 & .4 & .9 \\ .5 & 1 & 1 & 1 & 1 \\ .7 & 0 & 1 & 0 & 0 \\ 0 & 0 & 0 & 1 & .6 \\ 1 & 0 & .7 & .4 & 1 \end{bmatrix}$$

$$\mathbf{M}_{12} = \begin{bmatrix} 1 & 0 & .8 & 0 & .6 & .8 & 0 \\ 0 & 1 & 0 & .6 & 0 & .5 & 0 \\ .8 & 0 & 1 & .8 & 0 & 0 & 0 \\ 0 & .6 & .8 & 1 & 0 & 0 & .8 \\ .6 & 0 & 0 & 0 & 1 & .6 & 0 \\ .8 & .5 & 0 & 0 & .6 & 1 & 0 \\ 0 & 0 & 0 & .8 & 0 & 0 & 1 \end{bmatrix} \quad \mathbf{M}_{13} = \begin{bmatrix} 0 & .8 & .7 & .1 & .2 & 0 & 0 \\ .3 & 0 & 1 & 0 & .7 & .3 & .2 \\ 0 & 0 & .5 & 1 & 0 & .9 & .2 \\ .1 & .2 & 0 & 1 & 1 & 0 & .7 \\ .3 & .2 & 1 & 0 & .5 & 1 & 1 \\ 0 & .7 & .6 & .2 & 0 & 1 & .5 \\ 1 & 1 & .5 & .7 & .3 & .2 & 1 \end{bmatrix}$$

3.5. Repeat Exercise 3.4 under the assumption that matrices $\mathbf{M}_4$–$\mathbf{M}_7$ are pages of a three-dimensional array that represents a fuzzy ternary relation.

3.6. The fuzzy binary relation R is defined on sets $X = \{1, 2, \ldots, 100\}$ and $Y = \{50, 51, \ldots, 100\}$ and represents the relation "x is much smaller than y." It is defined by the membership function

$$\mu_R(x, y) = \begin{cases} 1 - \dfrac{x}{y}, & x \leq y, \\ \\ 0, & \text{otherwise,} \end{cases}$$

where $x \in X$ and $y \in Y$.

(a) What is the domain of R?

(b) What is the range of R?

(c) What is the height of R?

(d) Is R normal or subnormal?

(e) Is R completely specified or incompletely specified?

(f) Is R a relation from X onto Y or from X into Y?

(g) Is R a function or one-to-many?

(h) Is R many-to-one, one-to-one, or neither?

3.7. For each of the following binary relations on a single set, state whether the relation is reflexive, irreflexive, or antireflexive, symmetric, asymmetric, antisymmetric, or strictly antisymmetric, and transitive, nontransitive, or antitransitive:

(a) "is a sibling of";

(b) "is a parent of";

(c) "is smarter than";

(d) "is the same height as";

(e) "is at least as tall as".

3.8. For some of the binary relations given in Table 3.4, determine:

(a) the domain, range, and height of the relation;

(b) the resolution form of the relation;

(c) the inverse of the relation.

3.9. Prove that the max-min composition and min join are associative operations on binary fuzzy relations.

3.10. For some of the binary relations given in Table 3.4, draw each of the following:

(a) a sagittal diagram of the relation;

(b) a simple diagram of the relation under the assumption that $X = Y$.

3.11. Assuming $X = Y$, perform the max-min composition of some sequences of comparable relations given in Table 3.4. For example:

(a) $\mathbf{M}_1 \circ \mathbf{M}_2,\ \mathbf{M}_4 \circ \mathbf{M}_5,\ \mathbf{M}_8 \circ \mathbf{M}_9,\ \mathbf{M}_{12} \circ \mathbf{M}_{13}$;

(b) $\mathbf{M}_1 \circ \mathbf{M}_2 \circ \mathbf{M}_3,\ \mathbf{M}_4 \circ \mathbf{M}_5 \circ \mathbf{M}_6 \circ \mathbf{M}_7$, etc.

3.12. Repeat Exercise 3.11 for the max-product composition.

3.13. Using binary relations given in Table 3.4, perform the following relational joins:

(a) $\mathbf{M}_1 * \mathbf{M}_2^T$;

(b) $\mathbf{M}_4 * \mathbf{M}_5^T$;

(c) $\mathbf{M}_8 * \mathbf{M}_9^T$;

(d) $\mathbf{M}_{12} * \mathbf{M}_{13}^T$.

3.14. Prove that the properties of symmetry, reflexivity, and transitivity (or the lack of these properties) are preserved under inversion for both crisp and fuzzy relations.

3.15. Given a crisp equivalence relation $R(X, X)$ and an arbitrary equivalence class A_x of $R(X, X)$, prove that no member of A_x is related to any element of X that is not also a member of A_x.

3.16. Given a crisp equivalence relation $R(X, X)$, prove that the family of all equivalence classes of $R(X, X)$ forms a partition on X.

3.17. Assuming that $X = Y$ for binary relations given in Table 3.4, determine for some of the relations whether they are reflexive, ϵ-reflexive (for some $\epsilon < 1$), irreflexive, antireflexive, symmetric, asymmetric, antisymmetric, strictly antisymmetric, transitive, nontransitive, or antitransitive (assuming max-min transitivity).

3.18. Assuming that $X = Y$, determine the max-min transitive closure for some relations given in Table 3.4.

3.19. Repeat Exercise 3.18 for max-product transitive closure.

3.20. Given a fuzzy similarity relation $R(X, X)$ and two partitions $\pi(R_\alpha)$ and $\pi(R_\beta)$, where R_α and R_β are α-cuts and $\alpha \geq \beta$, prove that each element of $\pi(R_\alpha)$ is contained in some element of $\pi(R_\beta)$.

3.21. The transitive closure of the relation defined by matrix $\mathbf{M}_{12}$ in Table 3.4 (Exercise 3.18) is a similarity relation. Determine its partition tree.

3.22. Relations defined by matrices $\mathbf{M}_3$ and $\mathbf{M}_{12}$ in Table 3.4 are compatibility relations (assuming $X = Y$). Determine:
(a) simple diagrams of the relations;
(b) all complete α-covers of the relations.

3.23. Prove the following propositions:
(a) When $R(X, X)$ is a strictly antisymmetric crisp relation, then $R \cap R^{-1} = \varnothing$.
(b) When $R(X, X)$ is max-min transitive, then $R \circ R \subseteq R$.

3.24. Show that for every fuzzy partial ordering on X, the sets of undominated and undominating elements of X are nonempty.

3.25. Construct some fuzzy binary relations that are orderings of the various types introduced in Sec. 3.6.

3.26. Assuming $X = Y$, construct simplifications of some of the relations given in Table 3.4 under appropriate homomorphic mappings defined by you.

3.27. Solve the following fuzzy relation equations.

(a)
$$\mathbf{p} \circ \begin{bmatrix} .9 & .6 & 1 \\ .8 & .8 & .5 \\ .6 & .4 & .6 \end{bmatrix} = [.6\ .6\ .5];$$

(b)
$$\mathbf{p} \circ \begin{bmatrix} .5 & .7 & 0 & .2 \\ .4 & .6 & 1 & 0 \\ .2 & .4 & .5 & .6 \\ 0 & .2 & 0 & .8 \end{bmatrix} = [.5\ .5\ .4\ .2];$$

(c)
$$\mathbf{P} \circ \begin{bmatrix} .5 & 0 & .3 & 0 \\ .4 & 1 & .3 & 0 \\ 0 & .1 & 1 & .1 \\ .4 & .3 & .3 & .5 \end{bmatrix} = \begin{bmatrix} .5 & .3 & .3 & .1 \\ .5 & .4 & .4 & .2 \end{bmatrix};$$

(d)
$$\mathbf{P} \circ \begin{bmatrix} .2 & .4 & .6 & .7 \\ .2 & 0 & .6 & .8 \\ .1 & .4 & .6 & .7 \\ 0 & .3 & 0 & 1 \end{bmatrix} = \begin{bmatrix} .2 & .4 & .6 & .7 \\ .1 & .2 & .2 & .2 \end{bmatrix}.$$

4

FUZZY MEASURES

4.1 GENERAL DISCUSSION

The fuzzy set provides us with an intuitively pleasing method of representing one form of uncertainty. Consider, however, the jury members for a criminal trial who are uncertain about the guilt or innocence of the defendant. The uncertainty in this situation seems to be of a different type; the set of people who are guilty of the crime and the set of innocent people are assumed to have very distinct boundaries. The concern, therefore, is not with the degree to which the defendant is guilty but with the degree to which the evidence proves his or her membership in either the crisp set of guilty people or in the crisp set of innocent people. We assume that perfect evidence would point to full membership in one and only one of these sets. Our evidence, however, is rarely, if ever, perfect, and some uncertainty usually prevails. In order to represent this type of uncertainty, we could assign a value to each possible crisp set to which the element in question might belong. This value would indicate the degree of evidence or certainty of the element's membership in the set. Such a representation of uncertainty is known as a *fuzzy measure*. Note how this method differs from the assignment of membership grades in fuzzy sets. In the latter case, a value is assigned to each element of the universal set signifying its degree of membership in a particular set with unsharp boundaries. The fuzzy measure, on the other hand, assigns a value to each crisp set of the universal set signifying the degree of evidence or belief that a particular element belongs in the set.

A *fuzzy measure* is thus defined by a function

$$g : \mathcal{P}(X) \rightarrow [0, 1],$$

which assigns to each *crisp subset* of X a number in the unit interval $[0, 1]$.* When this number is assigned to a subset $A \in \mathcal{P}(X)$, $g(A)$ represents the degree of available evidence or our belief that a given element of X (a priori nonlocated in any subset of X) belongs to the subset A. The subset to which we assign the highest value represents our best guess concerning the particular element in question. For instance, suppose we are trying to diagnose an ill patient. In simplified terms, we may be trying to determine whether this patient belongs to the set of people with, say, pneumonia, bronchitis, emphysema, or a common cold. A physical examination may provide us with helpful yet inconclusive evidence. Therefore, we might assign a high value, say .75, to our best guess, bronchitis, and a lower value to the other possibilities, such as .45 to pneumonia, .3 to a common cold, and 0 to emphysema. These values reflect the degree to which the patient's symptoms provide evidence for one disease rather than another, and the collection of these values constitutes a fuzzy measure representing the uncertainty or ambiguity associated with several well-defined alternatives. It is important to understand how this type of uncertainty is distinct from the vagueness or lack of sharp boundaries that is represented by the fuzzy set.

The difference between these two types of uncertainty is also exhibited in the context of scientific observation or measurement. Observing attributes such as a type of cloud formation in meteorology, a characteristic posture of an animal in ethology, or a degree of defect of a tree in forestry clearly involves situations in which it is not possible to draw sharp boundaries; such observations or measurements are inherently vague and, consequently, their connection with the concept of the fuzzy set is suggestive. In most measurements in physics, on the other hand, such as the measurement of length, weight, electric current, or light intensity, we define classes with sharp boundaries. Given a measurement range, usually represented by an interval of real numbers $[a, b]$, we partition this interval into n disjoint subintervals

$$[a, a_1), [a_1, a_2), [a_2, a_3), \ldots, [a_{n-1}, b]$$

according to the desired (or feasible) accuracy. Then, theoretically, each observed magnitude fits exactly into one of the intervals. In practice, however, this would be warranted only if no observational errors were involved. Since measurement errors are unavoidable in principle, each observation that coincides with or is in close proximity to one of the boundaries $a_1, a_2, \ldots, a_{n-1}$ between two neighboring intervals involves uncertainty regarding its membership in the two crisp intervals (crisp subsets of the set of real numbers). This uncertainty clearly has all the characteristics of a fuzzy measure.

In order to qualify as a fuzzy measure, the function g must have certain properties. These required properties were traditionally assumed to be the usual axioms of probability theory (or probability measures). It turns out, however, that this assumption is not warranted. Fuzzy measures are defined by weaker axioms,

---

* Note that the domain of the function g is the power set $\mathcal{P}(X)$ of *crisp* subsets of X and not the power set $\tilde{\mathcal{P}}(X)$ of fuzzy subsets of X.

thus subsuming probability measures as a special type of fuzzy measure.* The following are *axioms of fuzzy measures*:

Axiom g1 (*boundary conditions*). $g(0) = 0$ and $g(X) = 1$.

Axiom g2 (*monotonicity*).   For every $A, B \in \mathcal{P}(X)$, if $A \subseteq B$, then $g(A) \le g(B)$.

Axiom g3 (*continuity*).   For every sequence $(A_i \in \mathcal{P}(X) \mid i \in \mathbb{N})$ of subsets of X, if either $A_1 \subseteq A_2 \subseteq \cdots$ or $A_1 \supseteq A_2 \supseteq \cdots$ (i.e., the sequence is monotonic), then

$$\lim_{i \to \infty} g(A_i) = g(\lim_{i \to \infty} A_i).$$

Axiom g1 states that despite our degree of evidence, we always know that the element in question definitely does not belong to the empty set and definitely does belong to the universal set. The empty set, by definition, does not contain any element and, hence, it cannot contain the element of our interest either; the universal set, on the other hand, contains all elements under consideration in each particular context and, therefore, it must contain our element as well.

Axiom g2 requires that the evidence of the membership of an element in a set must be at least as great as the evidence that the element belongs to any subset of that set. Indeed, when we know with some degree of certainty that the element belongs to a set, then our degree of certainty that it belongs to a larger set containing the former set can be greater or equal, but it cannot be smaller.

Axiom g3 is clearly applicable only to an infinite universal set. It can, therefore, be disregarded when we are dealing with a finite universal set. The axiom requires that for every infinite sequence $A_1, A_2, \ldots$ of nested (monotonic) subsets of X that converge to the set

$$A = \lim_{i \to \infty} A_i,$$

the sequence of numbers $g(A_1), g(A_2), \ldots$ must converge to the number $g(A)$. That is, the axiom requires that g is a continuous function. This axiom can also be viewed as a requirement of *consistency*: calculation of $g(A)$ in two different ways, either as the limit of $g(A_i)$ for $i \to \infty$ or by application of the function g to the limit of A_i for $i \to \infty$, is required to yield the same value.

A fuzzy measure is often defined more generally as a function

$$g : \mathcal{B} \to [0, 1],$$

where $\mathcal{B} \subset \mathcal{P}(X)$ is a family of subsets of X such that:

1. $\varnothing \in \mathcal{B}$ and $X \in \mathcal{B}$;
2. If $A \in \mathcal{B}$, then $\overline{A} \in \mathcal{B}$;

---

* The relationship between fuzzy measures and probability measures is discussed in Sec. 4.3.

3. $\mathcal{B}$ is closed under the operation of set union, that is, if $A \in \mathcal{B}$ and $B \in \mathcal{B}$, then also $A \cup B \in \mathcal{B}$.

The set $\mathcal{B}$ is usually called a *Borel field* or *σ-field*.

Since $A \cup B \supseteq A$ and $A \cup B \supseteq B$, we have

$$\max[g(A), g(B)] \leq g(A \cup B) \tag{4.1}$$

due to the required monotonicity of the function g (Axiom g2). Similarly, since $A \cap B \subseteq A$ and $A \cap B \subseteq B$, we have

$$g(A \cap B) \leq \min[g(A), g(B)]. \tag{4.2}$$

In the remainder of this chapter, we examine two important and well-developed special types of fuzzy measures, which are usually referred to as plausibility and belief measures. These two measures are complementary in the sense that one of them can be uniquely derived from the other. In addition, we also examine three important special types of plausibility and belief measures—the well-known probability measures and a pair of complementary measures referred to as possibility and necessity measures.

4.2 BELIEF AND PLAUSIBILITY MEASURES

A *belief measure* is a function

$$Bel: \mathcal{P}(X) \rightarrow [0, 1]$$

that satisfies the Axioms g1 through g3 of fuzzy measures and the following additional axiom:*

$$Bel(A_1 \cup A_2 \cup \cdots \cup A_n) \geq \sum_i Bel(A_i) - \sum_{i<j} Bel(A_i \cap A_j)$$

$$+ \cdots + (-1)^{n+1} Bel(A_1 \cap A_2 \cap \cdots \cap A_n) \tag{4.3}$$

for every $n \in \mathbb{N}$ and every collection of subsets of X. For $n = 2$, for example, this axiom has the form

$$Bel(A_1 \cup A_2) \geq Bel(A_1) + Bel(A_2) - Bel(A_1 \cap A_2); \tag{4.4}$$

for $n = 3$, we have

$$Bel(A_1 \cup A_2 \cup A_3) \geq Bel(A_1) + Bel(A_2) + Bel(A_3)$$

$$- Bel(A_1 \cap A_2) - Bel(A_1 \cap A_3)$$

$$- Bel(A_2 \cap A_3) \tag{4.5}$$

$$+ Bel(A_1 \cap A_2 \cap A_3).$$

---

* As previously mentioned, $\mathcal{P}(X)$ in this definition can be replaced, in general, by a Borel field $\mathcal{B}$ defined on X.

For each $A \in \mathcal{P}(X)$, $Bel(A)$ is interpreted as the *degree of belief* (based on available evidence) that a given element of X belongs to the set A. We may also view the subsets of X as answers to a particular question. We assume that some of the answers are correct, but we do not know with full certainty which ones they are.

When the sets $A_1, A_2, \ldots, A_n$ in Axiom (4.3) are pairwise disjoint ($A_i \cap A_j = \emptyset$ for all $i, j \in \mathbb{N}_n$ such that $i \neq j$), the axiom requires that the degree of belief associated with the union of the sets is not smaller than the sum of the degrees of belief pertaining to the individual sets. The basic axiom of belief measures is thus a weaker version of the additivity axiom of probability theory. This implies that probability measures are special cases of belief measures for which the equality in (4.3) is required.

We can easily show that Axiom (4.3) of belief measures implies the monotonicity axiom of fuzzy measures (Axiom g2). Let $A \subset B$ and let $C = B - A$. Then, $A \cup C = B$ and $A \cap C = \emptyset$. Now applying A and C to Axiom (4.3) for $n = 2$, as expressed by (4.4), we obtain

$$Bel(A \cup C) = Bel(B) \geq Bel(A) + Bel(C) - Bel(A \cap C).$$

Since $A \cap C = \emptyset$ and $Bel(\emptyset) = 0$ by Axiom g1, we have

$$Bel(B) \geq Bel(A) + Bel(C)$$

and, consequently, $Bel(B) \geq Bel(A)$.

Let $A_1 = A$ and $A_2 = \overline{A}$ in Eq. (4.4). Then, we can immediately derive the following fundamental property of belief measures:

$$Bel(A) + Bel(\overline{A}) \leq 1.$$

Associated with each belief measure is a plausibility measure Pl, defined by the equation

$$Pl(A) = 1 - Bel(\overline{A}) \tag{4.6}$$

for all $A \in \mathcal{P}(X)$. Similarly,

$$Bel(A) = 1 - Pl(\overline{A}). \tag{4.7}$$

Belief measures and plausibility measures are therefore mutually dual. Plausibility measures, however, can also be defined independently of belief measures.

A *plausibility measure* is a function

$$Pl : \mathcal{P}(X) \to [0, 1]$$

that satisfies the Axioms g1 through g3 of fuzzy measures and the following additional axiom:

$$Pl(A_1 \cap A_2 \cap \cdots \cap A_n) \leq \sum_i Pl(A_i) - \sum_{i<j} Pl(A_i \cup A_j) + \cdots$$

$$+ (-1)^{n+1} Pl(A_1 \cup A_2 \cup \cdots \cup A_n) \tag{4.8}$$

for every $n \in \mathbb{N}$ and every collection of subsets of X.

Let $n = 2$, $A_i = A$, and $A_2 = \overline{A}$ in Eq. (4.8). Then, we immediately obtain the following basic inequality of plausibility measures:

$$Pl(A) + Pl(\overline{A}) \geq 1.$$

Using Eq. (4.7), we can convert the axiom of belief measures, expressed by the inequality (4.3), into the axiom of plausibility measures, expressed by the inequality (4.8). We should realize, however, that the exact counterpart of (4.3) is a modification of (4.8) in which all sets are complemented. For example, when we apply Eq. (4.7) to the special form of the axiom of belief functions for $n = 3$, expressed by the inequality (4.5), we obtain

$$1 - Pl(\overline{A}_1 \cap \overline{A}_2 \cap \overline{A}_3) \geq 1 - Pl(\overline{A}_1) + 1 - Pl(\overline{A}_2) + 1 - Pl(\overline{A}_3)$$
$$- [1 - Pl(\overline{A}_1 \cup \overline{A}_2)] - [1 - Pl(\overline{A}_1 \cup \overline{A}_3)]$$
$$- [1 - Pl(\overline{A}_2 \cup \overline{A}_3)]$$
$$+ 1 - Pl(\overline{A}_1 \cup \overline{A}_2 \cup \overline{A}_3).$$

Canceling the 1s and multiplying the inequality by -1 results in

$$Pl(\overline{A}_1 \cap \overline{A}_2 \cap \overline{A}_3) \leq Pl(\overline{A}_1) + Pl(\overline{A}_2) + Pl(\overline{A}_3)$$
$$- Pl(\overline{A}_1 \cup \overline{A}_2) - Pl(\overline{A}_1 \cup \overline{A}_3) - Pl(\overline{A}_2 \cup \overline{A}_3)$$
$$+ Pl(\overline{A}_1 \cup \overline{A}_2 \cup \overline{A}_3).$$

This inequality is thus the plausibilistic counterpart of the inequality (4.5) for belief measures.

Every belief measure and its dual plausibility measure can be expressed in terms of a function

$$m : \mathscr{P}(X) \rightarrow [0, 1]$$

such that $m(\varnothing) = 0$ and

$$\sum_{A \in \mathscr{P}(X)} m(A) = 1, \tag{4.9}$$

where $m(A)$ is interpreted either as the *degree of evidence* supporting the claim *that a specific element of X belongs to the set A but not to any special subset of A*, or as the degree to which we *believe* that such a claim is warranted.

Consider, for example, that x is the age of a person and the universe of discourse is the set of all possible ages (in years) represented by an appropriate subset of nonnegative integers. Then $m(A)$ may represent the degree of evidence (or our belief) that a particular person is a teenager (when $A = \{13, 14, \ldots, 19\}$) or a person in his or her thirties (when $A = \{30, 31, \ldots, 39\}$), and so on.

Since Eq. (4.9) resembles a similar equation for probability distributions, the function m is usually called a *basic probability assignment*. It is crucial to realize, however, that there is a fundamental difference between probability distribution functions and basic probability assignments. Although we discuss this difference in more detail in Sec. 4.3, we can immediately see one obvious aspect of this difference: probability distribution functions are defined on X, whereas basic probability assignments are defined on the power set $\mathcal{P}(X)$ of X. In order to avoid confusion between these two concepts, we prefer to use the shorter term *basic assignment* instead of the usual full term *basic probability assignment*.

The requirement that $m(\varnothing) = 0$ has been questioned on some occasions. It has been argued that there might be cases in which it is not meaningful to assign any value of the basic assignment to some elements x of X. This may occur, for example, if x represents the ages of cars belonging to individuals in a population and it happens that some persons in the population own no cars. Such cases, it has been argued, can be conveniently handled by allowing $m(\varnothing) > 0$. These undesirable cases, however, can always be avoided by an appropriate choice of the universal set. Hence, for the sake of simplicity, we assume in this book that $m(\varnothing) = 0$. Basic assignments that possess this property are called *normal*.

Upon careful examination of the definition of the basic assignment, we observe the following:

1. It is not required that $m(X) = 1$.
2. It is not required that $m(A) \leq m(B)$ when $A \subset B$.
3. No relationship between $m(A)$ and $m(\overline{A})$ is required.

It follows from these observations that the basic assignments are not fuzzy measures. However, given a basic assignment m, a belief measure and a plausibility measure are uniquely determined by the formulas

$$Bel(A) = \sum_{B \subseteq A} m(B) \tag{4.10}$$

and

$$Pl(A) = \sum_{B \cap A \neq \varnothing} m(B), \tag{4.11}$$

which are applicable for all $A \in \mathcal{P}(X)$.

The relationship between $m(A)$ and $Bel(A)$, expressed by Eq. (4.10), has the following meaning: whereas $m(A)$ characterizes the degree of evidence or belief that the element in question belongs to the set A alone (i.e., exactly to set A), $Bel(A)$ represents the total evidence or belief that the element belongs to A as well as to the various special subsets of A. The plausibility measure $Pl(A)$, as defined by Eq. (4.11), has a different meaning: it represents not only the total evidence or belief that the element in question belongs to set A or to any of its subsets but also the additional evidence or belief associated with sets that overlap with A. Hence,

$$Pl(A) \geq Bel(A) \tag{4.12}$$

for all $A \in \mathcal{P}(X)$.

Every set $A \in \mathcal{P}(X)$ for which $m(A) > 0$ is usually called a *focal element* of m. As this name suggests, focal elements are subsets of X on which the available evidence focuses. When X is finite, m can be fully characterized by a list of its focal elements A with the corresponding values $m(A)$. The pair $(\mathcal{F}, m)$, where $\mathcal{F}$ and m denote a set of focal elements and the associated basic assignment, respectively, is often called a *body of evidence*.

Total ignorance is expressed in terms of the basic assignment by $m(X) = 1$ and $m(A) = 0$ for all $A \neq X$. That is, we know that the element is in the universal set, but we have no evidence about its location in any subset of X. It follows from Eq. (4.10) that the expression of total ignorance in terms of the corresponding belief measure is exactly the same: $Bel(X) = 1$ and $Bel(A) = 0$ for all $A \neq X$. However, the form of total ignorance in terms of the associated plausibility measure (obtained from the belief measure by Eq. (4.6)) is quite different: $Pl(\emptyset) = 0$ and $Pl(A) = 1$ for all $A \neq \emptyset$. This form follows directly from Eq. (4.11).

A basic assignment m is said to be a *simple support function* focused at A if $m(A) = s$, $m(X) = 1 - s$, and $m(B) = 0$ for all other sets B in $\mathcal{P}(X)$.

Given a belief measure Bel, the corresponding basic assignment m is determined by the formula

$$m(A) = \sum_{B \subseteq A} (-1)^{|A-B|} \, Bel(B) \tag{4.13}$$

for all $A \in \mathcal{P}(X)$, where $|A - B|$ denotes the cardinality of the set $A - B$. Since the derivation of this formula is rather lengthy, it is omitted here. To derive the basic assignment from the associated plausibility measure, we can employ Eqs. (4.13) and (4.7).

Evidence obtained in the same context from two independent sources (for example, from two experts in the field of inquiry) and expressed by two basic assignments m_1 and m_2 on some power set $\mathcal{P}(X)$ must be appropriately combined to obtain a joint basic assignment $m_{1,2}$. In general, evidence can be combined in various ways, some of which may take into consideration the reliability of the sources and other relevant aspects. The standard way of combining evidence is expressed by the formula

$$m_{1,2}(A) = \frac{\displaystyle\sum_{B \cap C = A} m_1(B) \cdot m_2(C)}{1 - K} \tag{4.14}$$

for $A \neq \emptyset$, where

$$K = \sum_{B \cap C = \emptyset} m_1(B) \cdot m_2(C); \tag{4.15}$$

$m_{1,2}(\emptyset) = 0$.

Formula (4.14) is referred to as *Dempster's rule of combination*. According to this rule, the degree of evidence $m_1(B)$ from the first source that focuses on set $B \in \mathcal{P}(X)$ and the degree of evidence $m_2(C)$ from the second source that focuses on set $C \in \mathcal{P}(X)$ are combined by taking the product $m_1(B) \cdot m_2(C)$, which focuses on the intersection $B \cap C$. This is exactly the same way in which the joint probability distribution is calculated from two independent marginal distributions and, consequently, it is justified on the same grounds. Since, however, some intersections of focal elements from the first and second source may result in the same set A, we must add the corresponding products to obtain $m_{1,2}(A)$. Moreover, some of the intersections may be empty. Since it is required that $m_{1,2}(\emptyset) = 0$, the value K expressed by Eq. (4.15) is not included in the definition of the joint basic assignment $m_{1,2}$. This means that the sum of products $m_1(B) \cdot m_2(C)$ for all focal elements B of m_1 and all focal elements C of m_2 such that $B \cap C \neq \emptyset$ is equal to $1 - K$. To obtain a normalized basic assignment $m_{1,2}$, as required (Eq. (4.9)), we must divide each of these products by this factor $1 - K$ as indicated in Eq. (4.14).

Example 4.1.

Assume that an old painting was discovered that strongly resembles paintings by Raphael. Such a discovery is likely to generate various questions regarding the status of the painting. Assume the following three questions:

1. Is the discovered painting a genuine painting by Raphael?
2. Is the discovered painting a product of one of Raphael's many disciples?
3. Is the discovered painting a counterfeit?

Let R, D, and C denote subsets of our universal set X—the set of all paintings— that contain the set of all paintings by Raphael, the set of all paintings by disciples of Raphael, and the set of all counterfeits of Raphael's paintings, respectively.

Assume now that two experts performed careful examinations of the painting and subsequently provided us with basic assignments m_1 and m_2 specified in Table 4.1. These are the degrees of evidence that each expert obtained by the examination and that support the various claims that the painting belongs to one of the sets of our concern. For example, $m_1(R \cup D) = .15$ is the degree of evidence obtained by the first expert that the painting was done by Raphael himself or that the painting was done by one of his disciples. Using Eq. (4.10), we can easily calculate the total evidence, Bel_1 and Bel_2, in each set, as shown in Table 4.1.

Applying Dempster's rule (Eq. (4.14)) to m_1 and m_2, we obtain the joint basic assignment $m_{1,2}$, which is also shown in Table 4.1. To determine the values of $m_{1,2}$,

TABLE 4.1. COMBINATION OF DEGREES OF EVIDENCE FROM TWO INDEPENDENT SOURCES (EXAMPLE 4.1).

Focal elements	Expert 1		Expert 2		Combined evidence	
	m_1	Bel_1	m_2	Bel_2	$m_{1,2}$	$Bel_{1,2}$
R	.05	.05	.15	.15	.21	.21
D	0	0	0	0	.01	.01
C	.05	.05	.05	.05	.09	.09
$R \cup D$	.15	.2	.05	.2	.12	.34
$R \cup C$	.1	.2	.2	.4	.2	.5
$D \cup C$	.05	.1	.05	.1	.06	.16
$R \cup D \cup C$	.6	1	.5	1	.31	1

we calculate the normalization factor $1 - K$ first. Applying Eq. (4.15), we obtain

$$K = m_1(R) \cdot m_2(D) + m_1(R) \cdot m_2(C) + m_1(R) \cdot m_2(D \cup C) + m_1(D) \cdot m_2(R)$$

$$+ m_1(D) \cdot m_2(C) + m_1(D) \cdot m_2(R \cup C) + m_1(C) \cdot m_2(R) + m_1(C) \cdot m_2(D)$$

$$+ m_1(C) \cdot m_2(R \cup D) + m_1(R \cup D) \cdot m_2(C) + m_1(R \cup C) \cdot m_2(D)$$

$$+ m_1(D \cup C) \cdot m_2(R)$$

$$= .03.$$

The normalization factor is then $1 - K = .97$. Values of $m_{1,2}$ are calculated by Eq. (4.14). For example

$$m_{1,2}(R) = [m_1(R) \cdot m_2(R) + m_1(R) \cdot m_2(R \cup D) + m_1(R) \cdot m_2(R \cup C)$$

$$+ m_1(R) \cdot m_2(R \cup D \cup C) + m_1(R \cup D) \cdot m_2(R)$$

$$+ m_1(R \cup D) \cdot m_2(R \cup C) + m_1(R \cup C) \cdot m_2(R)$$

$$+ m_1(R \cup C) \cdot m_2(R \cup D) + m_1(R \cup D \cup C) \cdot m_2(R)]/.97$$

$$= .21,$$

$$m_{1,2}(D) = [m_1(D) \cdot m_2(D) + m_1(D) \cdot m_2(R \cup D) + m_1(D) \cdot m_2(D \cup C)$$

$$+ m_1(D) \cdot m_2(R \cup D \cup C) + m_1(R \cup D) \cdot m_2(D)$$

$$+ m_1(R \cup D) \cdot m_2(D \cup C) + m_1(D \cup C) \cdot m_2(D)$$

$$+ m_1(D \cup C) \cdot m_2(R \cup D) + m_1(R \cup D \cup C) \cdot m_2(D)]/.97$$

$$= .01,$$

$$m_{1,2}(R \cup C) = [m_1(R \cup C) \cdot m_2(R \cup C) + m_1(R \cup C) \cdot m_2(R \cup D \cup C)$$

$$+ m_1(R \cup D \cup C) \cdot m_2(R \cup C)]/.97$$

$$= .2,$$

$$m_{1,2}(R \cup D \cup C) = [m_1(R \cup D \cup C) \cdot m_2(R \cup D \cup C)]/.97$$

$$= .31,$$

and similarly for the remaining focal elements C, $R \cup D$, and $D \cup C$. The joint basic assignment can now be used to calculate the joint belief $Bel_{1,2}$ (Table 4.1) and joint plausibility $Pl_{1,2}$.

Consider now a basic assignment m defined on the Cartesian product $Z = X \times Y$, that is,

$$m : \mathcal{P}(X \times Y) \rightarrow [0, 1].$$

Each focal element of m is in this case a binary relation R on $X \times Y$. Let R_x denote the projection of R on X, that is,

$$R_x = \{x \in X \mid (x, y) \in R \quad \text{for some} \quad y \in Y\}.$$

Similarly,

$$R_y = \{y \in Y \mid (x, y) \in R \text{ for some } x \in X\}$$

defines the projection of R on Y. We can now define the projection m_x of m on X by the formula

$$m_x(A) = \sum_{R:A=R_x} m(R) \qquad \text{for all } A \in \mathcal{P}(X). \qquad (4.16)$$

To calculate $m_x(A)$ according to this formula, we add the values of $m(R)$ for all focal elements R whose projection on X is A. This is appropriate since all the focal elements R whose projection on X is A are represented in m_x by A. Similarly,

$$m_y(B) = \sum_{R:B=R_y} m(R) \qquad \text{for all } B \in \mathcal{P}(Y) \qquad (4.17)$$

defines the projection of m on Y. Let m_x and m_y be called *marginal basic assignments*, and let $(\mathcal{F}_x, m_x)$ and $(\mathcal{F}_y, m_y)$ be the associated *marginal bodies of evidence*.

Two marginal bodies of evidence $(\mathcal{F}_x, m_x)$ and $(\mathcal{F}_y, m_y)$ are said to be *noninteractive* if and only if for all $A \in \mathcal{F}_x$ and all $B \in \mathcal{F}_y$

$$m(A \times B) = m_x(A) \cdot m_y(B) \qquad (4.18)$$

and

$$m(R) = 0 \qquad \text{for all } R \neq A \times B.$$

That is, two marginal bodies of evidence are noninteractive if and only if the only focal elements of the joint body of evidence are Cartesian products of focal elements of the marginal bodies and if the joint basic assignment is determined by the product of the marginal basic assignments. It is clear that this concept of noninteraction is based on Dempster's rule of combination. We may say that two marginal bodies of evidence are noninteractive if the joint basic assignment is obtained from the marginal basic assignment by Dempster's rule.

Example 4.2.

Consider the body of evidence given in Table 4.2(a). Focal elements are subsets of the Cartesian product $X \times Y$, where $X = \{1, 2, 3\}$ and $Y = \{a, b, c\}$; they are defined in the table by their characteristic functions. To emphasize that each focal element is, in fact, a binary relation on $X \times Y$, they are labeled $R_1, R_2, \ldots, R_{12}$. Employing Eqs. (4.16) and (4.17), we obtain the marginal bodies of evidence shown in Table 4.2(b). For example,

$$m_x(\{2, 3\}) = m(R_1) + m(R_2) + m(R_3) = .25,$$

$$m_x(\{1, 2\}) = m(R_5) + m(R_8) + m(R_{11}) = .3,$$

$$m_y(\{a\}) = m(R_2) + m(R_7) + m(R_8) + m(R_9) = .25,$$

$$m_y(\{a, b, c\}) = m(R_3) + m(R_{10}) + m(R_{11}) + m(R_{12}) = .5,$$

and similarly for the remaining sets A and B. We can easily verify that the joint basic

TABLE 4.2. JOINT AND MARGINAL BODIES OF EVIDENCE: ILLUSTRATION OF INDEPENDENCE (EXAMPLE 4.2).

(a) Joint body of evidence

$X \times Y$ 1a 1b 1c 2a 2b 2c 3a 3b 3c	$m(R_i)$
R_1 = 0 0 0 0 1 1 0 1 1	.0625
R_2 = 0 0 0 1 0 0 1 0 0	.0625
R_3 = 0 0 0 1 1 1 1 1 1	.125
R_4 = 0 1 1 0 0 0 0 1 1	.0375
R_5 = 0 1 1 0 1 1 0 0 0	.075
R_6 = 0 1 1 0 1 1 0 1 1	.075
R_7 = 1 0 0 0 0 0 1 0 0	.0375
R_8 = 1 0 0 1 0 0 0 0 0	.075
R_9 = 1 0 0 1 0 0 1 0 0	.075
R_{10} = 1 1 1 0 0 0 1 1 1	.075
R_{11} = 1 1 1 1 1 1 0 0 0	.15
R_{12} = 1 1 1 1 1 1 1 1 1	.15

$$m: \mathcal{P}(X \times Y) \to [0, 1]$$

(b) Marginal bodies of evidence

X 1 2 3	$m_x(A)$
A = 0 1 1	.25
1 0 1	.15
1 1 0	.3
1 1 1	.3

$$m_x: \mathcal{P}(X) \to [0, 1]$$

Y a b c	$m_y(B)$
B = 0 1 1	.25
1 0 0	.25
1 1 1	.5

$$m_y: \mathcal{P}(Y) \to [0, 1]$$

assignment m is uniquely determined in this case by the marginal basic assignments through Eq. (4.18). The marginal bodies of evidence are thus noninteractive. For example,

$$m(R_1) = m_x(\{2, 3\}) \cdot m_y(\{b, c\})$$

$$= .25 \times .25 = .0625.$$

Observe that $\{2, 3\} \times \{b, c\} = \{2b, 2c, 3b, 3c\} = R_1$. Similarly,

$$m(R_{10}) = m_x(\{1, 3\}) \cdot m_y(\{a, b, c\})$$

$$= .15 \times .5 = .075,$$

where $R_{10} = \{1, 3\} \times \{a, b, c\} = \{1a, 1b, 1c, 3a, 3b, 3c\}$.

4.3 PROBABILITY MEASURES

When Axiom (4.3) for belief measures is replaced with a stronger axiom

$$Bel(A \cup B) = Bel(A) + Bel(B) \quad \text{whenever} \quad A \cap B = \varnothing, \quad (4.19)$$

we obtain a special type of belief measures—the classical *probability measures* (sometimes also referred to as *Bayesian belief measures*).

The following theorem expresses the most fundamental property of probability measures as special types of belief measures.

Theorem 4.1. A belief measure Bel on a finite power set $\mathcal{P}(X)$ is a probability measure if and only if its basic assignment m is given by $m(\{x\}) = Bel(\{x\})$ and $m(A) = 0$ for all subsets of X that are not singletons.

Proof: Assume that Bel is a probability measure. For the empty set $\varnothing$, the theorem trivially holds, since $m(\varnothing) = 0$ by definition of m. Let $A \neq \varnothing$ and assume $A = \{x_1, x_2, \ldots, x_n\}$. Then, by repeated application of Axiom (4.19), we obtain

$$Bel(A) = Bel(\{x_1\}) + Bel(\{x_2, x_3, \ldots, x_n\})$$

$$= Bel(\{x_1\}) + Bel(\{x_2\}) + Bel(\{x_3, x_4, \ldots, x_n\})$$

$$= \cdots = Bel(\{x_1\}) + Bel(\{x_2\}) + \cdots + Bel(\{x_n\}).$$

Since $Bel(\{x\}) = m(\{x\})$ for any $x \in X$ by Eq. (4.10), we have

$$Bel(A) = \sum_{i=1}^{n} m(\{x_i\}).$$

Hence, Bel is defined in terms of a basic assignment that focuses only on singletons.

Assume now that a basic assignment m is given such that

$$\sum_{x \in X} m(\{x\}) = 1.$$

Then, for any sets $A, B \in \mathcal{P}(X)$ such that $A \cap B = \varnothing$, we have

$$Bel(A) + Bel(B) = \sum_{x \in A} m(\{x\}) + \sum_{x \in B} m(\{x\})$$

$$= \sum_{x \in A \cup B} m(\{x\}) = Bel(A \cup B)$$

and, consequently, Bel is a probability measure. This completes the proof. ∎

According to Theorem 4.1, probability measures on finite sets are thus fully represented by a function

$$p : X \to [0, 1]$$

such that $p(x) = m(\{x\})$. This function is usually called a *probability distribution function*. Let $\mathbf{p} = (p(x) \mid x \in X)$ be referred to as a *probability distribution* on X.

When the basic assignment focuses only on singletons, as required for probability measures, then the right hand sides of Eqs. (4.10) and (4.11) become equal. Hence,

$$Bel(A) = Pl(A) = \sum_{x \in A} m(\{x\})$$

for all $A \in \mathcal{P}(X)$; this can also be written, by utilizing the notion of a probability distribution function, as

$$Bel(A) = Pl(A) = \sum_{x \in A} p(x).$$

This means that the dual belief and plausibility measures merge under Axiom (4.19) of probability measures. It is therefore convenient in this case to denote them by a single symbol. Let us denote probability measures by P. Then,

$$P(A) = \sum_{x \in A} p(x) \tag{4.20}$$

for all $A \in \mathcal{P}(X)$.

Within probability measures, total ignorance is expressed by the uniform probability distribution (or basic assignment)

$$p(x) = \frac{1}{|X|} (= m(\{x\}))$$

for all $x \in X$. This follows directly from the fact that basic assignments of probability measures are required to focus only on singletons (Theorem 4.1).

Probability measures, which are the subject of probability theory, have been studied at length. The literature of probability theory, including textbooks at various levels, is abundant. We therefore assume that the reader is familiar with the fundamentals of probability theory and, consequently, we do not attempt a full coverage of probability theory in this book. However, we briefly review a few concepts from probability theory that are employed later in the text.

When a probability distribution **p** is defined on the Cartesian product $X \times Y$, it is called a *joint probability distribution*. Projections $\mathbf{p}_X$ and $\mathbf{p}_Y$ of **p** on X and Y, respectively, are called *marginal probability distributions*; they are defined by the formulas

$$p_X(x) = \sum_{y \in Y} p(x, y) \tag{4.21}$$

for each $x \in X$ and

$$p_Y(y) = \sum_{x \in X} p(x, y) \tag{4.22}$$

for each $y \in Y$. Sets X and Y are called *noninteractive* (in the probabilistic sense) with respect to p if

$$p(x, y) = p_X(x) \cdot p_Y(y) \tag{4.23}$$

for all $x \in X$ and all $y \in Y$. This definition is a special case of the general definition of noninteractive bodies of evidence expressed by Eq. (4.18) and based on Dempster's rule of combination.

Two conditional probability distributions $\mathbf{p}_{X|Y}$ and $\mathbf{p}_{Y|X}$ are defined in terms of a joint distribution **p** by the formulas

$$p_{X|Y}(x \mid y) = \frac{p(x, y)}{p_Y(y)} \tag{4.24}$$

and

$$p_{Y|X}(y \mid x) = \frac{p(x, y)}{p_X(x)} \tag{4.25}$$

for all $x \in X$ and all $y \in Y$. The value $p_{X|Y}(x, y)$ represents the probability that x occurs provided that y is known; similarly, $p_{Y|X}(y \mid x)$ designates the probability of y given x.

Set X is called *independent* of Y (in the probabilistic sense) if

$$p_{X|Y}(x \mid y) = p_X(x) \tag{4.26}$$

for all $x \in X$ and all $y \in Y$; similarly, set Y is called independent of X if

$$p_{Y|X}(y \mid x) = p_Y(y) \tag{4.27}$$

for all $x \in X$ and all $y \in Y$. Since the joint probability distribution is defined by

$$p(x, y) = p_{X|Y}(x \mid y) \cdot p_Y(y) = p_{Y|X}(y \mid x) \cdot p_X(x), \tag{4.28}$$

we can immediately see that the sets X and Y are independent of each other if and only if they are noninteractive.

4.4 POSSIBILITY AND NECESSITY MEASURES

We say that a family of subsets of a universal set is nested if these subsets can be ordered in such a way that each is contained within the next. Thus, $A_1 \subset A_2 \subset \cdots \subset A_n$ are nested sets.

When it is required that the focal elements of a body of evidence $(\mathcal{F}, m)$ be nested, the associated belief and plausibility measures are called *consonant*. This name appropriately reflects the fact that degrees of evidence allocated to focal elements that are nested do not conflict with each other; that is, nested focal elements are free of dissonance in evidence.

The requirement that focal elements are nested restricts belief and plausibility measures in a way that is characterized by the following theorem.

Theorem 4.2.   Given a consonant body of evidence $(\mathcal{F}, m)$, the associated consonant belief and plausibility measures possess the following properties:

(i) $Bel(A \cap B) = \min[Bel(A), Bel(B)]$ for all $A, B \in \mathcal{P}(X)$;
(ii) $Pl(A \cup B) = \max[Pl(A), Pl(B)]$ for all $A, B \in \mathcal{P}(X)$.

Proof: (i) Since the focal elements in $\mathcal{F}$ are nested, they may be linearly ordered by the subset relationship. Let $\mathcal{F} = \{A_1, A_2, \ldots, A_n\}$ and assume that $A_i \subset A_j$ whenever $i < j$. Consider now arbitrary subsets A and B of X. Let i_1 be the largest integer i such that $A_i \subset A$ and let i_2 be the largest integer i such that $A_i \subset B$. Then, $A_i \subset A$ and $A_i \subset B$ if and only if $i \leq i_1$ and $i \leq i_2$, respectively. Moreover, $A_i \subset A \cap B$ if and only if $i \leq \min(i_1, i_2)$. Hence,

$$Bel(A \cap B) = \sum_{i=1}^{\min(i_1, i_2)} m(A_i) = \min\left[\sum_{i=1}^{i_1} m(A_i), \sum_{i=1}^{i_2} m(A_i)\right]$$

$$= \min[Bel(A), Bel(B)].$$

(ii) Assume that (i) holds. Then, by Eq. (4.6),

$$Pl(A \cup B) = 1 - Bel(\overline{A \cup B}) = 1 - Bel(\overline{A} \cap \overline{B})$$

$$= 1 - \min[Bel(\overline{A}), Bel(\overline{B})]$$

$$= \max[1 - Bel(\overline{A}), 1 - Bel(\overline{B})]$$

$$= \max[Pl(A), Pl(B)]$$

for all $A, B \in \mathcal{P}(X)$. ■

Consonant belief and plausibility measures are usually referred to as *necessity measures* and *possibility measures*, respectively. Let η and π denote a necessity measure and a possibility measure on $\mathcal{P}(X)$, respectively. Then,

$$\eta(A \cap B) = \min[\eta(A), \eta(B)] \tag{4.29}$$

and

$$\pi(A \cup B) = \max[\pi(A), \pi(B)] \tag{4.30}$$

for all $A, B \in \mathcal{P}(X)$. Necessity and possibility measures are often defined axiomatically with these properties, which can then be used to derive as a theorem the fact that these measures are based on nested focal elements.

Necessity and possibility measures are related to each other by the equation

$$\eta(A) = 1 - \pi(\overline{A}) \tag{4.31}$$

for all $A \in \mathcal{P}(X)$, which is only a special version of Eq. (4.6). Because of this property, we need examine only one of these measures. Any property derived for this measure can then be easily converted by Eq. (4.31) for application to the other measure. Since possibility measures are more prominent in the literature than necessity measures, we focus on the former.

Recall that probability measures can conveniently be represented by probability distribution functions. It turns out that possibility measures can be represented in this way as well, as specified in the following theorem.

Theorem 4.3.   Every possibility measure π on $\mathcal{P}(X)$ can be uniquely determined by a *possibility distribution function*

$$r : X \to [0, 1]$$

via the formula

$$\pi(A) = \max_{x \in A} r(x) \tag{4.32}$$

for each $A \in \mathcal{P}(X)$.

Proof: We prove the theorem by induction on the cardinality of set A. Let $|A| = 1$. Then, $A = \{x\}$, where $x \in X$, and Eq. (4.32) is trivially satisfied. Assume now that Eq. (4.32) is satisfied for $|A| = n - 1$, and let $A = \{x_1, x_2, \ldots, x_n\}$.

Then, by Eq. (4.30),

$$\pi(A) = \max[\pi(\{x_1, x_2, \ldots, x_{n-1}\}), \pi(\{x_n\})]$$

$$= \max[\max[\pi(\{x_1\}), \pi(\{x_2\}), \ldots, \pi(\{x_{n-1}\})], \pi(\{x_n\})]$$

$$= \max[\pi(\{x_1\}), \pi(\{x_2\}), \ldots, \pi(\{x_n\})]$$

$$= \max_{x \in A} r(x). \quad \blacksquare$$

Let a possibility distribution function r be defined on the universal set $X = \{x_1, x_2, \ldots, x_n\}$. Then,

$$\mathbf{r} = (\rho_1, \rho_2, \ldots, \rho_n),$$

where $\rho_i = r(x_i)$ for all $x_i \in X$, is called a *possibility distribution* associated with the function r. The number of components in a possibility distribution is called its *length*.

It is convenient to order possibility distributions in such a way that $\rho_i \geq \rho_j$ when $i < j$. Let $^n\mathcal{R}$ denote the set of all ordered possibility distributions of length n and let

$$\mathcal{R} = \bigcup_{n \in \mathbb{N}} {}^n\mathcal{R}.$$

Given two possibility distributions

$$^1\mathbf{r} = (^1\rho_1, {}^1\rho_2, \ldots, {}^1\rho_n) \in {}^n\mathcal{R}$$

and

$$^2\mathbf{r} = (^2\rho_1, {}^2\rho_2, \ldots, {}^2\rho_n) \in {}^n\mathcal{R}$$

for some $n \in \mathbb{N}$, we define

$$^1\mathbf{r} \leq {}^2\mathbf{r} \text{ if and only if } {}^1\rho_i \leq {}^2\rho_i \quad \text{for all } i \in \mathbb{N}_n.$$

This ordering on $^n\mathcal{R}$ is partial and forms a lattice whose join, $\vee$, and meet, $\wedge$, are defined, respectively, as

$$^i\mathbf{r} \vee {}^j\mathbf{r} = (\max(^i\rho_1, {}^j\rho_1), \max(^i\rho_2, {}^j\rho_2), \ldots, \max(^i\rho_n, {}^j\rho_n))$$

and

$$^i\mathbf{r} \wedge {}^j\mathbf{r} = (\min(^i\rho_1, {}^j\rho_1), \min(^i\rho_2, {}^j\rho_2), \ldots, \min(^i\rho_n, {}^j\rho_n))$$

for all $^i\mathbf{r}, {}^j\mathbf{r} \in {}^n\mathcal{R}$. For each $n \in \mathbb{N}$, let $(^n\mathcal{R}, \leq)$ be called a *lattice of possibility distributions of length n*.

Consider again $X = \{x_1, x_2, \ldots, x_n\}$ and assume that a possibility measure π is defined on $\mathcal{P}(X)$ in terms of its basic assignment m. This requires (by the definition of possibility measures) that all focal elements be nested. Assume, without any loss of generality, that the focal elements are some or all of the subsets in the complete sequence of nested subsets

$$A_1 \subset A_2 \subset \cdots \subset A_n \, (= X)$$

where $A_i = \{x_1, \ldots, x_i\}$, $i \in \mathbb{N}_n$, as illustrated in Fig. 4.1(a). That is, $m(A) = 0$ for each $A \neq A_i$ ($i \in \mathbb{N}_n$) and

$$\sum_{i=1}^{n} m(A_i) = 1. \tag{4.33}$$

It is not required, however, that $m(A_i) \neq 0$ for all $i \in \mathbb{N}_n$, as illustrated by an example in Fig. 4.1(b).

It follows from the previous discussion that every possibility measure can be uniquely characterized by the *n*-tuple

$$\mathbf{m} = (\mu_1, \mu_2, \ldots, \mu_n)$$

for some $n \in \mathbb{N}$, where $\mu_i = m(A_i)$ for all $i \in \mathbb{N}_n$. Clearly,

$$\sum_{i=1}^{n} \mu_i = 1 \tag{4.34}$$

and $\mu_i \in [0, 1]$ for all $i \in \mathbb{N}_n$. Let $\mathbf{m}$ be called a *basic distribution*. Let $^n\mathcal{M}$ denote the set of all basic distributions with *n* components (distributions of length *n*), and let

$$\mathcal{M} = \bigcup_{n \in \mathbb{N}} {}^n\mathcal{M}.$$

Using the introduced notation, let us demonstrate now that each basic distribution $\mathbf{m} \in \mathcal{M}$ represents exactly one possibility distribution $\mathbf{r} \in \mathcal{R}$, and vice versa. First, it follows from Eq. (4.32) and from the definition of possibility measures as consonant plausibility measures that

$$\rho_i = r(x_i) = \pi(\{x_i\}) = Pl(\{x_i\})$$

for all $x \in X$. Hence, we can apply Eq. (4.11) to $Pl(\{x_i\})$ and thus obtain a set of equations

$$\rho_i = Pl(\{x_i\}) = \sum_{k=i}^{n} m(A_k) = \sum_{k=i}^{n} \mu_k, \tag{4.35}$$

with one equation for each $i \in \mathbb{N}_n$. Written more explicitly, these equations are

$$\begin{aligned}
\rho_1 &= \mu_1 + \mu_2 + \mu_3 + \cdots + \mu_i + \mu_{i+1} + \cdots + \mu_n \\
\rho_2 &= \phantom{\mu_1 + {}} \mu_2 + \mu_3 + \cdots + \mu_i + \mu_{i+1} + \cdots + \mu_n \\
&\vdots \\
\rho_i &= \phantom{\mu_1 + \mu_2 + \mu_3 + \cdots + {}} \mu_i + \mu_{i+1} + \cdots + \mu_n \\
&\vdots \\
\rho_n &= \phantom{\mu_1 + \mu_2 + \mu_3 + \cdots + \mu_i + \mu_{i+1} + \cdots + {}} \mu_n
\end{aligned} \tag{4.35'}$$

Solving these equations for μ_i ($i \in \mathbb{N}_n$), we obtain

$$\mu_i = \rho_i - \rho_{i+1} \tag{4.36}$$

for all $i \in \mathbb{N}_n$, where $\rho_{n+1} = 0$ by convention.

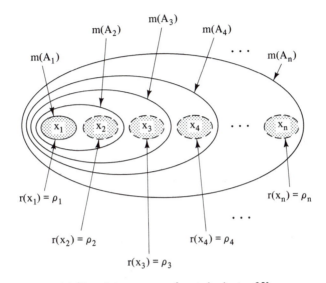

(a) Complete sequence of nested subsets of X

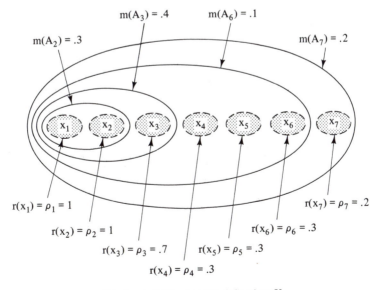

(b) A possibility measure defined on X

Figure 4.1.   Nested focal elements of possibility measure on $\mathscr{P}(X)$, where $X = \{x_1, x_2, \ldots, x_n\}$.

Equations (4.35) and (4.36) define a one-to-one correspondence

$$t : \mathcal{R} \leftrightarrow \mathcal{M}$$

between possibility distributions and the underlying basic distributions. Given

$$\mathbf{r} = (\rho_1, \rho_2, \ldots, \rho_n) \in \mathcal{R} \quad \text{and} \quad \mathbf{m} = (\mu_1, \mu_2, \ldots, \mu_n) \in \mathcal{M}$$

for some $n \in \mathbb{N}$, $t(\mathbf{r}) = \mathbf{m}$ if and only if Eq. (4.36) is satisfied; similarly, $t^{-1}(\mathbf{m}) = \mathbf{r}$ if and only if Eq. (4.35) is satisfied.

Function t enables us to define a partial ordering on set $\mathcal{M}$ in terms of the partial ordering defined on set $\mathcal{R}$. For all $^1\mathbf{m}, ^2\mathbf{m} \in \mathcal{M}$, we define

$$^1\mathbf{m} \le {}^2\mathbf{m} \quad \text{if and only if} \quad t^{-1}(^1\mathbf{m}) \le t^{-1}(^2\mathbf{m}).$$

Example 4.3.

Consider the basic assignment m specified in Fig. 4.1(b). We observe that Eq. (4.33) is satisfied, but $m(A_i) = 0$ in this case for some subsets of the complete sequence

$$A_1 \subset A_2 \subset \cdots \subset A_7$$

of nested subsets. For example, $m(A_1) = m(\{x_1\}) = 0$ and $m(A_4) = m(\{x_1, x_2, x_3, x_4\}) = 0$. The basic distribution is

$$\mathbf{m} = (0, .3, .4, 0, 0, .1, .2).$$

Applying Eq. (4.35) to this basic distribution for all $i \in \mathbb{N}_7$, we obtain the possibility distribution

$$\mathbf{r} = (1, 1, .7, .3, .3, .3, .2),$$

as shown in Fig. 4.1(b). For instance,

$$\rho_3 = \sum_{k=3}^{7} \mu_k = .4 + 0 + 0 + .1 + .2 = .7$$

or

$$\rho_5 = \sum_{k=5}^{7} \mu_k = 0 + .1 + .2 = .3.$$

Degrees of possibility $\pi(A)$ can now be calculated for any subset A of $X = \{x_1, x_2, \ldots, x_7\}$ from components of the possibility distribution $\mathbf{r}$ by Eq. (4.32). For instance,

$$\pi(\{x_1, \ldots, x_k\}) = \max(\rho_1, \ldots, \rho_k) = \max(1, \ldots, \rho_k) = 1$$

for each $k \in \mathbb{N}_7$; similarly,

$$\pi(\{x_3, x_4, x_5\}) = \max(\rho_3, \rho_4, \rho_5) = \max(.7, .3, .3) = .7.$$

It follows from Eqs. (4.35) and (4.9) that $\rho_1 = 1$ for each possibility distribution

$$\mathbf{r} = (\rho_1, \rho_2, \ldots, \rho_n) \in {}^n\mathcal{R}.$$

Hence, the *smallest possibility distribution* $\check{\mathbf{r}}_n$ of length n has the form

$$\check{\mathbf{r}}_n = (1, 0, 0, \ldots, 0)$$

with $n - 1$ zeros. This possibility distribution, whose basic assignment

$$t(\check{\mathbf{r}}_n) = (1, 0, 0, \ldots, 0)$$

has the same form, represents *perfect evidence* with no uncertainty involved. The *largest possibility distribution* $\hat{\mathbf{r}}_n$ of length n consists of all 1's, and

$$t(\hat{\mathbf{r}}_n) = (0, 0, \ldots, 0, 1)$$

with $n - 1$ zeros. This distribution represents *total ignorance*, that is, a situation in which no relevant evidence is available. In general, the larger the possibility distribution, the less specific the evidence and, consequently, the more ignorant we are.

Let us consider now *joint possibility distributions* $\mathbf{r}$ defined on the Cartesian product $X \times Y$. Projections $\mathbf{r}_X$ and $\mathbf{r}_Y$ of $\mathbf{r}$, which are called *marginal possibility distributions*, are defined by the formulas

$$r_X(x) = \max_{y \in Y}[r(x, y)] \qquad (4.37)$$

for each $x \in X$ and

$$r_Y(y) = \max_{x \in X}[r(x, y)] \qquad (4.38)$$

for each $y \in Y$. These formulas follow directly from Eq. (4.32). To see this, note that each particular element x of X, for which the marginal distribution $\mathbf{r}_X$ is defined, stands for the set $\{(x, y) \mid y \in Y\}$ of pairs in $X \times Y$ for which the joint distribution is defined. Hence, it must be that

$$\pi_X(\{x\}) = \pi(\{x, y\} \mid y \in Y),$$

where π_X and π are possibility measures corresponding to r_X and r, respectively. Applying Eq. (4.32) to the left-hand side of this equation, we obtain

$$\pi_X(\{x\}) = r_X(x);$$

and applying it to the right-hand side of the equation yields

$$\pi(\{(x, y) \mid y \in Y\}) = \max_{y \in Y} r(x, y).$$

Hence, Eq. (4.37) is obtained. We can argue similarly for Eq. (4.38).

Sets X and Y are called *noninteractive* (in the possibilistic sense) if and only if

$$r(x, y) = \min[r_X(x), r_Y(y)] \qquad (4.39)$$

for all $x \in X$ and all $y \in Y$. An example of noninteractive consonant bodies of evidence is given in Fig. 4.2, which is self-explanatory. Observe that for each joint possibility ρ_{ij} in Fig. 4.2(b), we have

$$\rho_{ij} = \min(\rho_i, \rho'_j),$$

where ρ_i, ρ_j' are the respective marginal possibilities given in Fig. 4.2(a). For instance,

$$\rho_{32} = \min(\rho_3, \rho_2') = \min(.6, 1) = .6$$

or

$$\rho_{43} = \min(\rho_4, \rho_3') = \min(.5, .3) = .3.$$

This definition of possibilistic noninteraction (Eq. (4.39)) is clearly not based upon Dempster's rule. Consequently, it does not conform to the general definition

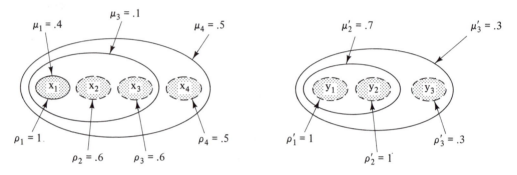

(a) Marginal consonant bodies of evidence

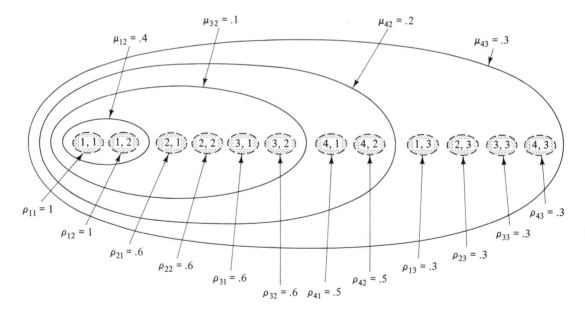

(b) Joint consonant body of evidence

Figure 4.2.   Example of noninteractive sets with possibility measure: (a) marginal consonant bodies of evidence; (b) joint consonant body of evidence.

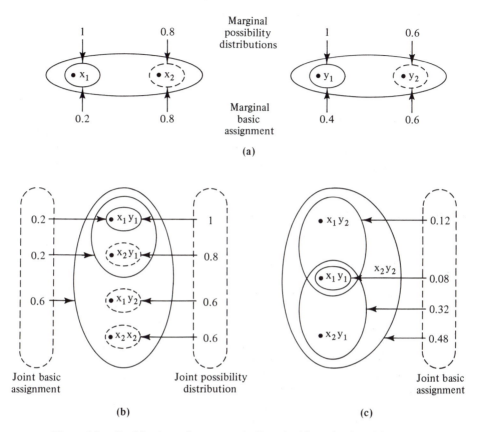

Figure 4.3. Combinations of consonant bodies of evidence by the minimum operator versus Dempster's rule (Example 4.4).

of noninteractive bodies of evidence as expressed by Eq. (4.18). Dempster's rule cannot be used for defining possibilistic noninteraction because it does not preserve consonance. This is illustrated by the following example.

Example 4.4

Consider the marginal possibility distributions and basic distributions specified in Fig. 4.3(a), which represent two consonant bodies of evidence. When we combine them by Eq. (4.39), we obtain the joint possibility distribution shown in Fig. 4.3(b). That is, the result of combining two nested (consonant) bodies of evidence by the min operator is again nested (consonant) so that we remain in the domain of possibility theory. In contrast to this, when Dempster's rule is employed we obtain the joint basic assignment shown in Fig. 4.3(c). This basic assignment does not represent a consonant (nested) body of evidence and, consequently, it is not a subject of possibility theory.

It follows directly from Eqs. (4.37) and (4.38) that

$$r(x, y) \leq r_X(x) \quad \text{and} \quad r(x, y) \leq r_Y(y).$$

This implies that, in general,

$$r(x, y) \le \min[r_X(x), r_Y(y)]. \tag{4.40}$$

The joint possibility distribution $r(x, y)$ defined by Eq. (4.39) is thus the largest joint distribution that satisfies the given marginal distributions. As such, it represents the least specific claim about the actual joint distribution. It is clearly desirable to define noninteraction in terms of this extreme condition.

The concept of conditional possibilities, which is required for defining possibilistic independence (analogous to probabilistic independence), is still a controversial issue. Some views on this issue are briefly overviewed in Note 4.8.

4.5 RELATIONSHIP AMONG CLASSES OF FUZZY MEASURES

The concept of a *fuzzy measure* provides us with a large framework within which various special classes of measures can be formulated, including the classical probability measures. The most fundamental property of fuzzy measures is their monotonicity with respect to the subset relationship.

The special classes of fuzzy measures that are introduced in this chapter are currently the most prominent types of measures in the literature. The inclusion relationship among them is summarized in Fig. 4.4.

Two large classes of fuzzy measures, which are covered in Sec. 4.2, are *belief measures and plausibility measures*. These are mutually dual in the sense that one of them can be uniquely determined from the other. Together, they form a theory that is usually referred to as a *mathematical theory of evidence*. This theory is fully characterized by a pair of *dual axioms of subadditivity*, expressed by Eqs. (4.3) and (4.8).

Each pair consisting of a belief measure and its dual plausibility measure is represented by a single function—the *basic probability assignment*, which assigns degrees of evidence or belief to certain specific subsets of the universal set. Subsets that are assigned nonzero degrees of evidence are called *focal elements*.

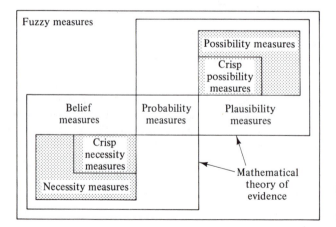

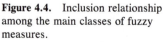

Figure 4.4.   Inclusion relationship among the main classes of fuzzy measures.

When all focal elements are *singletons*, the corresponding belief measure is equal to its dual plausibility measure. In these cases, belief and plausibility measures are not distinguished. The dual properties of subadditivity merge into a single property of *additivity*, expressed by Eq. (4.19), which is a fundamental property of the classical *probability measures*. That is, probability measures are the merging of belief and plausibility measures, as visually expressed in Fig. 4.4. An important property of probability measures is that each of them can be uniquely represented by a function defined on elements of the universal set rather than on its subsets. This function is called a *probability distribution function*.

When focal elements are *nested*, we obtain a special subclass of plausibility measures and its dual subclass of belief measures. These measures, which are covered in Sec. 4.4, are called *possibility measures* and *necessity measures*, respectively. Given a possibility measure, the dual necessity measure is uniquely determined. As is the case with probability measures, possibility and necessity measures can also be uniquely characterized by functions defined on the universal set. These functions are called a *possibility distribution function* and a *necessity distribution function*, respectively. When the values of these functions are restricted to 0s and 1s, the measures are referred to as *crisp possibility or necessity measures*.

Possibility, necessity, and probability measures do not overlap with one another except for one very special measure, which is characterized by only one focal element that is a singleton. The three distribution functions that represent probabilities, possibilities, and necessities become equal for this measure: one element of the universal set is assigned the value of 1, with all other elements being assigned a value of 0. This is clearly the only measure that represents perfect evidence.

NOTES

4.1. The concept of *fuzzy measures* was introduced by Sugeno [1977]. Good surveys of various types of measures subsumed under this broad concept were prepared by Dubois and Prade [1980a], Bacon [1981], and Wierzchon [1982a].

4.2. The *mathematical theory of evidence* that is based on the complementary *belief and plausibility measures* was originated and developed by Glenn Shafer [1976a, b]. It was motivated by previous work on upper and lower probabilities by Dempster [1967], as well as by Shafer's historical reflection upon the concept of probability [Shafer, 1978] and his critical examination of the Bayesian approach to evidence [Shafer, 1976a, 1981a]. Although Shafer's book [1976a] is still the most comprehensive presentation of the theory, new developments regarding the theory have been made after the book was published: Bacon [1981], Dubois and Prade [1982a, c, 1986], De Feriet [1982], and Wierzchon [1982] clarified the relationship of belief and plausibility measures to other types of fuzzy measures; Shafer [1979, 1981b], Smets [1981a], and Shafer and Tversky [1985] made new contributions to the theory, Zadeh [1979b] discussed the validity of Dempster's rule of combination, whereas Oblow [1987a, b] investigated the rules of combination of evidence in a comprehensive fashion and derived Dempster's rule as a special and well-justified rule for combining evidence from independent sources.

4.3. The derivation of formula (4.13) is presented in the book by Shafer [1976a]. The validity of the formula is stated in the book as Theorem 2.2 (p. 39); its proof is covered in Lemmas 2.1 through 2.3 (pp. 47–49). Proofs of Theorems 4.1 through 4.3 are also due to Shafer [1976a].

4.4. In addition to the basic assignment, belief measure, and plausibility measure, it is sometimes also convenient to use the function

$$Q(A) = \sum_{X \neq B \supset A} m(B), \qquad (4.41)$$

which is called a *commonality number*. For each $A \in \mathcal{P}(X)$, the value $Q(A)$ represents the total portion of belief that can move freely to every point of A. Given commonality numbers $Q(A)$ for all $A \in \mathcal{P}(X)$, the corresponding plausibility measure can be calculated by the formula

$$Pl(A) = \sum_{\varnothing \neq B \subset A} (-1)^{|B|+1} Q(B). \qquad (4.42)$$

A derivation of this formula is given in the book by Shafer [1976a, pp. 44, 52, 53].

4.5. Given a body of evidence $(\mathcal{F}, m)$, the corresponding functions Pl and Bel can be viewed as *upper and lower probabilities* that characterize a set of probability measures. Individual probability measures in the set can be defined by the following allocation procedure:

  1. Choose a particular $x_A \in A$ for each $A \in \mathcal{F}$.
  2. Set $P(\{x\}) = p(x) = \displaystyle\sum_{A:x_A = x} m(A)$ for all $x \in X$.

For all probability measures P that can be assigned in this way, the inequalities

$$Bel(A) \leq P(A) \leq Pl(A) \qquad (4.43)$$

are known to hold for all $A \in \mathcal{F}$. This explains the interpretation of Pl and Bel as upper and lower probabilities, respectively. A body of evidence thus represents for each $A \in \mathcal{P}(X)$ the range $[Bel(A), Pl(A)]$ of feasible probabilities; this range is clearly $[0, 1]$ for all $A \neq \varnothing$ in the face of total ignorance.

Lower and upper probabilities were first investigated by Dempster [1967] independently of the concepts of belief and plausibility measures; they are well discussed in a paper by De Feriet [1982].

4.6. From the copious literature on *probability theory*, let us identify a few references that are particularly suitable as supplementary readings to this text. The classical book by Kolmogorov [1950] is still considered the best axiomatic treatment of probability theory. One of the most popular textbooks on probability theory is a two-volume book by Feller [1950, 1966]; a more elementary text with many examples was prepared by Lipschutz [1968]. A comprehensive as well as thorough survey of both mathematical and interpretative aspects of probability theory can be found in the book by Fine [1973] and a paper by Walley and Fine [1979]. A history of ideas leading to probability theory is traced in the book by Hacking [1975] and the previously mentioned paper by Shafer [1978]. Among other notable books on probability theory, we recommend the books by Reichenbach [1949], Rényi [1970], Savage [1972], and De Finetti [1974].

4.7. Although we present *possibility theory* as a theory of consonant plausibility measures, it was originally introduced in the context of fuzzy sets by Zadeh [1978a]. Given a fuzzy subset A of X with membership grade function μ_A, Zadeh defines a possibility distribution function r associated with A as numerically equal to μ_A, that is,

$$r(x) = \mu_A(x) \tag{4.44}$$

for all $x \in X$; then, he defines the corresponding possibility measure π by the equation

$$\pi(A) = \sup_{x \in A} r(x) \tag{4.45}$$

for each $A \in \mathcal{P}(X)$. This equation becomes Eq. (4.32) when X is finite. In this approach to possibility theory, focal elements are represented by α-cuts of the associated fuzzy set. The relationship between these two alternative approaches to possibility theory is discussed by Dubois and Prade on several occasions [Dubois and Prade, 1982a, b, 1985d, 1986]. As supplementary readings on possibility theory, we also recommend papers by Giles [1982], Nguyen [1979], and Yager [1980d, e, 1983c].

4.8. In analogy with probability theory, *possibilistic independence* can be defined by the requirement that*

$$r(x \mid y) = r(x) \tag{4.46}$$

and

$$r(y \mid x) = r(y), \tag{4.47}$$

where $r(x \mid y)$ and $r(y \mid x)$ denote *conditional possibilities*. It is not obvious, however, how to define these conditional possibilities. Two views on this issue are expressed in the literature by Hisdal [1978, 1979, 1980] and Nguyen [1978]. Hisdal argues that the equations

$$r(x, y) = \min[r(y), r(x \mid y)] \tag{4.48}$$

$$r(x, y) = \min[r(x), r(y \mid x)] \tag{4.49}$$

must be satisfied for any two sets characterized by marginal possibility distributions $r(x)$ and $r(y)$ and joint possibility distribution $r(x, y)$. Then, it is clear that Eqs. (4.46) and (4.48) as well as Eqs. (4.47) and (4.49) imply Eq. (4.39). Hence, the possibilistic independence implies possibilistic noninteraction. The converse, however, is not true. Indeed, from Eqs. (4.39) and (4.48), we obtain

$$r(x \mid y) = \begin{cases} r(x) & \text{if } r(x) < r(y) \\ [r(x), 1] & \text{otherwise} \end{cases} \tag{4.50}$$

and similarly for the other conditional possibility. Hence, possibilistic noninteraction does not imply possibilistic independence. Nguyen [1978] takes a radically different approach to the meaning of the conditional possibilities. He defines "normalized" conditional possibilities in such a way that, by analogy to probability theory, possibilistic noninteraction is required to be equivalent to possibilistic independence.

---

* In this note, the meaning of each possibility function is uniquely determined by the argument shown in the parentheses and, consequently, we do not identify it by a subscript.

This requirement leads to the formula

$$r(x \mid y) = \begin{cases} r(x, y) & \text{if } r(x) \le r(y) \\ r(x, y) \cdot \dfrac{r(x)}{r(y)} & \text{otherwise.} \end{cases} \qquad (4.51)$$

where $r(x)/r(y)$ is a normalization factor.

4.9. It was shown by Puri and Ralescu [1982] that possibility measures do not satisfy the axiom of continuity of fuzzy measures except in some special cases. This result has no relevance, of course, to possibility measures that are defined on finite sets.

4.10. The need for a consonant theory of evidence was perceived by a British economist Shackle [1961] well before possibility theory was proposed by Zadeh [1978a] or emerged from the mathematical theory of evidence [Shafer, 1976a].

4.11. A fuzzy measure g is called a *Sugeno measure* if it satisfies, in addition to Axioms g1 through g3 of fuzzy measures, the following special axiom: for all $A, B \in \mathcal{P}(X)$, if $A \cap B = \varnothing$, then

$$g(A \cup B) = g(A) + g(B) + \lambda g(A)g(B), \qquad (4.52)$$

where $\lambda > -1$ is a parameter by which different types of Sugeno measures are distinguished. Sugeno measures were investigated by Wierzchon [1982]. Among other results, he shows that Sugeno measures are belief measures for $\lambda > 0$, plausibility measures for $\lambda < 0$, and probability measures for $\lambda = 0$. It is known, however, that Sugeno measures are a subset of belief and plausibility measures [Bacon, 1981].

4.12. Various other types of fuzzy measures have been studied. The following are useful references in this regard: Bacon [1981], Dubois and Prade [1980a, 1982a, c, 1986], Goodman [1982], Höhle [1984], Höhle and Klement [1984], and Klement [1982].

EXERCISES

4.1. Compare the concepts of fuzzy sets and fuzzy measures in some situations connected with, for example, the degree of education of a person, the age of a painting or of a collector's coin, the size of a city or a country, the degree of inflation or unemployment in a country, the distance between two cities, and so on.

4.2. Let $X = \{a, b, c, d\}$. Given the basic assignment $m(\{a, b, c\}) = .5$, $m(\{a, b, d\}) = .2$, and $m(X) = .3$, determine the corresponding belief and plausibility measures.

4.3. Repeat Exercise 4.2 for some of the basic assignments given in Table 4.3, where subsets of X are defined by their characteristic functions.

4.4. Show that the function Bel determined by Eq. (4.10) for any given basic assignment m is a belief measure.

4.5. Show that the function Pl determined by Eq. (4.11) for any given basic assignment m is a plausibility measure.

4.6. Let $X = \{a, b, c, d\}$. Given the belief measure $Bel(\{b\}) = .1$, $Bel(\{a, b\}) = .2$, $Bel(\{b, c\}) = .3$, $Bel(\{b, d\}) = .1$, $Bel(\{a, b, c\}) = .4$, $Bel(\{a, b, d\}) = .2$, $Bel(\{b, c, d\}) = .6$, and $Bel(X) = 1$, determine the corresponding basic assignment.

4.7. Using Eqs. (4.7) and (4.13), derive a formula by which the basic assignment for a given plausibility measure can be determined.

TABLE 4.3. BASIC ASSIGNMENTS EMPLOYED IN EXERCISES.

a	b	c	d	m_1	m_2	m_3	m_4	m_5	m_6	m_7	m_8	m_9	m_{10}
0	0	0	0	0	0	0	0	0	0	0	0	0	0
0	0	0	1	.2	0	0	.2	.2	.3	0	.1	0	.7
0	0	1	0	0	.4	0	0	.2	0	0	.1	0	.1
0	0	1	1	0	0	0	.1	0	.3	0	.1	0	0
0	1	0	0	0	.5	0	0	.3	0	0	.1	0	0
0	1	0	1	0	0	0	0	0	0	.4	.1	0	0
0	1	1	0	.3	0	0	0	0	0	.2	.1	.1	0
0	1	1	1	0	0	0	0	0	.3	.1	.1	.1	.2
1	0	0	0	.1	.1	.2	0	.3	0	.3	.1	.1	0
1	0	0	1	0	0	0	0	0	0	0	.1	.1	0
1	0	1	0	.1	0	.3	0	0	0	0	.1	.1	0
1	0	1	1	0	0	0	0	0	0	0	0	.1	0
1	1	0	0	0	0	0	0	0	0	0	0	.1	0
1	1	0	1	.2	0	0	0	0	0	0	0	.1	0
1	1	1	0	.1	0	.4	0	0	0	0	0	.1	0
1	1	1	1	0	0	.1	.7	0	.1	0	0	.1	0

4.8. Calculate the joint basic assignment $m_{1,2}$ for the focal elements C, $R \cup D$, and $D \cup C$ in Example 4.1. Also determine $Bel_{1,2}$ for these focal elements.

4.9. For each of the focal elements in Example 4.1 (Raphael's paintings), determine the ranges $[Bel_1(A), Pl_1(A)]$, $[Bel_2(A), Pl_2(A)]$, $[Bel_{1,2}(A), Pl_{1,2}(A)]$, which can be viewed as ranges of feasible probabilities corresponding to these bodies of evidence (Note 4.5).

4.10. Repeat Exercise 4.9 for some of the basic assignments given in Table 4.3.

4.11. Calculate $m_x(\{1, 3\})$, $m_x(X)$, and $m_y(\{b, c\})$ in Example 4.2.

4.12. Verify completely that the marginal bodies of evidence given in Table 4.2(b), which were obtained as projections of the joint body of evidence given in Table 4.2(a), are noninteractive.

4.13. Determine whether each of the basic assignments given in Table 4.3 represents a probability measure, possibility measure, or neither of these.

4.14. Given two noninteractive marginal possibility distributions $\mathbf{r}_X = (1, .8, .5)$ and $\mathbf{r}_y = (1, .7)$ on sets $X = \{a, b, c\}$ and $Y = \{\alpha, \beta\}$, respectively, determine the corresponding basic distributions. Then, calculate the joint basic distribution in two different ways:
   (a) by the min operator;
   (b) by Dempster's rule.
Show visually (as in Fig. 4.3) the focal elements of the marginal and joint distributions.

4.15. Repeat Exercise 4.14 for the following marginal possibility distributions:
   (a) $\mathbf{r}_X = (1, .7, .2)$ on $X = \{a, b, c\}$ and $\mathbf{r}_Y = (1, 1, .4)$ on $Y = \{\alpha, \beta, \gamma\}$;
   (b) $\mathbf{r}_X = (1, .9, .6, .2)$ on $X = \{a, b, c, d\}$ and $\mathbf{r}_Y = (1, .6)$ on $Y = \{\alpha, \beta\}$.

4.16. Express one of the conditional probabilities in terms of the other conditional probability and the marginal probabilities.

4.17. Let $X = \{a, b, c, d, e, f, g\}$ and $Y = N_7$. Using joint probability distributions on $X \times Y$, given in terms of the matrix

$$
\begin{array}{c}
 \\ a \\ b \\ c \\ d \\ e \\ f \\ g
\end{array}
\begin{array}{cccccccc}
1 & 2 & 3 & 4 & 5 & 6 & 7 \\
\left[\begin{array}{ccccccc}
.08 & 0 & .02 & 0 & 0 & .01 & 0 \\
0 & .05 & 0 & 0 & .05 & 0 & 0 \\
0 & 0 & 0 & 0 & 0 & 0 & .03 \\
.03 & 0 & 0 & .3 & 0 & 0 & 0 \\
0 & 0 & .01 & .01 & .2 & .03 & 0 \\
0 & .05 & 0 & 0 & 0 & .1 & 0 \\
0 & 0 & 0 & .02 & 0 & .01 & 0
\end{array}\right]
\end{array}
$$

determine:

(a) marginal probabilities;
(b) both conditional probabilities;
(c) hypothetical joint probability distribution based on the assumption of non-interaction.

4.18. Show that a belief measure that satisfies Eq. (4.29) is based on nested focal elements.

4.19. Show that a plausibility measure that satisfies Eq. (4.30) is based on nested focal elements.

4.20. Determine the basic assignment, possibility measure, and necessity measure for each of the following possibility distributions defined on $X = \{x_i \mid i \in N_n\}$ for appropriate values of n:

(a) $r_1 = (1, .8, .8, .5, .2)$;
(b) $r_2 = (1, 1, 1, .7, .7, .7, .7)$;
(c) $r_3 = (1, .9, .8, .6, .5, .3, .3)$;
(d) $r_4 = (1, .5, .4, .3, .2, .1)$;
(e) $r_5 = (1, 1, .8, .8, .5, .5, .5, .1)$.
Assume in each case that $\rho_i = r(x_i)$, $i \in N_n$.

4.21. Determine the possibility distribution, possibility measure, and necessity measure for each of the following basic distributions defined on $X = \{x_i \mid i \in N_n\}$ for appropriate values of n:

(a) $m_1 = (0, 0, .3, .2, 0, .4, .1)$;
(b) $m_2 = (.1, .1, .1, 0, .1, .2, .2, .2)$;
(c) $m_3 = (0, 0, 0, 0, .5, .5)$;
(d) $m_4 = (0, .2, 0, .2, 0, .3, 0, .3)$;
(e) $m_5 = (.1, .2, .3, .4)$;
(f) $m_6 = (.4, .3, .2, 1)$.
Assume in each case that $\mu_i = m(\{x_1, x_2, \ldots, x_i\})$, $i \in N_n$.

4.22. Calculate the commonality numbers for some of the basic assignments given in Table 4.3 (Note 4.4).

4.23. Let $X = \{a, b, c, d, e\}$ and $Y = N_8$. Using a joint possibility distribution on $X \times Y$ given in terms of the matrix

$$
\begin{array}{c c}
 & \begin{array}{cccccccc} 1 & 2 & 3 & 4 & 5 & 6 & 7 & 8 \end{array} \\
\begin{array}{c} a \\ b \\ c \\ d \\ e \end{array} &
\left[\begin{array}{cccccccc}
1 & 0 & 0 & .3 & .5 & .2 & .4 & .1 \\
0 & .7 & 0 & .6 & 1 & 0 & .4 & .3 \\
0 & .5 & 0 & 0 & 1 & 0 & 1 & .5 \\
1 & 1 & 1 & .5 & 0 & 0 & 1 & .4 \\
.8 & 0 & .9 & 0 & 1 & .7 & 1 & .2
\end{array}\right]
\end{array}
$$

determine:

(a) marginal possibilities;

(b) joint and marginal basic assignments;

(c) both conditional possibilities of the two forms introduced in Note 4.8;

(d) hypothetical joint possibility distributions based on the assumption of non-interaction.

5

UNCERTAINTY

AND INFORMATION

5.1 TYPES OF UNCERTAINTY

Upon consulting a common dictionary about the term *uncertain*, we find that the word has a broad semantic content. For example, *Webster's New Twentieth Century Dictionary* gives six clusters of meanings for the term:

1. Not certainly known; questionable; problematical.
2. Vague; not definite or determined.
3. Doubtful; not having certain knowledge; not sure.
4. Ambiguous.
5. Not steady or constant; varying.
6. Liable to change or vary; not dependable or reliable.

When we further examine these various meanings, again using the dictionary, two categories of uncertainty emerge quite naturally; they are captured quite well by the terms *vagueness* and *ambiguity,* respectively.

In general, *vagueness* is associated with the difficulty of making sharp or precise distinctions in the world; that is, some domain of interest is vague if it cannot be delimited by sharp boundaries. *Ambiguity,* on the other hand, is associated with one-to-many relations, that is, situations in which the choice between two or more alternatives is left unspecified.

Each of these two distinct forms of uncertainty—vagueness and ambiguity—is connected with a set of kindred concepts. Some of the concepts connected with vagueness are fuzziness, haziness, cloudiness, unclearness, indistinctiveness, and sharplessness; some of the concepts connected with ambiguity are nonspecificity, one-to-many relation, variety, generality, diversity, and divergence.

It is easy to see that the concept of a fuzzy set provides a basic mathematical framework for dealing with vagueness. The concept of a fuzzy measure, on the other hand, provides a general framework for dealing with ambiguity. Indeed, a fuzzy measure specifies the degrees to which an arbitrary element of the universal set X (which is a priori nonlocated) belongs to the individual crisp subsets of X. That is, the measure specifies a set of alternative subsets of X that are associated with any given element of X to various degrees according to the available evidence.

Three types of ambiguity are readily recognizable within this framework. The first is connected with the size of the subsets that are designated by a fuzzy measure as prospective locations of the element in question. The larger the subsets, the less specific the characterization. This type of ambiguity therefore has the meaning of *nonspecificity in evidence,* as expressed by the given fuzzy measure.

The second type of ambiguity is exhibited by disjoint subsets of X that are designated by the given fuzzy measure as prospective locations of the element of concern. In this case, evidence focusing on one subset conflicts with evidence focusing on the other subsets. This type of ambiguity thus has the meaning of conflict or *dissonance in evidence.*

The third type of ambiguity is associated with the number of subsets of X that are designated by a fuzzy measure as prospective locations of the element under consideration and that do not overlap or overlap only partially. The multitude of partially or totally conflicting evidence is a source of confusion. This type of ambiguity therefore characterizes *confusion in evidence.*

It follows from this preliminary discussion that fuzzy sets and fuzzy measures reflect two fundamentally different types of uncertainty—vagueness and ambiguity. Consequently, each of them constitutes a distinct framework within which appropriate measures of the corresponding type of uncertainty must be formulated. In accordance with a current terminological trend in the literature, measures of uncertainty related to vagueness are referred to in this text as *measures of fuzziness.* They are discussed in Sec. 5.2. Measures related to ambiguity are further divided into three types referred to as *measures of nonspecificity, measures of dissonance,* and *measures of confusion* in evidence. These are discussed in Secs. 5.4 through 5.6.

Prior to the entry of fuzzy set theory into our mathematical repertory, the only well-developed mathematical apparatus for dealing with uncertainty was *probability theory.* Although useful and successful in many applications, probability theory is, in fact, appropriate for dealing with only a very special type of uncertainty. Its limitations, some of which are discussed later in this chapter, have increasingly been recognized.

It is well known that a measure of uncertainty can also be used for measuring information. That is, the amount of uncertainty regarding some situation represents the total amount of potential information in this situation. According to this view, the reduction of uncertainty by a certain amount (due to new evidence from, for instance, the outcome of an experiment or a received message) indicates the gain of an equal amount of information.

A measure of uncertainty, when adopted as a *measure of information,* does not include semantic and pragmatic aspects of information. As such, it is not adequate for dealing with information in human communication. However, when we are dealing with structural (syntactic) aspects of systems, such a measure is not only adequate but highly desirable. It can be used for measuring the degree of constraint among variables of interest, thus comprising a powerful tool for dealing with systems problems such as systems modeling, analysis, or design.

5.2 MEASURES OF FUZZINESS

The question of how to measure vagueness or fuzziness has been one of the issues associated with the development of the theory of fuzzy sets. In general, a *measure of fuzziness* is a function

$$f : \tilde{\mathcal{P}}(X) \to \mathbb{R},$$

where $\tilde{\mathcal{P}}(X)$ denotes the set of all fuzzy subsets of X. That is, the function f assigns a value $f(A)$ to each fuzzy subset A of X that characterizes the degree of fuzziness of A.

In order to qualify as a meaningful measure of fuzziness, f must satisfy certain axiomatic requirements. Although not necessarily unique, these requirements must fully capture the meaning of an intuitively acceptable characterization of the concept *degree of fuzziness*.

There are three axiomatic requirements that every meaningful measure of fuzziness must satisfy. Only one of them is unique; the remaining two depend on the meaning given to the concept of the degree of fuzziness.

The unique requirement states that the degree of fuzziness must be zero for all crisp sets in $\tilde{\mathcal{P}}(X)$ and only for these. Formally,

Axiom f1. $f(A) = 0$ if and only if A is a crisp set.

It is obvious that any function f violating this requirement would be totally unacceptable as a measure of fuzziness.

The second requirement is based upon a particular definition of the relation "sharper than" (i.e., "less fuzzy than") on the set $\tilde{\mathcal{P}}(X)$. If, according to a particular meaning given to the concept of the degree of fuzziness, set A is viewed (perceived) as sharper (less fuzzy) than set B, it is required that $f(A) \le f(B)$. Formally,

Axiom f2. If $A < B$, then $f(A) \le f(B)$;

here $A < B$ denotes that A is sharper than B. Clearly, each particular definition of the sharpness relation must properly capture the underlying conception of the degree of fuzziness.

The third requirement states that the degree of fuzziness must attain the maximum value only for a fuzzy set in $\tilde{\mathcal{P}}(X)$ that is viewed (perceived) as maximally fuzzy. Formally,

Axiom f3. $f(A)$ assumes the maximum value if and only if A is maximally fuzzy.

Of course, for each particular conception of the degree of fuzziness, the term *maximally fuzzy* attains a unique meaning.

Several measures of fuzziness have been proposed in the literature. One of them, perhaps the best known, is based on the following concepts:

1. The sharpness relation $A < B$ in Axiom f2 is defined by

$$\mu_A(x) \le \mu_B(x) \qquad \text{for } \mu_B(x) \le \tfrac{1}{2}$$

and

$$\mu_A(x) \ge \mu_B(x) \qquad \text{for } \mu_B(x) \ge \tfrac{1}{2}$$

for all $x \in X$.

2. The term *maximally fuzzy* in Axiom f3 is defined by the membership grade $\tfrac{1}{2}$ for all $x \in X$.

This measure of fuzziness is defined by the function

$$f(A) = - \sum_{x \in X} (\mu_A(x)\log_2\mu_A(x) + [1 - \mu_A(x)]\log_2[1 - \mu_A(x)]). \qquad (5.1)$$

Its normalized version, $\hat{f}(A)$, for which

$$0 \le \hat{f}(A) \le 1,$$

is clearly given by

$$\hat{f}(A) = \frac{f(A)}{|X|},$$

where $|X|$ denotes the cardinality of the universal set X.

Another measure of fuzziness, referred to as an *index of fuzziness*, is defined in terms of a metric distance (Hamming or Euclidean) of A from any of the nearest crisp sets, say crisp set C, for which

$$\mu_C(x) = 0 \qquad \text{if } \mu_A(x) \le \tfrac{1}{2}$$

and

$$\mu_C(x) = 1 \qquad \text{if } \mu_A(x) > \tfrac{1}{2}.$$

When the Hamming distance is used, the measure of fuzziness is expressed by the function

$$f(A) = \sum_{x \in X} |\mu_A(x) - \mu_C(x)|; \qquad (5.2)$$

for the Euclidean distance,

$$f(A) = (\sum_{x \in X} [\mu_A(x) - \mu_C(x)]^2)^{1/2}. \qquad (5.3)$$

It is clear that other metric distances may be used as well. For example, the Minkowski class of distances yields a class of fuzzy measures

$$f_w(A) = (\sum_{x \in X} | \mu_A(x) - \mu_C(x) |^w)^{1/w}, \tag{5.4}$$

where $w \in [1, \infty]$; obviously, (5.2) and (5.3) are special cases of (5.4) for $w = 1$ and $w = 2$, respectively. It follows directly from Theorem 2.6 that

$$f_\infty(A) = \max_{x \in X} | \mu_A(x) - \mu_C(x) |. \tag{5.5}$$

It is easy to see that (5.4), for every $w \in [1, \infty]$, is based on the same definition of sharpness in Axiom f2 and maximum fuzziness in Axiom f3 as (5.1). In fact, it is known that (5.1) and (5.4) are only special cases of a larger class of measures of fuzziness, which are all based on these meanings given to requirements Axiom f2 and Axiom f3. This larger class can be expressed by the form

$$f(A) = h(\sum_{x \in X} g_x(\mu_A(x))), \tag{5.6}$$

where g_x ($x \in X$) are functions

$$g_x : [0, 1] \to \mathbb{R}^+$$

(different, in general, for different x), which are all monotonically increasing in $[0, \frac{1}{2}]$, monotonically decreasing in $[\frac{1}{2}, 1]$, and satisfy the requirements that $g_x(0) = g_x(1) = 0$ and that $g_x(\frac{1}{2})$ be a unique maximum of g_x; h is a monotonically increasing function from $\mathbb{R}^+$ to $\mathbb{R}^+$.

For example, when

$$g_x(\mu_A(X)) = -\mu_A(x)\log_2\mu_A(x) - [1 - \mu_A(x)]\log_2[1 - \mu_A(x)]$$

for all $x \in X$ and h is an identity function on $\mathbb{R}^+$, measure (5.1) is obtained. When, for a given $w \in [1, \infty]$,

$$g_x(\mu_A(x)) = \begin{cases} \mu_A^w(x) & \text{for } \mu_A(x) \in [0, \frac{1}{2}] \\ [1 - \mu_A^w(x)] & \text{for } \mu_A(x) \in (\frac{1}{2}, 1] \end{cases}$$

for all $x \in X$ and

$$h(b) = b^{1/w},$$

measures of the class (5.4) are obtained.

It has been more recently argued that the degree of fuzziness of a fuzzy set can be expressed, in the most natural way, in terms of the lack of distinction between the set and its complement. Indeed, it is precisely the lack of distinction between sets and their complements that distinguishes fuzzy sets from crisp sets. The less a set differs from its complement, the fuzzier it is.

The formulation of measures of fuzziness based upon this approach depends on the fuzzy complement selected. When we employ this approach within the general class of fuzzy complements c, whose properties are discussed in Sec. 2.2, the term *sharpness* in Axiom f2 is defined by

$$A < B \quad \text{if and only if} \quad | \mu_A(x) - c(\mu_A(x)) | \geq | \mu_B(x) - c(\mu_B(x)) |$$

for all $x \in X$, and the term *maximum fuzziness* in Axiom f3 means that $\mu_A(x) = e_c$ for all $x \in X$, provided that the complement employed has an equilibrium e_c.

It has been established that a general class of measures of fuzziness based upon the lack of distinction between a set and its complement is exactly the same as the class of measures of fuzziness in which this lack of distinction is expressed in terms of a metric distance that is based on some form of aggregating the individual differences

$$| \mu_A(x) - c(\mu_A(x)) | = \delta_{c,A}(x)$$

for all $x \in X$. For example, using any metric distance from the Minkowski class, we obtain for each $w \in [1, \infty]$ one particular distance

$$D_{c,w}(A, A^c) = [\sum_{x \in X} \delta_{c,A}^w(x)]^{1/w},   \qquad (5.7)$$

where A^c denotes the complement of A produced by function c.

For a general metric distance D, the measure of fuzziness has the form

$$f_{c,D}(A) = D_c(Z, Z^c) - D_c(A, A^c),   \qquad (5.8)$$

where Z denotes any arbitrary crisp subset of X so that $D_c(Z, Z^c)$ is the largest possible distance in $\tilde{\mathcal{P}}(X)$ for a given c. Clearly,

$$0 \le f_{c,D}(A) \le D_{c,w}(Z, Z^c).   \qquad (5.9)$$

The normalized version $\hat{f}_{c,D}$ of this measure of fuzziness is given by the formula

$$\hat{f}_{c,D}(A) = 1 - \frac{D_c(A, A^c)}{D_c(Z, Z^c)} ;   \qquad (5.10)$$

that is,

$$0 \le \hat{f}_{c,D}(A) \le 1.$$

When the Minkowski class of metric distances, expressed by Eq. (5.7), is used, let $f_{c,w}$ denote the measure of fuzziness for the distance $D_{c,w}$. Then,

$$D_{c,w}(Z, Z^c) = | X |^{1/w},   \qquad (5.11)$$

Eq. (5.8) becomes

$$f_{c,w}(A) = | X |^{1/w} - D_{c,w}(A, A^c),   \qquad (5.12)$$

and Eq. (5.10) becomes

$$\hat{f}_{c,w} = 1 - \frac{D_{c,w}(A, A^c)}{| X |^{1/w}}.   \qquad (5.13)$$

Example 5.1

Consider the fuzzy set A on $X = \mathbb{N}_9$ specified in Table 5.1. To illustrate the effect of the complement and distance employed, Table 5.1 shows calculations of the measure of fuzziness (regular and normalized) for two complements of the Yager class and for four distances of the Minkowski class ($w = 1, 2, 3, 5$). For each case, we first calculate the local differences $\delta_{c,A}(x)$ for all $x \in X$. Then, using Eq. (5.7), we calculate for each case the distance $D_{c,w}(A, A^c)$ between the given set A and its complement A^c, as well as the distance $D_{c,w}(Z, Z^c)$ between an arbitrary crisp set on X and its complement; by Eq. (5.11), we have $D_{c,w}(Z, Z^c) = |X|^{1/w} = 9^{1/w}$, which is independent of the complement employed. Finally, using Eqs. (5.12) and (5.13), we calculate the measure of fuzziness $f_{c,w}(A)$ and its normalized version $\hat{f}_{c,w}(A)$, respectively.

Observe that the measure of fuzziness (regular as well as normalized) decreases with increasing w for both of the complements. The rate of decrease, however, is smaller for the classical fuzzy complement.

Measures of fuzziness defined in terms of different distance functions are based upon different measurement units. Although the choice of a unit is not a critical issue, it is often desirable to use a unit that is intuitively appealing in the sense that it has a simple interpretation in terms of some significant canonical situation. For instance, it seems intuitively pleasing for one unit of fuzziness to indicate that the membership in the fuzzy set of one element of the universal set is maximally uncertain. A measure of fuzziness equal to two of these units would therefore indicate that we cannot determine set membership or nonmembership at all for two elements of the universal set. Obviously, the maximally fuzzy set is one for which we cannot make this determination for any of the elements of the universal set; the fuzziness in this case is equal to the cardinality of the (finite) universal set. To define this unit formally, we require that the degree of fuzziness is 1 for every set A (defined on a finite universal set X) for which $\mu_A(\dot{x})$ is equal to the equilibrium of the complement employed for one particular $\dot{x} \in X$ and that

TABLE 5.1. THE EFFECT OF THE COMPLEMENT AND DISTANCE (VALUE OF w) EMPLOYED ON THE MEASURE OF FUZZINESS EXPRESSED BY EQS. (5.8) AND (5.10).

		$c(a) = 1 - a$				$c(a) = (1 - a^2)^{1/2}$			
X	$\mu_A(x)$	$\delta^1_{c,A}(x)$	$\delta^2_{c,A}(x)$	$\delta^3_{c,A}(x)$	$\delta^5_{c,A}(x)$	$\delta^1_{c,A}(x)$	$\delta^2_{c,A}(x)$	$\delta^3_{c,A}(x)$	$\delta^5_{c,A}(x)$
1	.2	.6	.36	.22	.08	.78	.61	.47	.29
2	.5	0	0	0	0	.37	.14	.05	.01
3	.7	.4	.16	.06	0	.01	0	0	0
4	.8	.6	.36	.22	.08	.20	.04	0	0
5	.9	.8	.64	.51	.11	.46	.21	.09	.02
6	1	1	1	1	1	1	1	1	1
7	.6	.2	.04	.01	0	.20	.04	.01	0
8	.4	.2	.04	.01	0	.52	.27	.14	.04
9	.1	.8	.64	.51	.11	.89	.79	.70	.56
		$D_{c,1}(A, A^c)$	$D_{c,2}(A, A^c)$	$D_{c,3}(A, A^c)$	$D_{c,5}(A, A^c)$	$D_{c,1}(A, A^c)$	$D_{c,2}(A, A^c)$	$D_{c,3}(A, A^c)$	$D_{c,5}(A, A^c)$
		4.6	1.8	1.36	1.05	4.43	1.76	1.35	1.14
		$|X|$	$|X|^{1/2}$	$|X|^{1/3}$	$|X|^{1/5}$	$|X|$	$|X|^{1/2}$	$|X|^{1/3}$	$|X|^{1/5}$
		9.0	3.0	2.08	1.55	9.0	3.0	2.08	1.55
		$f_{c,1}(A)$	$f_{c,2}(A)$	$f_{c,3}(A)$	$f_{c,5}(A)$	$f_{c,1}(A)$	$f_{c,2}(A)$	$f_{c,3}(A)$	$f_{c,5}(A)$
		4.4	1.2	.72	.5	4.57	1.24	.73	.41
		$\hat{f}_{c,1}(A)$	$\hat{f}_{c,2}(A)$	$\hat{f}_{c,3}(A)$	$\hat{f}_{c,5}(A)$	$\hat{f}_{c,1}(A)$	$\hat{f}_{c,2}(A)$	$\hat{f}_{c,3}(A)$	$\hat{f}_{c,5}(A)$
		.49	.4	.35	.32	.51	.41	.35	.26

$\mu_A(x) \in \{0, 1\}$ for all $x \in X$ that are different from $\dot{x}$. (Note that the equilibrium is that membership value which is the same in a fuzzy set and its complement; it therefore indicates maximum uncertainty concerning set membership.) Then,

$$\delta_{c,A}(x) = \begin{cases} 0 & \text{when } x = \dot{x} \\ 1 & \text{otherwise} \end{cases}$$

and our requirement leads to the equation

$$f_{c,w}(A) = |X|^{1/w} - (|X| - 1)^{1/w} = 1.$$

This equation is satisfied only for $w = 1$, thus implying that our requirement is satisfied only by the Hamming distance.

This unit of fuzziness characterizes the maximum fuzziness (vagueness) regarding the truth value of a single proposition, namely, the proposition that an element x belongs in fuzzy set A. Since truth values of propositions are often denoted by binary digits 0 and 1, the unit that characterizes full uncertainty (represented in our case by vagueness) for a single proposition is usually called a *bit*, which is an abbreviation for binary digit. The same reasoning is used to develop the notion of a bit in the context of the Shannon entropy discussed in Sec. 5.3, as well as any other measure of uncertainty.

When we accept bits as units of fuzziness, Eqs. (5.12) and (5.13) become more specific:

$$f_c(A) = |X| - \sum_{x \in X} |\mu_A(x) - c(\mu_A(x))| \qquad \text{(in bits)}, \qquad (5.14)$$

$$\hat{f}_c(A) = \frac{f_c(A)}{|X|}. \qquad (5.15)$$

The two values of $f_{c,1}$ in Table 5.1 provide us with the degrees of fuzziness, as measured in bits, of the set A for the two complements.

The previous definitions of measures of fuzziness are based on the assumption that the universal set X is finite. These definitions can be readily extended to infinite sets. Consider, for example, that $X = [a, b]$. Then, Eq. (5.7) becomes

$$D_{c,w}(A, A^c) = \left(\int_a^b \delta_{c,A}^w(x) \, dx \right)^{1/w} \qquad (5.16)$$

Since

$$D_{c,w}(Z, Z^c) = \left(\int_a^b dx \right)^{1/w} = (b - a)^{1/w}, \qquad (5.17)$$

we obtain

$$f_{c,w}(A) = (b - a)^{1/w} - \left(\int_a^b \delta_{c,A}^w(x) \, dx \right)^{1/w} \qquad (5.18)$$

and

$$\hat{f}_{c,w}(A) = \frac{f_{c,w}(A)}{(b - a)^{1/w}} .$$ (5.19)

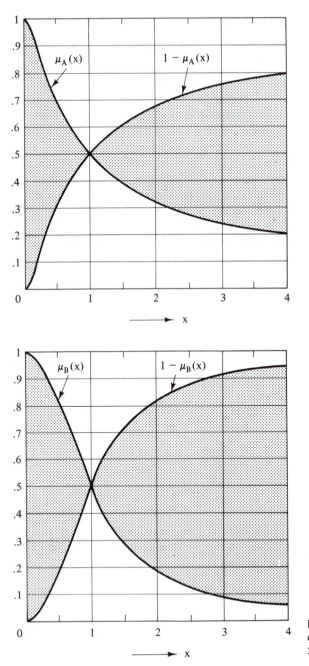

Figure 5.1. Comparison of the degree of fuzziness of two fuzzy sets (Example 5.2).

Example 5.2

Consider two fuzzy sets, A and B, defined on the set of real numbers $X = [0, 4]$ by the membership grade functions

$$\mu_A(x) = \frac{1}{1 + x}$$

and

$$\mu_B(x) = \frac{1}{1 + x^2} .$$

Plots of these functions and their standard classical complements are shown in Fig. 5.1. By inspecting these plots, we observe that set A is less distinct from its complement than is set B. To calculate the actual degrees of fuzziness of these sets, let us use the Hamming distance ($w = 1$). Then,

$$\delta_{c,A}(x) = |2\mu_A(x) - 1| = \left| \frac{2}{1 + x} - 1 \right|$$

and

$$
\begin{aligned}
D_{c,1}(A, A^c) &= \int_0^4 \left| \frac{2}{1 + x} - 1 \right| dx \\
&= \int_0^1 \left(\frac{2}{1 + x} - 1 \right) dx + \int_1^4 \left(1 - \frac{2}{1 + x} \right) dx \\
&= [2 \ln(1 + x) - x]_0^1 + [x - 2 \ln(1 + x)]_1^4 \\
&= 1.55.
\end{aligned}
$$

Now applying Eq. (5.18), we obtain

$$f_{c,1}(A) = 4 - 1.55 = 2.45;$$

after normalization, expressed by Eq. (5.19), we have

$$\hat{f}_{c,1}(A) = \frac{2.45}{4} = .61.$$

The following is the same calculation for the set B:

$$\delta_{c,B}(x) = |2\mu_B(x) - 1| = \left| \frac{2}{1 + x^2} - 1 \right|,$$

$$
\begin{aligned}
D_{c,1}(B, B^c) &= \int_0^4 \left| \frac{2}{1 + x^2} - 1 \right| dx \\
&= \int_0^1 \left(\frac{2}{1 + x^2} - 1 \right) dx + \int_1^4 \left(1 - \frac{2}{1 + x^2} \right) dx \\
&= [2 \tan^{-1}x - x]_0^1 + [x - 2 \tan^{-1}x]_1^4 \\
&= 4 \tan^{-1}1 - 2 \tan^{-1}4 + 2 = 2.49,
\end{aligned}
$$

$$f_{c,1}(B) = 4 - 2.49 = 1.51,$$

$$\hat{f}_{c,1}(B) = \frac{1.51}{4} = .38.$$

Thus, as we anticipated, the degree of fuzziness of the set B is smaller than that of the set A. The difference between the two degrees of fuzziness is .94 bits.

5.3 CLASSICAL MEASURES OF UNCERTAINTY

Prior to the theory of fuzzy sets, two principal measures of uncertainty were recognized. One of them, proposed by Hartley [1928], is based solely on the classical set theory. The other, introduced by Shannon [1948], is formulated in terms of probability theory. Both of these measures pertain to some aspects of ambiguity, as opposed to vagueness or fuzziness. As shown later in this chapter, however, each measures a different aspect of ambiguity: Hartley's measure pertains to nonspecificity, Shannon's measure to conflict or dissonance in evidence.

Both Hartley and Shannon introduced their measures for the purpose of measuring information in terms of uncertainty. Therefore, these measures are often referred to as *measures of information*. It has been more common, however, to refer to the measure invented by Shannon as the *Shannon entropy*. The name *entropy* was suggested by Shannon himself, presumably because of a similarity in the mathematical form between his measure and that of physical entropy as defined in certain formulations of statistical mechanics.

In this section, we introduce these classical measures of uncertainty and information and review their most fundamental properties. We refer to them by names that are predominant in the literature: the *Hartley information* and *Shannon entropy*.

Hartley Information

Consider a finite set X of n elements that are viewed as symbols that convey certain meanings to the parties involved in a certain context. For example, the set may consist of all possible measurements of a physical variable (defined in a specific way and subject to a specific accuracy), all possible states of an investigated system, all possible primitive messages, and the like.

Sequences can be formed from elements of set X by successive selections. These are determined, for example, by the variable measured, by the dynamic properties of the system under investigation, or by the sender of a message. At each selection, all possible elements that might have been chosen are eliminated except one. Similarly, as the selection proceeds, all possible sequences of elements of each particular length are eliminated except one.

Since all elements of X represent possible alternatives prior to a selection, we experience ambiguity, the amount of which is proportional to the number of alternatives. This ambiguity is totally resolved when one of the alternatives is selected. The amount of information conveyed (to the experimenter, observer, message receiver, and the like) can thus be meaningfully defined as the amount of ambiguity eliminated by the selection.

The number of all possible sequences of s selections ($s \in \mathbb{N}$) from set X is n^s, where $n = |X|$. The amount of information $I(n^s)$ associated with s selections from X should be proportional to s, that is,

$$I(n^s) = K(n) \cdot s, \tag{5.20}$$

where $K(n)$ is a constant, as yet unknown, that depends on n.

Consider two sets X_1 and X_2 such that $|X_1| = n_1$ and $|X_2| = n_2$. When the numbers s_1 and s_2 of selections from sets X_1 and X_2, respectively, are such that they yield the same number of sequences, then the amount of information associated with these sequences should also be the same. Formally, when

$$n_1^{s_1} = n_2^{s_2}, \tag{5.21}$$

then

$$K(n_1)s_1 = K(n_2)s_2. \tag{5.22}$$

From Eqs. (5.21) and (5.22), we obtain

$$\frac{s_2}{s_1} = \frac{\log_b n_1}{\log_b n_2}$$

and

$$\frac{s_2}{s_1} = \frac{K(n_1)}{K(n_2)},$$

respectively. Hence,

$$\frac{\log_b n_1}{\log_b n_2} = \frac{K(n_1)}{K(n_2)}.$$

This equation can be satisfied only by $K(n) = K_0 \log_b n$, where K_0 is a common constant. The amount of information conveyed by a sequence of s selections from a set with n elements is thus given by the formula

$$I(n^s) = K_0 s \log_b n. \tag{5.23}$$

By making a particular choice of values for K_0 and b in this formula, we define a unit by which information is measured. When we choose $K_0 = 1$ and $b = 2$, the information is measured in bits.* Then,

$$I(n^s) = s \log_2 n = \log_2 n^s$$

or

$$I(N) = \log_2 N, \tag{5.24}$$

where N denotes the total number of alternatives regardless of whether they are realized by one selection from a set or by a sequence of selections. One bit of information is obtained when one of two possible alternatives ($N = 2$) is eliminated. This is equivalent to obtaining the truth value of a single proposition, which was not known prior to the observation or receipt of the message. Because of

---

* This unit is introduced in Sec. 5.2 in the context of measures of fuzziness.

this intuitively appealing interpretation of the amount of information measured in bits, Hartley information is usually presented in the form (5.24).

For each $N \in \mathbb{N}$, the value of $I(N)$ can also be viewed as the amount of information needed to characterize one of N alternatives. When measured in bits, $I(N)$ expresses the number of single propositions whose truth values must be determined in order to characterize one of the alternatives.

Hartley information can also be characterized by the following axioms:

Axiom I1 (*additivity*). $I(N \cdot M) = I(N) + I(M)$ for all $N, M \in \mathbb{N}$.

Axiom I2 (*monotonicity*). $I(N) \leq I(N + 1)$ for all $N \in \mathbb{N}$.

Axiom I3 (*normalization*). $I(2) = 1$.

Axiom I1 involves a set with $N \cdot M$ elements, which can be partitioned into N subsets each with M elements. A characterization of an element from the full set requires the amount $I(N \cdot M)$ of information. However, we can also proceed in two steps to characterize the elements by taking advantage of the partition of the set. First, we characterize the subset to which the element belongs; the required amount of information is $I(N)$. Then, we characterize the element within the subset; here the required amount of information is $I(M)$. These two amounts of information completely characterize an element of the full set and, hence, their sum should be equal to $I(N \cdot M)$. This is exactly what the axiom requires.

Axiom I2 represents an essential and rather obvious requirement: the larger the number of alternatives, the more information is gained by eliminating all of them except one. Axiom I3 is needed only to define the unit of information. In our case, the defined unit is the bit.

As expressed by the following uniqueness theorem, the Hartley information (5.24) is the only function that satisfies these axioms.

Theorem 5.1.   Function $I(N) = \log_2 N$ is the only function that satisfies Axioms I1 through I3.

Proof: Let N be an integer greater than 2. For every integer i, define the integer $q(i)$ such that

$$2^{q(i)} \leq N^i < 2^{q(i)+1}. \tag{5.25}$$

These inequalities can be written as

$$q(i)\log_2 2 \leq i \log_2 N < [q(i) + 1]\log_2 2.$$

When we divide the inequalities by i and replace $\log_2 2$ with 1, we obtain

$$\frac{q(i)}{i} \leq \log_2 N < \frac{q(i) + 1}{i} \tag{5.26}$$

and, consequently,

$$\lim_{i \to \infty} \frac{q(i)}{i} = \log_2 N. \tag{5.27}$$

Let I denote a function that satisfies Axioms I1 through I3. Then, by Axiom I2,

$$I(a) \leq I(b) \tag{5.28}$$

for $a < b$. Combining (5.28) and (5.25), we obtain

$$I(2^{q(i)}) \leq I(N^i) \leq I(2^{q(i)+1}). \tag{5.29}$$

By Axiom I1, we have

$$I(a^k) = kI(a). \tag{5.30}$$

Hence,

$$I(N^i) = iI(N), \tag{5.31}$$

$$I(2^{q(i)}) = q(i)I(2),$$

and

$$I(2^{q(i)+1}) = (q(i) + 1)I(2).$$

Since $I(2) = 1$ by Axiom I3, we can rewrite the last two equations as

$$I(2^{q(i)}) = q(i)$$

and

$$I(2^{q(i)+1}) = q(i) + 1.$$

Applying these two equations and (5.31) to (5.29), we obtain

$$q(i) \leq iI(N) \leq q(i) + 1$$

and, consequently,

$$\lim_{i \to \infty} \frac{q(i)}{i} = I(N). \tag{5.32}$$

Comparing (5.32) with (5.27), we conclude that $I(N) = \log_2 N$ for $N > 2$. Since $\log_2 2 = 1$ and $\log_2 1 = 0$, function $\log_2 N$ clearly satisfies the axioms for $N = 1$, 2 as well. This concludes the proof. ■

Consider now two sets X and Y that are interrelated in the sense that selections from one of the sets are constrained by selections from the other. Assume that the constraint is expressed by a relation $R \subset X \times Y$. Then, three types of Hartley information can be defined on these sets:

- *Simple information*

$$I(X) = \log_2 |X|$$

$$I(Y) = \log_2 |Y|$$

- *Joint Information*

$$I(X, Y) = \log_2 |R|$$

• *Conditional information*

$$I(X \mid Y) = \log_2 \frac{|R|}{|Y|} = \log_2 |R| - \log_2 |Y|$$

$$I(Y \mid X) = \log_2 \frac{|R|}{|X|} = \log_2 |R| - \log_2 |X|$$

Observe that $|R \mid/\mid Y|$ in $I(X \mid Y)$ represents the average number of elements of X that can be selected under the condition that an element of Y has already been selected. Similarly, $|R \mid/\mid X|$ in $I(Y \mid X)$ characterizes the average number of elements of Y that can be selected provided that an element of X has been selected. Observe also that

$$I(X \mid Y) = I(X, Y) - I(Y) \tag{5.33}$$

and

$$I(Y \mid X) = I(X, Y) - I(X). \tag{5.34}$$

If selections from X do not depend on selections from Y, then sets X and Y are called *noninteractive*. Then, $R = X \times Y$ and we can readily obtain the following equations:

$$I(X, Y) = \log_2 |X \times Y| = \log_2 (|X| \cdot |Y|)$$

$$= \log_2 |X| + \log_2 |Y| = I(X) + I(Y),$$

$$I(X \mid Y) = I(X),$$

$$I(Y \mid X) = I(Y).$$

The following symmetric function, which is usually referred to as *information transmission,* is a useful indicator of the strength of constraint between sets X and Y:

$$T(X, Y) = I(X) + I(Y) - I(X, Y). \tag{5.35}$$

When the sets are noninteractive, we have $T(X, Y) = 0$; otherwise, $T(X, Y) > 0$.

Information transmission can be generalized to express the constraint among more than two sets. It is always expressed as the difference between the total information based on the individual sets and the joint information. Formally,

$$T(X_1, X_2, \ldots, X_n) = \sum_{i=1}^{n} I(X_i) - I(X_1, X_2, \ldots, X_n). \tag{5.36}$$

Example 5.3

Consider two variables x and y whose values are taken from sets $X = \{$low, medium, high$\}$ and $Y = \{1, 2, 3, 4\}$, respectively. It is known that the variables are constrained by the relation R expressed by the matrix

$$\begin{array}{c c} & \begin{array}{c c c c} 1 & 2 & 3 & 4 \end{array} \\ \begin{array}{r} \text{Low} \\ \text{Medium} \\ \text{High} \end{array} & \begin{bmatrix} 1 & 1 & 1 & 1 \\ 1 & 0 & 1 & 0 \\ 0 & 1 & 0 & 0 \end{bmatrix} \end{array}$$

We can see that the low value of x does not constrain y at all, the medium value of x constrains y partially, and the high value constrains it totally. The following types of Hartley information can be calculated in this example:

$$I(X) = \log_2 |X| = \log_2 3 = 1.6,$$

$$I(Y) = \log_2 |Y| = \log_2 4 = 2,$$

$$I(X, Y) = \log_2 |R| = \log_2 7 = 2.8,$$

$$I(X \mid Y) = I(X, Y) - I(Y) = 2.8 - 2 = .8,$$

$$I(Y \mid X) = I(X, Y) - I(X) = 2.8 - 1.6 = 1.2,$$

$$T(X, Y) = I(X) + I(Y) - I(X, Y) = 1.6 + 2 - 2.8 = .8.$$

The Hartley information is based on uncertainty associated with a choice among a certain number of alternatives. The larger the number of alternatives, the larger the uncertainty and, consequently, the more information is measured by the uncertainty. The degree of uncertainty in this case can be viewed, most naturally, as the degree of nonspecificity. Indeed, the fewer alternatives we have, the more specific is our choice, and vice versa.

The Hartley information can thus be viewed as a simple *measure of nonspecificity* based solely on classical set theory. As discussed in Sec. 5.6, it is a special case of more general nonspecificity measures based upon fuzzy set theory.

Shannon Entropy

The Shannon entropy, which is a measure of uncertainty and information formulated in terms of probability theory, is expressed by the function

$$H(p(x) \mid x \in X) = -\sum_{x \in X} p(x)\log_2 p(x), \qquad (5.37)$$

where $(p(x) \mid x \in X)$ is a probability distribution on a finite set X. It is thus a function of the form

$$H : \mathscr{P} \to [0, \infty),$$

where $\mathscr{P}$ denotes the set of all probability distributions on finite sets.

Shannon entropy was considered for many years to be the only feasible basis for information theory. It has certainly dominated the literature on information theory since it was proposed by Shannon in 1948. Hartley information, which is in fact a predecessor of Shannon entropy, is rarely mentioned in the current literature. When it is mentioned, it is almost always given one of two probabilistic interpretations. In the first, it is viewed as a measure that only distinguishes between zero and nonzero probabilities in the given probability distribution, that is, a measure that is totally insensitive to the actual values of the probabilities. It is derived from Shannon entropy by replacing any nonzero probability in the probability distribution with one.

The second probabilistic interpretation views Hartley information as equivalent to Shannon entropy under the assumption that all elements of the set X are equally probable. In this case, the equal probabilities are $1/|X|$. When we substitute them for $p(x)$ in formula (5.37), we readily obtain the Hartley information $\log_2 |X|$.

Although the probabilistic interpretations try to subsume the Hartley information under the framework of information theory based on the Shannon entropy, such attempts are ill-conceived. This should be obvious from the axiomatic treatment of the Hartley information earlier in this section, which is totally independent of any probabilistic assumptions. In fact, the Shannon entropy and Hartley information measure quite different aspects of uncertainty and information, as discussed in Secs. 5.4 through 5.6.

Why is the Shannon entropy significant as a measure of uncertainty and information? First, let us justify it on simple intuitive grounds. Suppose a particular element x of our universal set X occurs with the probability $p(x)$. When the probability of x is very high, say $p(x) = .99$, then the actual occurrence of x is taken almost for granted and, consequently, its occurrence does not surprise us very much. That is, our uncertainty in anticipating x is quite small and, therefore, our observation that x has actually occurred contains very little information content. When the probability is very small, on the other hand, say $p(x) = .01$, then we are greatly surprised by the occurrence of x. This means that we are highly uncertain in our anticipation of x and, hence, the observation of x has a very large information content. The information content of observing x, expressed by our anticipatory uncertainty prior to the observation, should therefore be characterized by a decreasing function of the probability $p(x)$: the more likely the occurrence of x, the less information the observation of it actually contains.

Let u denote a function that for each $x \in X$ with probability $p(x)$ characterizes the anticipatory uncertainty of x. Since $p(x) \in [0, 1]$, we have

$$u : [0, 1] \to [0, \infty),$$

where

$$u(a) > u(b) \qquad \text{for } a < b.$$

In addition, function u should be additive with respect to joint observations of elements from two sets that are independent in the probabilistic sense. That is, for each $x \in X$ and each $y \in Y$, if

$$p(x, y) = p(x) \cdot p(y),$$

then u should satisfy the equation

$$u(p(x) \cdot p(y)) = u(p(x)) + u(p(y)).$$

This functional equation is known as the Cauchy equation. Its solution is

$$u(a) = K \log_b a,$$

where K is a constant ($K \in \mathbb{R}$). Since u is required to be a decreasing function on $[0, 1]$ and the logarithmic function is increasing, K must be negative. When

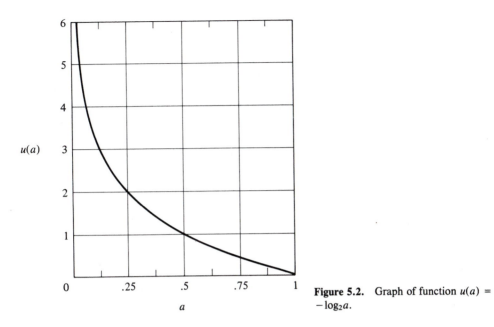

Figure 5.2. Graph of function $u(a) = -\log_2 a$.

we take $b = 2$ and add a normalization requirement that $u(\tfrac{1}{2}) = 1$ (to measure the anticipatory uncertainty in bits), we obtain $K = -1$ and

$$u(a) = -\log_2 a.$$

A graph of this function is shown in Fig. 5.2.

Consider now a set X of alternatives with probabilities $p(x)$ for each $x \in X$. It is certain that exactly one of them must occur under some relevant scenario; for example, exactly one must be observed as an outcome of an experiment or be received as a message. When we know that x occurs (by actually observing it as the outcome of our experiment or by receiving it as a message), the information content of this fact is $-\log_2 p(x)$ bits. Prior to the occurrence of the particular x, this information content is not known. However, we can calculate the expected information content as the weighted arithmetic mean

$$-\sum_{x \in X} p(x)\log_2 p(x),$$

which is exactly the Shannon entropy.

When only two alternatives whose probabilities are a and $1 - a$ are available, the expected information $H(a, 1 - a)$ depends on a, as illustrated in Fig. 5.3(a); graphs of its components $-a \log_2 a$ and $-(1 - a)\log_2(1 - a)$ are shown in Fig. 5.3(b).

Probabilistic measures of uncertainty and information have also been treated axiomatically in various ways. To illustrate this more rigorous treatment, assume that

$$X = \{x_1, x_2, \ldots, x_n\},$$

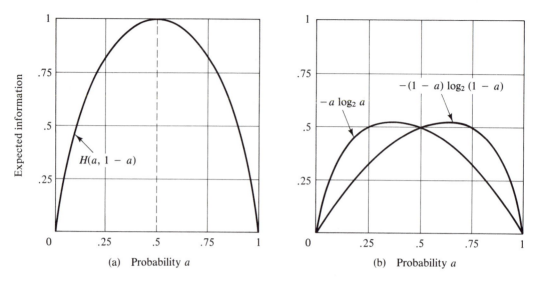

Figure 5.3. Graphs of $H(a, 1 - a)$ and its components $-a \log_2 a$ and $-(1 - a)\log_2(1 - a)$.

and let p_i denote the probability of $x_i \in X$ for all $i \in \mathbb{N}_n$. In addition, let

$$^n\mathscr{P} = \left\{ (p_1, p_2, \ldots, p_n) \mid p_i \geq 0 \text{ for all } i \in \mathbb{N}_n \text{ and } \sum_{i=1}^{n} p_i = 1 \right\}$$

denote the set of all probability distributions with n components and let

$$\mathscr{P} = \bigcup_{n \in \mathbb{N}} {}^n\mathscr{P}.$$

Then, every measure of uncertainty based on probability distributions is characterized by a function

$$H : \mathscr{P} \to [0, \infty)$$

that satisfies some requirements considered desirable for such a measure.

     Different subsets of the following requirements, which are universally considered essential for a probabilistic measure of uncertainty and information, are usually taken as axioms of probabilistic information theory:

     (H1) Expansibility—when a component with zero probability is added to a probability distribution, the uncertainty should not change; formally,

$$H(p_1, p_2, \ldots, p_n) = H(p_1, p_2, \ldots, p_n, 0)$$

for all $(p_1, p_2, \ldots, p_n) \in \mathscr{P}$.

     (H2) Symmetry—the uncertainty should be invariant with respect to permutations of probabilities of a given probability distribution; formally,

$$H(p_1, p_2, \ldots, p_n) = H(\text{perm}(p_1, p_2, \ldots, p_n))$$

for all $(p_1, p_2, \ldots, p_n) \in \mathscr{P}$ and all permutations $\text{perm}(p_1, p_2, \ldots, p_n)$ of $(p_1, p_2, \ldots, p_n)$.

(H3) Continuity—function H should be continuous in all its arguments p_1, $p_2, \ldots, p_n$ $(n \in \mathbb{N})$. This requirement is often replaced with a weaker requirement: $H(p, 1 - p)$ is a continuous function of p in the interval $[0, 1]$.

(H4) Maximum—for every $n \in \mathbb{N}$, the maximum uncertainty should be obtained when all probabilities are equal to $1/n$. Formally,

$$H(p_1, p_2, \ldots, p_n) \leq H\left(\frac{1}{n}, \frac{1}{n}, \ldots, \frac{1}{n}\right)$$

for all $n \in \mathbb{N}$.

(H5) Subadditivity—the uncertainty of a joint probability distribution should not be greater than the sum of the uncertainties of the corresponding marginal probability distributions; formally,

$$H(p_{11}, p_{12}, \ldots, p_{1n}, p_{21}, p_{22}, \ldots, p_{2n}, \ldots, p_{s1}, p_{s2}, \ldots, p_{sn})$$

$$\leq H\left(\sum_{i=1}^{s} p_{i1}, \sum_{i=1}^{s} p_{i2}, \ldots, \sum_{i=1}^{s} p_{in}\right) \tag{5.38}$$

$$+ H\left(\sum_{j=1}^{n} p_{1j}, \sum_{j=1}^{n} p_{2j}, \ldots, \sum_{j=1}^{n} p_{sj}\right)$$

for any joint probability distribution in $\mathscr{P}$.

(H6) Additivity—for any two marginal probability distributions that are noninteractive, the uncertainty of the associated joint distribution should be equal to the sum of the uncertainties of the marginal distributions. Formally,

$$H(p_1 q_1, p_1 q_2, \ldots, p_1 q_s, p_2 q_1, p_2 q_2, \ldots, p_2 q_s, \ldots, p_n q_1, p_n q_2, \ldots, p_n q_s)$$

$$= H(p_1, p_2, \ldots, p_n) + H(q_1, q_2, \ldots, q_s) \tag{5.39}$$

for any two probability distributions $(p_1, p_2, \ldots, p_n) \in \mathscr{P}$ and $(q_1, q_2, \ldots, q_s)$ $\in \mathscr{P}$. This requirement is sometimes replaced with a *weaker (special) additivity requirement* defined only for marginal distributions with equal probabilities $1/n$ and $1/s$, that is,

$$H\left(\frac{1}{ns}, \frac{1}{ns}, \ldots, \frac{1}{ns}\right) = H\left(\frac{1}{n}, \frac{1}{n}, \ldots, \frac{1}{n}\right) + H\left(\frac{1}{s}, \frac{1}{s}, \ldots, \frac{1}{s}\right)$$

for all $n, s \in \mathbb{N}$. When we introduce a convenient function

$$f(n) = H\left(\frac{1}{n}, \frac{1}{n}, \ldots, \frac{1}{n}\right)$$

for all $n \in \mathbb{N}$, this requirement can be expressed in a simpler way as

$$f(ns) = f(n) + f(s) \tag{5.40}$$

for all $n, s \in \mathbb{N}$. Let this requirement be called a *weak additivity*.

(H7) Monotonicity—for probability distributions with equal probabilities $1/n$ ($n \in \mathbb{N}$), the uncertainty should increase with increasing n. Formally, the function

$$f(n) = H\left(\frac{1}{n}, \frac{1}{n}, \ldots, \frac{1}{n}\right)$$

should be a monotonically increasing function of n, that is,

$$\text{if } n < s, \text{ then } f(n) < f(s)$$

for all $n, s \in \mathbb{N}$.

(H8) Branching—given a probability distribution on a universal set X, we should be able to measure the uncertainty associated with it either directly or indirectly by adding uncertainties involved in a two-stage measurement process: in the first stage, we measure the uncertainty of two disjoint subsets of X that form a partition of X; in the second stage, we measure the uncertainty associated with the conditional probability distributions on these subsets and weight each of them by the probability of the subset. To formalize this requirement, let $A = \{x_1, x_2, \ldots, x_s\}$ and $B = \{x_{s+1}, x_{s+2}, \ldots, x_n\}$ be two disjoint subsets of the universal set $X = \{x_1, x_2, \ldots, x_n\}$. Furthermore, given a probability distribution $(p_1, p_2, \ldots, p_n)$ on X, where $p_i = p(x_i)$ for all $i \in \mathbb{N}_n$, let

$$p_A = \sum_{i=1}^{s} p_i \quad \text{and} \quad p_B = \sum_{i=s+1}^{n} p_i$$

denote the probabilities of A and B, respectively. Then, the branching requirement means that

$$H(p_1, p_2, \ldots, p_n) = H(p_A, p_B) + p_A H\left(\frac{p_1}{p_A}, \frac{p_2}{p_A}, \ldots, \frac{p_s}{p_A}\right)$$

$$+ p_B H\left(\frac{p_{s+1}}{p_B}, \frac{p_{s+2}}{p_B}, \ldots, \frac{p_n}{p_B}\right) \tag{5.41}$$

should be satisfied for any probability distribution $(p_1, p_2, \ldots, p_n) \in \mathscr{P}$ and any subsets A and B of X that form a partition on X. This requirement, which is also called a *grouping requirement,* is sometimes presented in various alternative forms. For example, one of its weaker forms is described by the equation

$$H(p_1, p_2, p_3) = H(p_1 + p_2, p_3) + (p_1 + p_2)H\left(\frac{p_1}{p_1 + p_2}, \frac{p_2}{p_1 + p_2}\right).$$

It matters little which of these forms is adopted since they can be derived from each other.

(H9) Normalization—to ensure (if desirable) that the uncertainty is measured in bits, we require that

$$H(\tfrac{1}{2}, \tfrac{1}{2}) = 1.$$

The listed requirements for probabilistic measures of uncertainty are extensively discussed and well justified in the plentiful literature on classical information theory. The following subsets of these requirements are the best known examples of axiomatic characterizations of the probabilistic measure of uncertainty:

1. Continuity, weak additivity, monotonicity, branching, and normalization.
2. Expansibility, continuity, maximum, branching, and normalization.
3. Symmetry, continuity, branching, and normalization.
4. Expansibility, symmetry, continuity, subadditivity, additivity, and normalization.

Any of these collections of requirements (as well as some additional collections) is, when taken as a set of axioms, sufficient to characterize the Shannon entropy uniquely. That is, it has been proven that the Shannon entropy is the only function that satisfies any of these sets of axioms. To illustrate in detail this important issue of uniqueness, which gives the Shannon entropy its great significance, we present the uniqueness proof for the first of the listed sets of axioms, that is, for continuity, weak additivity, monotonicity, branching, and normalization. Since the proof is rather lengthy, it is placed in Appendix A.1.

The literature dealing with information theory based on the Shannon entropy is extensive. We do not attempt to give a comprehensive coverage of the theory in this book. In the rest of this section, however, we briefly overview the most fundamental properties of the Shannon entropy. In addition, some notes at the end of this chapter provide the reader with key literature resources for a deeper study of various aspects of this classical information theory.

First, let us show that

$$0 \le H(p_1, p_2, \ldots, p_n) \le \log_2 n. \tag{5.42}$$

To derive the lower bound, we observe that $-p_i \log_2 p_i \ge 0$ for all $p_i \in (0, 1]$. For $p_i = 0$, the function $-p_i \log_2 p_i$ is not defined. However, employing l'Hospital's rule for indeterminate forms, we can calculate its limit for $p_i \to 0$:

$$\lim_{p_i \to 0} -p_i \log_2 p_i = \lim_{p_i \to 0} \frac{-\log_2 p_i}{\dfrac{1}{p_i}} = \lim_{p_i \to 0} \frac{\dfrac{-1}{p_i \ln 2}}{\dfrac{-1}{p_i^2}} = \lim_{p_i \to 0} \frac{p_i}{\ln 2} = 0.$$

Clearly, the lower bound of (5.42) is obtained only when $p_i = 1$ for some particular i and $p_j = 0$ for all $j \neq i$ $(i, j \in \mathbb{N}_n)$; this probability distribution indeed represents no uncertainty.

To derive the upper bound of (5.42), we must determine the maximum of function H for all probability distributions in $^n\mathcal{P}$. In this case, H is a function of n variables, but one of these is dependent on the other variables due to the requirement that the probabilities must add to one. Without any loss of generality, let p_n be the dependent variable. Then,

$$p_n = 1 - (p_1 + p_2 + \cdots + p_{n-1}),$$

and the necessary conditions for an extremal value of H are

$$\frac{\partial H}{\partial p_i} = 0 \qquad \text{for all } i \in \mathbb{N}_{n-1}.$$

The partial derivatives are

$$\frac{\partial H}{\partial p_i} = \sum_{k=1}^{n} \frac{\partial H}{\partial p_k} \frac{\partial p_k}{\partial p_i} = \frac{d(-p_i \log_2 p_i)}{dp_i} + \frac{d(-p_n \log_2 p_n)}{dp_n} \frac{\partial p_n}{\partial p_i}$$

since

$$\frac{\partial H}{\partial p_k} \frac{\partial p_k}{\partial p_i} = 0 \qquad \text{for all } k \neq i, n.$$

Clearly,

$$\frac{\partial p_n}{\partial p_i} = \frac{\partial[1 - (p_1 + p_2 + \cdots + p_i + \cdots + p_{n-1})]}{\partial p_i} = -1$$

and, consequently,

$$\frac{\partial H}{\partial p_i} = -\log_2 p_i - \frac{1}{\ln 2} + \log_2 p_n + \frac{1}{\ln 2}$$

$$= -\log_2 p_i + \log_2 p_n.$$

By setting the derivatives to zero, we obtain

$$p_i = p_n \quad \text{for all} \quad i \in \mathbb{N}_{n-1}.$$

Hence, an extremal value of H exists for the distribution with equal probabilities $1/n$. Since

$$\left. \frac{\partial^2 H}{\partial p_i^2} \right|_{p_i = 1/n} = \left. \frac{-1}{p_i \ln 2} \right|_{p_i = 1/n} = \frac{-n}{\ln 2} < 0$$

for all $i \in \mathbb{N}_n$,

$$H\left(\frac{1}{n}, \frac{1}{n}, \ldots, \frac{1}{n}\right) = \log_2 n$$

is the maximum of H in $^n\mathcal{P}$.

In some applications, it is desirable to define the ratio

$$\hat{H}(p_1, p_2, \ldots, p_n) = \frac{H(p_1, p_2, \ldots, p_n)}{\log_2 n}$$

of the actual entropy and its upper bound, which is called a *normalized Shannon entropy*. Clearly,

$$0 \le \hat{H}(p_1, p_2, \ldots, p_n) \le 1.$$

Before examining additional properties of the Shannon entropy, let us illustrate the meaning of its branching property, which is employed as one of the axioms in our uniqueness proof (Appendix A.1). In principle, the branching property allows us to calculate uncertainty either directly, in terms of the probability distribution on the universal set, or in stages, using probability distributions on various subsets of the universal set. These alternative ways of calculating uncertainty can be best illustrated by an example.

Example 5.4

Let the set $X = \{x_1, x_2, x_3, x_4\}$ with the probability distribution

$$p = (p_1 = .25, p_2 = .5, p_3 = .125, p_4 = .125)$$

be given where p_i denotes the probability of x_i for all $i \in \mathbb{N}_4$. Consider the four branching schemes specified in Fig. 5.4 for calculating the uncertainty of this probability distribution. Employing the branching property of Shannon entropy, the resulting uncertainty should be the same regardless of which of the branching schemes we use. Let us perform and compare the four schemes of calculating the uncertainty.

Scheme I. According to this scheme, we calculate the uncertainty directly: $H(p) = -.25 \log_2 .25 - .5 \log_2 .5 - 2 \times .125 \log_2 .125 = .5 + .5 + .375 + .375 = 1.75.$
Scheme II. $H(p) = H(p_A, p_B) + p_A H(p_1/p_A, p_2/p_A) + p_B H(p_3/p_B, p_4/p_B) = H(\frac{3}{4}, \frac{1}{4}) + .75H(\frac{1}{3}, \frac{2}{3}) + .25H(\frac{1}{2}, \frac{1}{2}) = .811 + .689 + .25 = 1.75.$
Scheme III. $H(p) = H(p_1, p_A) + p_A H(p_2/p_A, p_3/p_A, p_4/p_A) = H(\frac{1}{4}, \frac{3}{4}) + .75H(\frac{2}{3}, \frac{1}{6}, \frac{1}{6}) = .811 + .939 = 1.75.$
Scheme IV. $H(p) = H(p_1, p_A) + p_A H(p_2/p_A, p_B/p_A) + p_B H(p_3/p_B, p_4/p_B) = H(\frac{1}{4}, \frac{3}{4}) + .75H(\frac{2}{3}, \frac{1}{3}) + .25H(\frac{1}{2}, \frac{1}{2}) = .811 + .689 + .25 = 1.75.$

These results thus demonstrate that the uncertainty can be calculated in terms of any branching scheme. There are, of course, many additional branching schemes in this example, each of which can be employed for calculating the uncertainty and each of which must lead to the same result.

We now present a theorem that plays an important role in classical information theory. This theorem is essential for proving some basic properties of Shannon entropy as well as for introducing some additional important concepts of information theory.

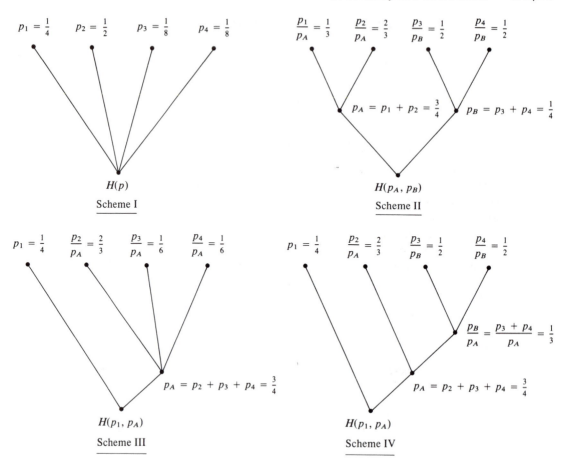

Figure 5.4. Application of the branching property of Shannon entropy (Example 5.4).

Theorem 5.2. The inequality

$$-\sum_{i=1}^{n} p_i\log_2 p_i \leq -\sum_{i=1}^{n} p_i\log_2 q_i \qquad (5.43)$$

is satisfied for all probability distributions $(p_i \mid i \in \mathbb{N}_n) \in {}^n\mathscr{P}$ and $(q_i \mid i \in \mathbb{N}_n) \in {}^n\mathscr{P}$ and for all $n \in \mathbb{N}$; the equality in (5.43) holds if and only if $p_i = q_i$ for all $i \in \mathbb{N}_n$.

Proof: Consider the function

$$h(p_i, q_i) = p_i(\log_2 p_i - \log_2 q_i) - p_i + q_i$$

for $p_i, q_i \in [0, 1]$. This function is finite and differentiable for all values of p_i and q_i except the pair $q_i = 0$ and $p_i \neq 0$. For each fixed $q_i \neq 0$, the partial derivative with respect to p_i is

$$\frac{\partial h(p_i, q_i)}{\partial p_i} = \frac{1}{\ln 2}\, (\log_2 p_i - \log_2 q_i).$$

That is,

$$\frac{\partial h(p_i, q_i)}{\partial p_i} \begin{cases} < 0 & \text{for } p_i < q_i \\ = 0 & \text{for } p_i = q_i \\ > 0 & \text{for } p_i > q_i \end{cases}$$

and, consequently, h is a convex function of p_i with its minimum at $p_i = q_i$. Hence, for any given i, we have

$$p_i(\log_2 p_i - \log_2 q_i) - p_i + q_i \geq 0,$$

where the equality holds if and only if $p_i = q_i$. This inequality is also satisfied for $q_i = 0$, since the expression on its left-hand side is $+\infty$ if $p_i \neq 0$ and $q_i = 0$, and it is zero if $p_i = 0$ and $q_i = 0$. Taking the sum of this inequality for all $i \in \mathbb{N}_n$, we obtain

$$\sum_{i=1}^{n} [p_i(\log_2 p_i - \log_2 q_i) - p_i + q_i] \geq 0,$$

which can also be written as

$$\sum_{i=1}^{n} p_i \log_2 p_i - \sum_{i=1}^{n} p_i \log_2 q_i - \sum_{i=1}^{n} p_i + \sum_{i=1}^{n} q_i \geq 0.$$

The last two terms on the left-hand side of this inequality cancel each other, since each of the sums is equal to one. Hence,

$$\sum_{i=1}^{n} p_i \log_2 p_i - \sum_{i=1}^{n} p_i \log_2 q_i \geq 0,$$

which is equivalent to (5.43). ∎

The equality (5.43) is usually referred to as the *Gibbs' inequality* or *Gibbs' theorem*.

Let us examine now Shannon entropies of joint, marginal, and conditional probability distributions defined on two sets X and Y. In agreement with a common practice in the literature dealing with Shannon entropy, we simplify the notation in the rest of this section by using $H(X)$ instead of $H(p(x) \mid x \in X)$ or $H(p_1, p_2, \ldots, p_n)$. Furthermore, assuming that $x \in X$ and $y \in Y$, we use the symbols $p(x)$ and $p(y)$ to denote marginal probabilities on sets X and Y, respectively, the symbol $p(x, y)$ for joint probabilities on $X \times Y$, and the symbols $p(x \mid y)$ and $p(y \mid x)$ for the corresponding conditional probabilities. In this simplified notation, the meaning of each symbol is uniquely determined by the argument shown in the parentheses.

Given two sets X and Y, we can recognize three types of entropies:

1. Two *simple entropies* based on the marginal probability distributions,

$$H(X) = -\sum_{x \in X} p(x)\log_2 p(x)$$

and

$$H(Y) = - \sum_{y \in Y} p(y)\log_2 p(y).$$

2. A *joint entropy* defined in terms of the joint probability distribution on $X \times Y$,

$$H(X, Y) = - \sum_{(x,y) \in X \times Y} p(x, y)\log_2 p(x, y).$$

3. Two *conditional entropies* defined in terms of weighted averages of local conditional entropies as

$$H(X \mid Y) = - \sum_{y \in Y} p(y) \sum_{x \in X} p(x \mid y)\log_2 p(x \mid y)$$

and

$$H(Y \mid X) = - \sum_{x \in X} p(x) \sum_{y \in Y} p(y \mid x)\log_2 p(y \mid x).$$

In addition to these entropies, the function

$$T(X, Y) = H(X) + H(Y) - H(X, Y) \tag{5.44}$$

is often used in the literature as a measure of the strength of relationship (in the probabilistic sense) between elements of sets X and Y. The function is called *information transmission*. It is analogous to the function defined by Eq. (5.35) for Hartley information; it can be generalized to more than two sets in the same way.

Our next subject is an examination of the relationship among the various types of entropies and the information transmission. The key properties of this relationship are expressed by the next several theorems.

Theorem 5.3.

$$H(X \mid Y) = H(X, Y) - H(Y). \tag{5.45}$$

Proof:

$$H(X \mid Y) = - \sum_{y \in Y} p(y) \sum_{x \in X} p(x \mid y)\log_2 p(x \mid y)$$

$$= - \sum_{y \in Y} p(y) \sum_{x \in X} \frac{p(x, y)}{p(y)} \log_2 \frac{p(x, y)}{p(y)}$$

$$= - \sum_{y \in Y} \sum_{x \in X} p(x, y)\log_2 \frac{p(x, y)}{p(y)} \quad *$$

$$= H(X, Y) + \sum_{y \in Y} \sum_{x \in X} p(x, y)\log_2 p(y)$$

---

* The following useful form of the conditional entropy follows immediately from this equation:

$$H(X \mid Y) = - \sum_{y \in Y} \sum_{x \in X} p(x, y)\log_2 p(x \mid y).$$

$$= H(X, Y) + \sum_{y \in Y} \log_2 p(y) \sum_{x \in X} p(x, y)$$

$$= H(X, Y) + \sum_{y \in Y} p(y)\log_2 p(y)$$

$$= H(X, Y) - H(Y). \quad \blacksquare$$

The same theorem can obviously be proven for the other conditional entropy as well:

$$H(Y \mid X) = H(X, Y) - H(X). \tag{5.46}$$

The theorem can also be generalized to more than two sets. The general form, which is derived from either

$$H(X, Y) = H(X \mid Y) - H(Y)$$

or

$$H(X, Y) = H(Y \mid X) - H(X),$$

is

$$H(X_1, X_2, \ldots, X_n) = H(X_1) + H(X_2 \mid X_1) + H(X_3 \mid X_1, X_2)$$

$$+ \cdots + H(X_n \mid X_1, X_2, \ldots, X_{n-1}). \tag{5.47}$$

This equation is valid for any permutation of the sets involved.

Theorem 5.4.

$$H(X, Y) \le H(X) + H(Y). \tag{5.48}$$

Proof:

$$H(X) = - \sum_{x \in X} p(x)\log_2 p(x)$$

$$= - \sum_{x \in X} \sum_{y \in Y} p(x, y)\log_2 \sum_{y \in Y} p(x, y),$$

$$H(Y) = - \sum_{y \in Y} p(y)\log_2 p(y)$$

$$= - \sum_{y \in Y} \sum_{x \in X} p(x, y)\log_2 \sum_{x \in X} p(x, y),$$

$$H(X) + H(Y) = - \sum_{x \in X} \sum_{y \in Y} p(x, y)[\log_2 \sum_{x \in X} p(x, y)$$

$$+ \log_2 \sum_{y \in Y} p(x, y)]$$

$$= - \sum_{(x,y) \in X \times Y} p(x, y)[\log_2 p(x) + \log_2 p(y)]$$

$$= - \sum_{(x,y) \in X \times Y} p(x, y)\log_2 [p(x) \cdot p(y)].$$

By Gibbs' inequality (Theorem 5.2), we have

$$H(X, Y) = - \sum_{(x,y) \in X \times Y} p(x, y) \log_2 p(x, y)$$

$$\leq - \sum_{(x,y) \in X \times Y} p(x, y) \log_2 [p(x) \cdot p(y)] = H(X) + H(Y).$$

Hence, $H(X, Y) \leq H(X) + H(Y)$; furthermore (again by Gibbs' inequality), the equality holds if and only if

$$p(x, y) = p(x) \cdot p(y),$$

that is, only if the sets X and Y are noninteractive in the probabilistic sense. ■

Theorem 5.4 can be easily generalized to more than two sets. Its general form is

$$H(X_1, X_2, \ldots, X_s) \leq \sum_{i=1}^{s} H(X_i), \qquad (5.49)$$

which holds for every $s \in \mathbb{N}$.

Theorem 5.5.

$$H(X) \geq H(X \mid Y). \qquad (5.50)$$

Proof: From Theorem 5.3,

$$H(X, Y) = H(X \mid Y) + H(Y),$$

and from Theorem 5.4,

$$H(X, Y) \leq H(X) + H(Y).$$

Hence,

$$H(X \mid Y) + H(Y) \leq H(X) + H(Y)$$

and the inequality

$$H(X \mid Y) \leq H(X)$$

follows immediately. ■

Exchanging X and Y in Theorem 5.5, we obtain

$$H(Y) \geq H(Y \mid X). \qquad (5.51)$$

Additional equations expressing the relationship among the various entropies and the information transmission can be obtained by simple formula manipulations with the aid of the key properties contained in Theorems 5.2 through 5.5. For example, when we substitute for $H(X, Y)$ from Eq. (5.45) into Eq. (5.44), we obtain

$$T(X, Y) = H(X) - H(X \mid Y); \qquad (5.52)$$

similarly, by substituting for $H(X, Y)$ from Eq. (5.46) into Eq. (5.44), we obtain

$$T(X, Y) = H(Y) - H(Y \mid X). \tag{5.53}$$

By comparing Eqs. (5.52) and (5.53), we also have

$$H(X) - H(Y) = H(X \mid Y) - H(Y \mid X). \tag{5.54}$$

We leave additional simple derivations of this kind to the reader as an exercise.

Boltzmann Entropy

One important aspect of Shannon entropy remains to be discussed. It is connected with its restriction to finite sets. Is this restriction necessary? It seems that the formula

$$B(q(x) \mid x \in [a, b]) = - \int_a^b q(x) \log_2 q(x)\, dx, \tag{5.55}$$

where q denotes a probability density function on the real interval $[a, b]$, is analogous to formula (5.37) for Shannon entropy and could thus be viewed as an extension of Shannon entropy to the domain of real numbers. Moreover, function B defined by Eq. (5.55) is usually referred to as the *Boltzmann entropy*. Although the analogy between the two functions is suggestive, the following question cannot be avoided: is the Boltzmann entropy a genuine extension of the Shannon entropy? To answer this nontrivial question, we must establish a connection between the two functions.

Let q be a probability density function on the interval $[a, b]$ of real numbers. That is, $q(x) \geq 0$ for all $x \in [a, b]$ and

$$\int_a^b q(x)\, dx = 1. \tag{5.56}$$

Consider a sequence of probability distributions $^n\mathbf{p} = (^np_1, ^np_2, \ldots, ^np_n)$ such that

$$^np_i = \int_{x_{i-1}}^{x_i} q(x)\, dx \tag{5.57}$$

for every $n \in \mathbb{N}$ and every $i \in \mathbb{N}_n$, where

$$x_i = a + i \frac{b - a}{n}$$

for each $i \in \mathbb{N}_n$. For convenience, let

$$\Delta_n = \frac{b - a}{n}$$

so that

$$x_i = a + i\Delta_n.$$

For each probability distribution $^n\mathbf{p} = (^np_1, {}^np_2, \ldots, {}^np_n)$, let $^n\mathbf{d}(x)$ denote a probability density function on $[a, b]$ such that

$$^n\mathbf{d}(x) = (^nd_i(x) \mid i \in \mathbb{N}_n),$$

where

$$^nd_i(x) = \frac{^np_i}{\Delta_n} \quad \text{for} \quad x \in [x_{i-1}, x_i) \tag{5.58}$$

and for all $i \in \mathbb{N}_n$. Then, due to continuity of $q(x)$, the sequence $(^n\mathbf{d}(x) \mid n \in \mathbb{N})$ converges to $q(x)$ uniformly on $[a, b]$.

Given the probability distribution $^n\mathbf{p}$ for some $n \in \mathbb{N}$, its Shannon entropy is

$$H(^n\mathbf{p}) = -\sum_{i=1}^{n} {}^np_i \log_2 {}^np_i$$

or, using the introduced probability density function $^n\mathbf{d}$,

$$H(^n\mathbf{p}) = -\sum_{i=1}^{n} {}^nd_i(x) \Delta_n \log_2[^nd_i(x)\Delta_n].$$

This equation can be modified as follows:

$$H(^n\mathbf{p}) = -\sum_{i=1}^{n} {}^nd_i(x)\Delta_n \log_2 {}^nd_i(x) - \sum_{i=1}^{n} {}^nd_i(x)\Delta_n \log_2 \Delta_n$$

$$= -\sum_{i=1}^{n} [^nd_i(x)\log_2 {}^nd_i(x)]\Delta_n - \log_2\Delta_n \sum_{i=1}^{n} {}^np_i.$$

Since probabilities ^np_i of the distribution $^n\mathbf{p}$ must add to one and by the definition of Δ_n, we obtain

$$H(^n\mathbf{p}) = -\sum_{i=1}^{n} [^nd_i(x)\log_2 {}^nd_i(x)]\Delta_n + \log_2 \frac{n}{b-a}. \tag{5.59}$$

When $n \to \infty$ (or $\Delta_n \to 0$), we have

$$\lim_{n\to\infty} -\sum_{i=1}^{n} [^nd_i(x)\log_2 {}^nd_i(x)]\Delta_n = -\int_a^b q(x)\log_2 q(x)\, dx$$

according to the introduced relation among $^n\mathbf{p}$, $q(x)$, and $^nd_i(x)$, in particular Eqs. (5.57) and (5.58). Equation (5.59) can thus be rewritten for $n \to \infty$ as

$$\lim_{n\to\infty} H(^n\mathbf{p}) = B(q(x)) + \lim_{n\to\infty} \frac{n}{b-a}. \tag{5.60}$$

The last term in this equation clearly diverges. This means that the Boltzmann entropy *is not* a limit of the Shannon entropy for $n \to \infty$ and, consequently, it is not a measure of uncertainty and information.

The discrepancy between Shannon and Boltzmann entropies can be reconciled, for example, when we take the difference of entropies for two probability distributions. Then, the last term in Eq. (5.60), which is the same for both entropies, is canceled and the difference of the Shannon entropies converges for $n \to \infty$ to the difference of the corresponding Boltzmann entropies. One implication of this fact is that

$$T_B(q(x, y) \mid x \in [a, b], y \in [\alpha, \beta]) = - \int_a^b q(x)\log_2 q(x) \, dx$$

$$- \int_\alpha^\beta q(y)\log_2 q(y) \, dy + \int_a^b \int_\alpha^\beta q(x, y)\log_2 q(x, y) \, dxdy \quad (5.61)$$

is a genuine *Boltzmann counterpart* of the information transmission T defined by Eq. (5.44). Also

$$B(q, x), q'(x) \mid x \in [a, b]) = - \int_a^b q(x)\log_2 q'(x) \, dx + \int_a^b q(x)\log_2 q(x) \, dx \quad (5.62)$$

is a genuine Boltzmann counterpart of the function

$$H(p(x), p'(x) \mid x \in X) = - \sum_{x \in X} p(x)\log_2 p'(x) + \sum_{x \in X} p(x)\log_2 p(x), \quad (5.63)$$

which is known in information theory as the *Shannon cross-entropy* or *relative entropy* (often also called a *directed divergence*). This function is usually employed as a measure of the degree to which an estimated probability distribution **p'** approximates the distribution **p**. Observe that the function is always non-negative, due to the Gibbs' inequality (Theorem 5.2), but it is applicable only when $p(x) \neq 0$ implies $p'(x) \neq 0$.

We return to the cross-entropy again in Sec. 5.8 and describe some of its applications in Chap. 6.

5.4 MEASURES OF DISSONANCE

Let us begin now to investigate the various aspects of uncertainty within the framework of belief and plausibility measures. In this section, we focus on one aspect of uncertainty that is connected with conflict or dissonance in evidence.

At the common-sense level, we associate the terms *conflict* and *dissonance* with properties such as sharp disagreement (between claims, beliefs, interests, etc.), opposition, incompatibility, contradiction, or inconsistency. When we apply the common-sense meaning of these terms to the domain of belief and plausibility measures (within which we deal with degrees of evidence regarding subsets of a universal set), we encounter conflict or dissonance in evidence whenever nonzero degrees of evidence are allocated to disjoint subsets of the universal set. This follows directly from the fact that the element of concern (a priori nonlocated) can belong to only one of several disjoint subsets of the universal set. Whenever our evidence suggests that the element may belong to either of two disjoint subsets, then we clearly have a conflict in the evidence.

In order to derive a measure of conflict or dissonance in evidence within the framework of belief and plausibility measures, let us consider two bodies of evidence,* $(\mathcal{F}_1, m_1)$ and $(\mathcal{F}_2, m_2)$. Assume that these bodies of evidence, which are obtained from two independent sources, are defined on the same universal set X and for the same purpose. They conflict with each other whenever $m_1(A) \neq 0$, $m_2(B) \neq 0$, and $A \cap B = \varnothing$. Hence, the total amount of their conflict or dissonance should be monotonic increasing with

$$K = \sum_{A \cap B = \varnothing} m_1(A) \cdot m_2(B), \qquad (5.64)$$

where $A \in \mathcal{F}_1$ and $B \in \mathcal{F}_2$. Observe that K is also the factor employed in normalizing the joint basic assignment obtained by Dempster's rule (4.14).

The total amount of conflict between two bodies of evidence is usually expressed in the literature by the function

$$\mathrm{con}(m_1, m_2) = -\log_2(1 - K). \qquad (5.65)$$

This function takes values from 1 to ∞ and it is clearly monotonic increasing with K; $\mathrm{con}(m_1, m_2) = 0$ only if m_1 and m_2 do not conflict at all ($K = 0$), and $\mathrm{con}(m_1, m_2) = \infty$ only if they conflict totally ($K = 1$). The choice of the logarithmic function of K is motivated by similar arguments encountered in the discussion of Shannon entropy in Sec. 5.3; logarithm base 2 is used here solely for consistency with our formulation of the Hartley information and Shannon entropy in Sec. 5.3.

Our aim in this section is to derive a function

$$E : \mathcal{M} \to [0, \infty)$$

such that $E(m)$ can be justified as a meaningful measure of conflict or dissonance in evidence represented by the basic assignment m; $\mathcal{M}$ denotes here the set of all basic assignments on power sets with 2^n elements for any $n \in \mathbb{N}$. For further reference, let E be called a *measure of dissonance*.

Function (5.65) is obviously relevant for defining a meaningful measure of dissonance, but it must be adjusted to express a conflict within a single body of evidence. To make an appropriate adjustment of the function, let m_A denote a basic assignment on X such that

$$m_A(B) = \begin{cases} 1 & \text{if } B = A \\ 0 & \text{otherwise} \end{cases}$$

for all $B \in \mathcal{P}(X)$. Then, given a body of evidence $(\mathcal{F}, m)$, the function

$$\mathrm{con}(m, m_A) = -\log_2[1 - \sum_{B \cap C = \varnothing} m(B) \cdot m_A(C)],$$

where $A, B, C \in \mathcal{F}$, expresses the amount of conflict of the focal element A with all the other focal elements in $\mathcal{F}$. Since $m_A(C) = 0$ for all $C \neq A$, this function

---

* The term *body of evidence* is used here in exactly the same sense as defined in Sec. 4.2.

can be defined by a simpler form

$$\text{con}(m, m_A) = -\log_2[1 - \sum_{B \cap A = \varnothing} m(B)]. \qquad (5.66)$$

Furthermore, observe that

$$1 - \sum_{B \cap A = \varnothing} m(B) = \sum_{B \cap A \neq \varnothing} m(B)$$

and, by Eq. (4.11),

$$\sum_{B \cap A \neq \varnothing} m(B) = Pl(A).$$

Hence,

$$\text{con}(m, m_A) = -\log_2 Pl(A). \qquad (5.67)$$

It is now easy to go one step further and define $E(m)$ as a weighted average of $\text{con}(m, m_A)$ for all $A \in \mathcal{F}$. That is, we define the *measure of dissonance* as

$$E(m) = \sum_{A \in \mathcal{F}} m(A)\text{con}(m, m_A)$$

or, using Eq. (5.67), as

$$E(m) = -\sum_{A \in \mathcal{F}} m(A)\log_2 Pl(A). \qquad (5.68)$$

Example 5.5

Let $m(\{x_1, x_2\}) = .4$, $m(\{x_3\}) = .1$, $m(\{x_1, x_3\}) = .3$, and $m(\{x_1, x_2, x_3\}) = .2$ be a basic assignment representing a body of evidence with four focal elements. In order to calculate $E(m)$, we must determine the degrees of plausibility for the focal elements. We have

$$Pl(\{x_1, x_2\}) = m(\{x_1, x_2\}) + m(\{x_1, x_3\}) + m(\{x_1, x_2, x_3\}) = .9;$$

$$Pl(\{x_3\}) = m(\{x_3\}) + m(\{x_1, x_3\}) + m(\{x_1, x_2, x_3\}) = .6;$$

$$Pl(\{x_1, x_3\}) = Pl(\{x_1, x_2, x_3\}) = 1.$$

Applying formula (5.68), we obtain

$$E(m) = -.4\log_2.9 - .1\log_2.6 - .3\log_2 1 - .2\log_2 1 = .06 + .07 + 0 + 0 = .14.$$

Before we investigate various mathematical properties of function E defined by Eq. (5.68), let us demonstrate that it reduces to the Shannon entropy when m represents a probability distribution.

Theorem 5.6.   If m represents a probability distribution on X, then the measure of dissonance in evidence, as defined by Eq. (5.68), becomes equivalent to the Shannon entropy (Eq. (5.37)).

Proof: For probability measures defined on X, $m(A) = 0$ for $A \neq \{x\}$, $x \in X$. Hence,

$$m(A)\log_2 Pl(A) = 0$$

for all $A \neq \{x\}$, $x \in X$. This implies that

$$E(m) = - \sum_{x \in X} m(\{x\})\log_2 Pl(\{x\}).$$

Since

$$m(\{x\}) = Pl(\{x\}) = p(x),$$

where $p(x)$ denotes the probability of x, we have

$$E(m) = - \sum_{x \in X} p(x)\log_2 p(x) = H(p(x) \mid x \in X). \quad \blacksquare$$

The significance of Theorem 5.6 lies in the new insight it gives us into the meaning of Shannon entropy. It is quite clear from the theorem that Shannon entropy measures one specific type of uncertainty, that of dissonance in evidence.

Some properties of the measure of dissonance E are rather obvious. Since $m(A) \in [0, 1]$ and $Pl(A) \in [0, 1]$ for all $A \in \mathcal{P}(X)$, clearly $E(m) \geq 0$ for all basic assignments m; since $m(A) \neq 0$ implies $Pl(A) \neq 0$, it is guaranteed that $E(m)$ is finite for every $m \in \mathcal{M}$. Moreover, it is also clear that E is expansible, continuous, and properly normalized so that its unit of measurement is the bit. Minima and maxima of E are characterized by the following three theorems. Proofs of these theorems are rather simple and we leave them to the reader as an exercise.

Theorem 5.7.   If $(\mathcal{F}, m)$ is a consonant body of evidence, then $E(m) = 0$.

Theorem 5.8. $E(m) = 0$ if and only if m has the following property: for all $A, B \in \mathcal{P}(X)$, if $m(A) \neq 0$ and $m(B) \neq 0$, then $A \cap B \neq \emptyset$.

Theorem 5.9.   For all basic assignments defined on any universal set X with n elements, $\log_2 n$ is the only maximum of function E, which is obtained for the uniformly distributed probability measure, that is, for $m(\{x\}) = 1/n$ for each $x \in X$.

Given a universal set X, function E thus has the range

$$0 \leq E(m) \leq \log_2 |X|. \tag{5.69}$$

The maximum, which is unique, is identical with the maximum of the Shannon entropy on X. The minimum is not unique; it consists of all possibility and necessity measures as well as some other belief and plausibility measures (as characterized by Theorem 5.8).

It is not surprising that possibility and necessity measures have no dissonance in evidence. These measures are based on nested focal elements and, consequently, no conflict in evidence is involved. This is even reflected in their name—consonant measures.

Function E is also additive with respect to marginal bodies of evidence that are noninteractive in the sense of Eq. (4.18). This is expressed by the next theorem.

Theorem 5.10.   Let m_X and m_Y be marginal basic assignments on set X and Y, respectively, and let m be a joint basic assignment on $X \times Y$ such that

$$m(A \times B) = m_X(A) \cdot m_Y(B)$$

for all $A \in \mathcal{P}(X)$ and $B \in \mathcal{P}(Y)$. Then,

$$E(m) = E(m_X) + E(m_Y). \tag{5.70}$$

Proof: First, we must determine the meaning of $Pl(A \times B)$ in this case. By definition,

$$Pl(A \times B) = \sum m(C \times D),$$

where the sum is taken over all sets C and D such that $(C \times D) \cap (A \times B) \neq \varnothing$. This equation can be rewritten, according to the assumption of the theorem, as

$$Pl(A \times B) = \sum m_X(C) \cdot m_Y(D),$$

where the sum is taken over the same sets C and D as before. This equation is clearly identical to

$$Pl(A \times B) = \sum_{C \cap A \neq \varnothing} m_X(C) \cdot \sum_{D \cap B \neq \varnothing} m_Y(D).$$

Hence,

$$Pl(A \times B) = Pl_X(A) \cdot Pl_Y(B). \tag{5.71}$$

Using this result, the rest of the proof consists of the following simple manipulations of the formula for $E(m)$, where all the sums are taken over $A \in \mathcal{P}(X)$ and $B \in \mathcal{P}(Y)$ or just over the corresponding focal elements:

$$
\begin{aligned}
E(m) &= -\sum_{A,B} m(A \times B)\log_2 Pl(A \times B) \\
&= -\sum_{A,B} m_X(A) \cdot m_Y(B)\log_2(Pl_X(A) \cdot Pl_Y(B)) \\
&= -\sum_{A,B} m_X(A) \cdot m_Y(B)\log_2 Pl_X(A) - \sum_{A,B} m_X(A) \cdot m_Y(B)\log_2 Pl_Y(B) \\
&= -\sum_{A} m_X(A)\log_2 Pl_X(A) \sum_{B} m_Y(B) - \sum_{B} m_Y(B)\log_2 Pl_Y(B) \sum_{A} m_X(A) \\
&= -\sum_{A} m_X(A)\log_2 Pl_X(A) - \sum_{B} m_Y(B)\log_2 Pl_Y(B) \\
&= E(m_X) + E(m_Y). \quad \blacksquare
\end{aligned}
$$

Example 5.6

To illustrate the meaning of Theorem 5.10, let us consider the noninteractive marginal basic assignments m_X and m_Y on sets $X = \{x_1, x_2, x_3\}$ and $Y = \{y_1, y_2, y_3\}$, respectively, that are specified in Table 5.2(a). Their joint basic assignment m, which is defined by

$$m(A \times B) = m_X(A) \cdot m_Y(B)$$

TABLE 5.2. NONINTERACTIVE MARGINAL AND JOINT BASIC ASSIGNMENTS, PLAUSIBILITIES, AND BELIEFS (EXAMPLES 5.6 AND 5.7).

(a) $m(A \times B) = m_X(A) \cdot m_Y(B)$

$M(A \times B)$	$m_Y(\{y_1\}) = .2$	$m_Y(\{y_2, y_3\}) = .5$	$m_Y(\{y_2\}) = .3$
$m_X(\{x_1, x_2\}) = .4$	.08	.2	.12
$m_X(\{x_3\}) = .1$	.02	.05	.03
$m_X(\{x_1, x_3\}) = .3$	.06	.15	.09
$m_X(\{x_1, x_2, x_3\}) = .2$	.04	.1	.06

(b) $Pl(A \times B) = Pl_x(A) \cdot Pl_Y(B)$

$Pl(A \times B)$	$Pl_Y(\{y_1\}) = .2$	$Pl_Y(\{y_2, y_3\}) = .8$	$Pl_Y(\{y_2\}) = .8$
$Pl_X(\{x_1, x_2\}) = .9$	.18	.72	.72
$Pl_X(\{x_3\}) = .6$	.12	.48	.48
$Pl_X(\{x_1, x_3\}) = 1$	.2	.8	.8
$Pl_X(\{x_1, x_2, x_3\}) = 1$	.2	.8	.8

(c) $Bel(A \times B) = Bel_x(A) \cdot Bel_y(B)$

$Bel(A \times B)$	$Bel_Y(\{y_1\}) = .2$	$Bel_Y(\{y_2, y_3\}) = .8$	$Bel_Y(\{y_2\}) = .3$
$Bel_X(\{x_1, x_2\}) = .4$	.08	.32	.12
$Bel_X(x_3) = .1$	.02	.08	.03
$Bel_X(x_1, x_3) = .4$	.08	.32	.12
$Bel_X(x_1, x_2, x_3) = 1$	.2	.8	.3

for all $A \in \mathcal{P}(X)$ and $B \in \mathcal{P}(Y)$, is also given in Table 5.2(a). To illustrate the additivity of the measure of dissonance E, we need the corresponding marginal and joint plausibilities for all focal elements; these are given in Table 5.2(b). The marginal plausibilities are calculated from their basic assignment counterparts by Eq. (4.11); the joint plausibilities can then be calculated by Eq. (5.71). We can now calculate the amount of dissonance in the marginal and joint bodies of evidence:

$$E(m_X) = .14 \quad \text{(from Example 5.5),}$$

$$E(m_Y) = -.2 \log_2 .2 - .5 \log_2 .8 - .3 \log_2 .8$$

$$= .46 + .16 + .1 = .72,$$

$$E(m) = -.08 \log_2 .18 - .2 \log_2 .72 - .12 \log_2 .72 - .02 \log_2 .12$$

$$- .05 \log_2 .48 - .03 \log_2 .48 - .06 \log_2 .2 - .15 \log_2 .8$$

$$- .09 \log_2 .8 - .04 \log_2 .2 - .1 \log_2 .8 - .06 \log_2 .8$$

$$= .2 + .1 + .06 + .06 + .05 + .03 + .14 + .05$$

$$+ .03 + .09 + .03 + .02 = .86.$$

We can see that $E(m_X) + E(m_Y) = .14 + .72 = .86 = E(m)$.

5.5. *MEASURES OF CONFUSION*

As explained in Sec. 4.2, plausibility and belief measures are dual in the sense that

$$Pl(A) = 1 - Bel(\overline{A}).$$

It is thus reasonable to define a natural companion of the dissonance measure E by replacing $Pl(A)$ in the definition of E (Eq. (5.68)) with $Bel(A)$. This results in the function

$$C(m) = - \sum_{A \in \mathscr{F}} m(A)\log_2 Bel(A), \tag{5.72}$$

where $\mathscr{F}$ is the set of focal elements of the basic assignment m. This function is usually called in the literature a *measure of confusion*.

Function C is clearly expansible, continuous, and normalized so that its unit of measurement is the bit. Since $Bel(A) \leq Pl(A)$, it follows immediately from Eqs. (5.68) and (5.72) that

$$C(m) \geq E(m) \tag{5.73}$$

for any basic assignment m. It is also obvious that $C(m) \geq 0$. In addition, since $m(A) \neq 0$ implies $Bel(A) \neq 0$, $C(m)$ is always finite.

The minimum of function C, $C(m) = 0$, can be attained if and only if $Bel(A) = 1$ for all focal elements. According to Eq. (4.10), this is possible only if $m(A) = 1$ for one particular $A \in \mathscr{P}(X)$ and $m(B) = 0$ for all $B \in \mathscr{P}(X)$ different from A. Unlike the dissonance measure E, the confusion measure C is not zero for consonant bodies of evidence that have at least two focal elements.

Since $Bel(A) \geq m(A)$, it follows from Eq. (5.72) that

$$C(m) \leq - \sum_{A \in \mathscr{F}} m(A)\log_2 m(A). \tag{5.74}$$

Hence, C attains its maximum (not necessarily unique) if and only if m is uniformly distributed among the largest possible number of subsets of X such that none of these subsets is included in any other. Families of subsets, none of which is included in any other, are known in combinatorial theory as *antichains*. It is well established that the largest antichains consist of all subsets with $n/2$ elements when n is even and with either $[n/2]$ or $[n/2] + 1$ elements when n is odd, where $[n/2]$ denotes the integer part of $n/2$. Hence, C has one maximum when n is even and two maxima when n is odd. For any n, the largest antichains contain subsets. The maximum of C is thus attained when

$$m(A) = \cfrac{1}{\left(\begin{array}{c} n \\ \left[\dfrac{n}{2} \right] \end{array} \right)}$$

for all subsets in one of the largest antichains. This maximum is equal to

$$\log_2 \left(\begin{array}{c} n \\ \left[\dfrac{n}{2} \right] \end{array} \right)$$

Hence,

$$0 \le C(m) \le \log_2 \left(\begin{array}{c} n \\ \left[\dfrac{n}{2} \right] \end{array} \right), \tag{5.75}$$

where $n = |X|$.

We can now see that the name *measure of confusion* captures quite well the nature of the function C defined by Eq. (5.72). Indeed, $C(m)$ characterizes the multitude of subsets supported by evidence as well as the uniformity of the distribution of strength of evidence among the subsets. Clearly, the greater the number of subsets involved and the more uniform the distribution, the more we tend to be confused by the presentation of evidence.

When m represents a probability measure, then $m(A) = Pl(A) = Bel(A)$ for all focal elements (singletons) and, consequently, the confusion measure becomes the Shannon entropy. The generalization of the Shannon entropy from probability measures to evidence measures is thus not unique. Functions E and C, both of which are generalizations of Shannon entropy, are sometimes called *entropy-like measures*.

Like the dissonance measure E, the confusion measure is also additive with respect to marginal bodies of evidence that are noninteractive. This property is formally stated by the following theorem.

Theorem 5.11.   Let m_X and m_Y be marginal basic assignments on sets X and Y, respectively, and let m be a joint basic assignment on $X \times Y$ such that

$$m(A \times B) = m_X(A) \cdot m_Y(B)$$

for all $A \in \mathcal{P}(X)$ and $B \in \mathcal{P}(Y)$. Then,

$$C(m) = C(m_X) + C(m_Y). \tag{5.76}$$

Proof: The proof is analogous to the proof of Theorem 5.10. ■

Example 5.7

This is a continuation of Example 5.6, again using the noninteractive marginal basic assignments given in Table 5.2(a). We want to illustrate the additivity of the confusion measure C. For this purpose, we need the corresponding marginal and joint beliefs. First, we calculate the marginal beliefs from their basic assignment counterparts by Eq. (4.10); the joint beliefs are then calculated by the formula

$$Bel(A \times B) = Bel_X(A) \cdot Bel_Y(B), \tag{5.77}$$

which is a direct analogy of Eq. (5.71). All the calculated beliefs are given in Table 5.2(c). Applying Eq. (5.72), we calculate the amount of confusion in the marginal and joint bodies of evidence.

$$C(m_X) = -.4 \log_2 .4 - .1 \log_2 .1 - .3 \log_2 .4 - .2 \log_2 1$$

$$= .53 + .33 + .4 + 0 = 1.26,$$

$$C(m_Y) = -.2 \log_2 .2 - .5 \log_2 .8 - .3 \log_2 .3$$

$$= .46 + .16 + .52 = 1.14,$$

$$C(m) = -.08 \log_2 .08 - .2 \log_2 .32 - .12 \log_2 .12 - .02 \log_2 .02$$

$$-.05 \log_2 .08 - .03 \log_2 .03 - .06 \log_2 .08 - .15 \log_2 .32$$

$$-.09 \log_2 .12 - .04 \log_2 .2 - .1 \log_2 .8 - .06 \log_2 .3$$

$$= .29 + .33 + .37 + .11 + .18 + .15 + .22 + .25$$

$$+ .28 + .09 + .03 + .01 = 2.4.$$

Hence, $C(m_X) + C(m_Y) = 1.26 + 1.14 = 2.4 = C(m)$.

5.6 MEASURES OF NONSPECIFICITY

As shown in Secs. 5.4 and 5.5, the measures of dissonance and confusion are generalizations of the Shannon entropy in the framework of the mathematical theory of evidence. A natural question arises in this context: can the Hartley information be generalized in a similar way?

While the literature on Shannon entropy is abundant, Hartley information has been relatively obscured and often ill conceived; in fact, it has usually been dismissed as a measure far less interesting and useful than the Shannon entropy. To reconcile the two measures of uncertainty and information, which are not directly comparable, the Hartley measure (which is not continuous) has frequently been viewed as a special case of the Shannon entropy under various probabilistic assumptions. However, as argued in Sec. 5.3, the Hartley measure in its genuine form is not based on any probabilistic assumptions at all.

Strictly speaking, the Hartley measure is based upon one concept only—the concept of a finite set of possible entities, which can be interpreted as experimental outcomes, states of a system, events, messages, and the like, or as sequences of these. In order to use this measure, possible entities must be distinguished, within a given universe of discourse, from those that are not possible. It is thus the *possibility* of each relevant entity that matters in the Hartley measure. In this sense, the Hartley measure can be meaningfully generalized only through broadening the notion of possibility. This avenue is now available in terms of possibility theory.

A measure of uncertainty and information formulated in terms of possibility theory should thus be a generalization of the Hartley measure. As argued in Sec. 5.3, the Hartley measure captures that aspect of uncertainty that is well characterized by the term *nonspecificity*. It is therefore reasonable to assume that the possibilistic measure of uncertainty derived from the Hartley measure is also connected with nonspecificity. Since possibility and probability theories do not overlap and are both subsumed under the mathematical theory of evidence, they are alternative (complementary) theories, neither of which is a generalization of the other. Hence, the possibilistic measure of uncertainty is on equal footing with its probabilistic counterpart—the Shannon entropy—even though they measure different aspects of uncertainty. It is thus reasonable to demand that the possibilistic measure of uncertainty satisfies appropriate possibilistic counterparts of the probabilistic requirements upon which the Shannon entropy is founded. In addition, of course, it must also satisfy appropriate generalizations of the requirements upon which the Hartley measure is founded.

Let

$$U : \mathfrak{R} \to [0, \infty),$$

where $\mathfrak{R}$ denotes the set of all ordered possibility distributions (as introduced in Sec. 4.4), be a function such that $U(\mathbf{r})$ is supposed to characterize the amount of uncertainty associated with the possibility distribution $\mathbf{r}$. To qualify as an acceptable possibilistic measure of uncertainty, the function U must satisfy appropriate requirements. If U is supposed to be a possibilistic counterpart of the Shannon entropy and, at the same time, a generalizaton of the Hartley measure, we must require that function U satisfy all properties of the Shannon entropy for which possibilistic counterparts are meaningful as well as appropriate generalizations of the properties of the Hartley measure. This results in the following requirements for function U:*

(U1) Expansibility—when components of zero are added to a given possibility distribution, the value of U should not change.

(U2) Subadditivity—when $\mathbf{r}_X$ and $\mathbf{r}_Y$ are marginal possibility distributions that are calculated from a joint possibility distribution $\mathbf{r}$ by Eqs. (4.37) and (4.38), then $U(\mathbf{r}) \leq U(\mathbf{r}_X) + U(\mathbf{r}_Y)$.

(U3) Additivity—if $\mathbf{r}_X$ and $\mathbf{r}_Y$ in (U2) are noninteractive (in the possibilistic sense), that is, $r(x, y) = \min[r_X(x), r_Y(y)]$ for all $x \in X$ and $y \in Y$, then $U(\mathbf{r}) = U(\mathbf{r}_X) + U((\mathbf{r}_Y)$.

---

* The meaning of each of these requirements is discussed after they are all stated.

(U4) Continuity—U should be a continuous function.

(U5) Monotonicity—for any pair $^1\mathbf{r}$, $^2\mathbf{r}$ of possibility distributions of the same length ($^1\mathbf{r}$, $^2\mathbf{r} \in \mathcal{R}$ for some $n \in \mathbb{N}$), if $^1\mathbf{r} \leq {}^2\mathbf{r}$, then $U(^1\mathbf{r}) \leq U(^2\mathbf{r})$.

(U6) Minimum—for all possibility distributions, $U(\mathbf{r}) = 0$ if and only if exactly one component of $\mathbf{r}$ is equal to 1 and all of its remaining components are 0s.

(U7) Maximum—among all possibility distributions of the same length n ($n \in \mathbb{N}$), function U should attain its maximum for the distribution for which all elements are 1s.

(U8) Branching—for every possibility distribution $\mathbf{r} = (\rho_1, \rho_2, \ldots, \rho_n)$ of length n ($n \in \mathbb{N}$),

$$U(\rho_1, \rho_2, \ldots, \rho_n) = U(\rho_1, \rho_2, \ldots, \rho_{k-2}, \rho_k, \rho_k, \rho_{k+1}, \ldots, \rho_n)$$

$$+ (\rho_{k-2} - \rho_k)U \left(\underbrace{1, 1, \ldots, 1}_{k-2}, \frac{\rho_{k-1} - \rho_k}{\rho_{k-2} - \rho_k}, \underbrace{0, 0, \ldots, 0}_{n-k+1} \right)$$

$$- (\rho_{k-2} - \rho_k)U(\underbrace{1, 1, \ldots, 1}_{k-2}, \underbrace{0, 0, \ldots, 0}_{n-k+2}).$$

$$(5.78)$$

for each $k \in \mathbb{N}_{3,n}$.

(U9) Normalization—$U(1, 1) = 1$.

The expansibility and continuity requirements are virtually the same as their probabilistic counterparts. The only difference is that they are applied here to possibility distributions rather than to probability distributions. The requirements of subadditivity and additivity differ more from their probabilistic counterparts, but the differences are rather obvious: marginal possibility distributions are obtained from their joint distributions by the max operator rather than by summation, and possibilistic noninteraction is expressed by the min operator rather than by the product. The normalization requirement defines again the unit of measurement; it requires, in possibilistic terms, that the uncertainty associated with the total ignorance regarding two alternatives (e.g., truth values of a proposition) be equal to one so that the possibilistic uncertainty is measured in bits.

The requirement of monotonicity (U5) is a generalization of the monotonicity requirement for the Hartley measure. Clearly, we obtain the latter (special) requirement when we restrict components of the possibility distributions $^1\mathbf{r}$ and $^2\mathbf{r}$ in (U5) only to 0s and 1s. Observe that components of these possibility distributions can also be viewed as membership grades by which fuzzy sets are defined on the universal set under consideration. When restricted to 0s and 1s, the distributions define crisp sets and, consequently, the Hartley measure becomes directly applicable to them.

The requirement of the minimum of U is exactly the same as its probabilistic counterpart. The possibility distribution in which all components are 0s except one, which represents perfect evidence focusing on a single element of the universal set, is the only possibility distribution that is also a probability distribution. Due to the normalization requirements for possibility and probability distributions, the nonzero element in both cases is equal to 1. Any uncertainty measure should clearly be required to yield a value of zero for this distribution, regardless of whether it is viewed as a possibility distribution or as a probability distribution.

The requirement of the maximum of U is based on the characterization of total ignorance. In possibilistic terms, it is expressed by the possibility distribution all of whose elements are 1s; its probabilistic counterpart is the probability distribution whose components are all equal to $1/|X|$.

The most difficult requirement to comprehend is undoubtedly the branching requirement (U8). Like its various probabilistic counterparts, it requires that the function U must be capable of measuring the uncertainty in two ways. It is measured either directly for the given possibility distribution or indirectly by adding uncertainties associated with a combination of possibility distributions that reflect a two-stage measuring process. In the first stage of measurement, the distinction between the possibility values assigned to any two neighboring components is ignored (either the larger component ρ_{k-1} is replaced with the smaller component ρ_k for some k, as in our formulation of branching, or vice versa) and the uncertainty of this less refined possibility distribution is calculated. In the second stage, the uncertainty is calculated in a local frame of reference, which is defined by a possibility distribution that distinguishes only between the two neighboring possibility values that are not distinguished in the first stage of measurement. The uncertainty calculated in the local frame must be scaled back to the original frame by a suitable weighting factor. This local uncertainty must be added when the higher possibility value was replaced with the lower value (which reduced the uncertainty), and it must be subtracted in the opposite case.

The first term on the right-hand side of Eq. (5.78) represents the uncertainty obtained in the first stage of measurement. The remaining two terms represent the second stage, associated with the local uncertainty. The first of these terms expresses the loss of uncertainty caused by ignoring the component ρ_{k-1} in the given possibility distribution $\mathbf{r}$, but it introduces some additional uncertainty equivalent to the uncertainty of a crisp possibility distribution with $k-1$ components (due to the fact that the coarsening occurs at the $(k-1)$st place of $\mathbf{r}$). This additional uncertainty is excluded by the last term in Eq. (5.78). The last two terms in Eq. (5.78) thus together represent the local uncertainty. The value of k in the branching requirement (U8) is restricted to $k \geq 3$ for the following reasons: the replacement for $k = 2$ would not guarantee that the resulting distribution is normalized; for $k = 1$, the replacement is not possible at all.

An alternative branching requirement can be formulated by replacing the lower possibility value ρ_{k+1} with the higher value ρ_k for some $k \in \mathbb{N}_{n-1}$ in the first stage of measurement. This results in the form

$$U(\rho_1, \rho_2, \ldots, \rho_n) = U(\rho_1, \rho_2, \ldots, \rho_{k-1}, \rho_k, \rho_k, \rho_{k+2}, \ldots, \rho_n)$$

$$+ (\rho_k - \rho_{k+2}) U\left(\underbrace{1, \ldots, 1}_{k}, \frac{\rho_{k+1} - \rho_{k+2}}{\rho_k - \rho_{k+2}}, \underbrace{0, 0, \ldots, 0}_{n-k-1}\right) \qquad (5.79)$$

$$- (\rho_k - \rho_{k+2}) U(\underbrace{1, 1, \ldots, 1}_{k+1}, \underbrace{0, 0, \ldots, 0}_{n-k-1}),$$

which is applicable to each $k \in \mathbb{N}_{n-1}$.

It has been established that the only function that satisfies the requirements (U1) through (U9) is

$$U(\mathbf{r}) = \sum_{i=1}^{n} (\rho_i - \rho_{i+1})\log_2 |A_i|, \qquad (5.80)$$

where $\rho_{n+1} = 0$ by convention and

$$A_i = \{x \in X \mid r(x) \geq \rho_i\}.$$

It is important to remember at this point that the function U is defined on *ordered* possibility distributions. That is, it is always the case that $\rho_{i+1} \leq \rho_i$ in Eq. (5.80).

Observe that the set A_i in Eq. (5.80) can be viewed as the ρ_i-cut of a fuzzy set A defined on X by

$$\mu_A(x) = r(x)$$

for all $x \in X$. In this sense, $U(\mathbf{r})$ can be viewed as a weighted average of the Hartley information of the ρ_i-cuts A_i of the fuzzy set A; since $\rho_i - \rho_{i+1} = m(A_i)$ by Eq. (4.36), the weights of sets A_i are the values $m(A_i)$ of the basic assignment corresponding to the possibility distribution $\mathbf{r}$. Function U is thus applicable not only to possibility measures but also to fuzzy sets.

Function U defined by Eq. (5.80) is usually called a *U-uncertainty*. Since possibility distributions on which U is defined are assumed to be ordered and the sets A_i are nested, it is obvious that $|A_i| = i$ for all $i \in \mathbb{N}_n$. Equation (5.80) can thus be rewritten in a simpler form,

$$U(\mathbf{r}) = \sum_{i=1}^{n} (\rho_i - \rho_{i+1})\log_2 i. \qquad (5.81)$$

Furthermore, using the one-to-one correspondence between possibility distributions and basic distributions, given by Eqs. (4.35) and (4.36), we can also express the U-uncertainty in terms of the basic distributions. One form, based upon Eq. (5.80), is

$$U(\mathbf{r}) = U(t^{-1}(\mathbf{m})) = \sum_{i=1}^{n} m(A_i)\log_2 |A_i|; \qquad (5.82)$$

another form, based upon Eq. (5.81), is

$$U(\mathbf{r}) = U(t^{-1}(\mathbf{m})) = \sum_{i=1}^{n} \mu_i \log_2 i, \qquad (5.83)$$

where $(\mu_i \mid i \in \mathbb{N}_n) = (t(r_i) \mid i \in \mathbb{N}_n)$.

Example 5.8

Calculate $U(\mathbf{r})$ for the possibility distribution

$$\mathbf{r} = (1, 1, .8, .7, .7, .7, .4, .3, .2, .2).$$

Let us use Eq. (5.83), which is particularly convenient for calculating the U-uncertainty. First, using Eq. (4.36), we determine

$$\mathbf{m} = t(\mathbf{r}) = (0, .2, .1, 0, 0, .3, .1, .1, 0, .2).$$

Now, we apply components of $\mathbf{m}$ to Eq. (5.83) and obtain

$$U(\mathbf{r}) = 0 \log_2 1 + .2 \log_2 2 + .1 \log_2 3 + 0 \log_2 4 + 0 \log_2 5$$

$$+ .3 \log_2 6 + .1 \log_2 7 + .1 \log_2 8 + 0 \log_2 9 + .2 \log_2 10$$

$$= .2 + .16 + .78 + .28 + .3 + .66 = 2.38.$$

The proof that the U-uncertainty expressed by any of the forms (5.80) through (5.83) is the only function that satisfies requirements (U1) through (U9) for a possibilistic measure of uncertainty and information is quite complex and, therefore, it is covered in Appendix A.2. The proof uses only the requirements of expansibility, additivity, monotonicity, branching, and normalization as axioms. The remaining requirements are then easily derived as properties of the U-uncertainty.

As we argued in Sec. 5.3, the Hartley measure focuses on one of the facets of uncertainty which is well characterized by the term *nonspecificity*. Since the U-uncertainty is a generalization of the Hartley measure, it is reasonable to expect that it measures the same aspect of uncertainty. That it does, indeed, measure nonspecificity can best be seen from its form (5.82). This form readily reveals that the U-uncertainty is basically the average value of the Hartley measure for all focal elements, weighted by values of the basic assignment. The U-uncertainty is thus clearly a *possibilistic measure of nonspecificity*.

The U-uncertainty can also be expressed in the form

$$U(\mathbf{r}) = \int_0^1 \log_2 |A_\rho| \, d\rho, \qquad (5.84)$$

where

$$A_\rho = \{x \in X \mid r(x) \geq \rho\}.$$

Let us illustrate the use of this form by an example.

Example 5.9

Consider a possibility distribution on $\mathbb{N}$ that is defined by the dots in the diagram in Fig. 5.5; we assume that $r(x) = 0$ for all $x > 15$. Applying formula (5.84) to this

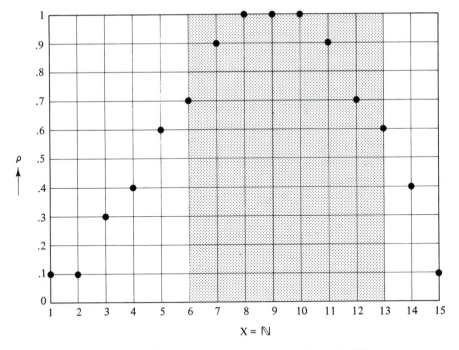

Figure 5.5. Nonspecificity versus fuzziness (Example 5.9).

distribution, we obtain

$$\int_0^1 \log_2 |A_\rho| \, d\rho = \int_0^{.1} \log_2 15 \, d\rho + \int_{.1}^{.3} \log_2 12 \, d\rho + \int_{.3}^{.4} \log_2 11 \, d\rho$$

$$+ \int_{.4}^{.6} \log_2 9 \, d\rho + \int_{.6}^{.7} \log_2 7 \, d\rho + \int_{.7}^{.9} \log_2 5 \, d\rho + \int_{.9}^1 \log_2 3 \, d\rho$$

$$= .1 \log_2 15 + .2 \log_2 12 + .1 \log_2 11$$

$$+ .2 \log_2 9 + .1 \log_2 7 + .2 \log_2 5 + .1 \log_2 3$$

$$= .39 + .72 + .35 + .63 + .28 + .46 + .16 = 3.05.$$

The amount of nonspecificity in the given possibility distribution is thus approximately 3 bits. This is equivalent to the nonspecificity of any possibility distribution in which eight components are equal to 1 and all remaining components are 0s, for example, the distribution

$$r'(x) = \begin{cases} 1 & \text{for } x \in \mathbb{N}_{6,13} \\ 0 & \text{otherwise,} \end{cases}$$

which is illustrated in Fig. 5.5 by the shaded area. This distribution is associated with the crisp set $\mathbb{N}_{6,13}$, whereas the given distribution **r** is associated with a fuzzy set A such that $\mu_A(x) = r(x)$ for all $x \in \mathbb{N}$. Using Eq. (5.14), we can calculate the degree of fuzziness of this set as follows:

$$f(A) = 15 - \sum_{i=1}^{15} |2r(x) - 1|$$

$$= 15 - (.8 + .8 + .4 + .2 + .2 + .4 + .8$$

$$+ 1 + 1 + 1 + .8 + .4 + .2 + .2 + .8)$$

$$= 15 - 9 = 6 \text{ bits}.$$

This result illustrates the fundamental difference between the two types of uncertainty—nonspecificity and fuzziness, both of which are applicable to possibility measures as well as to fuzzy sets. While the nonspecificity of $\mathbf{r}$ is virtually the same as the nonspecificity of $\mathbf{r}'$ (3 bits), their degrees of fuzziness are quite different: $\mathbf{r}$ is associated with 6 bits of fuzziness, whereas $\mathbf{r}'$ does not involve any fuzziness at all.

Let us now calculate the nonspecificity of the possibility distribution defined in Fig. 5.5 by Eq. (5.83). First, we express $\mathbf{r}$ in its ordered form:

$$\mathbf{r} = (1, 1, 1, .9, .9, .7, .7, .6, .6, .4, .4, .3, .1, .1, .1).$$

Then, we determine the corresponding basic distribution by Eq. (4.36):

$$\mathbf{m} = t(\mathbf{r}) = (0, 0, .1, 0, .2, 0, .1, 0, .2, 0, .1, .2, 0, 0, .1).$$

Now, we are ready to use Eq. (5.83):

$$U(\mathbf{r}) = U(t^{-1}(\mathbf{m})) = .1 \log_2 3 + .2 \log_2 5 + .1 \log_2 7 + .2 \log_2 9$$

$$+ .1 \log_2 11 + .2 \log_2 12 + .1 \log_2 15 = 2.99.$$

Let us consider now U-uncertainties associated with joint and marginal possibility distributions defined on X and Y. In analogy to Shannon entropy, let us simplify the notation by using $U(X)$ instead of $U(r(x) \mid x \in X)$. Then, we have

$$U(X) = \sum_{A \in \mathscr{F}_X} m_X(A) \log_2 |A|,$$

$$U(Y) = \sum_{B \in \mathscr{F}_Y} m_Y(B) \log_2 |B|,$$

$$U(X, Y) = \sum_{A \times B \in \mathscr{F}} m(A \times B) \log_2 |A \times B|,$$

where $\mathscr{F}_X$, $\mathscr{F}_Y$, and $\mathscr{F}$ are sets of focal elements of $\mathbf{m}_X$, $\mathbf{m}_Y$, and $\mathbf{m}$, respectively. In addition, we define

$$U(X \mid Y) = \sum_{A \times B \in \mathscr{F}} m(A \times B) \log_2 \frac{|A \times B|}{|B|}$$

and

$$U(Y \mid X) = \sum_{A \times B \in \mathscr{F}} m(A \times B) \log_2 \frac{|A \times B|}{|A|}$$

as generalizations of the two forms of the conditional Hartley measure.

Observe that $|A \times B|/|B|$ represents for each focal element $A \times B$ in $\mathscr{F}$ the average number of elements of A that can be selected provided that an element of B has already been selected. Expressing $U(X \mid Y)$ in the form of Eq. (5.84), we have

$$
\begin{aligned}
U(X \mid Y) &= \int_0^1 \log_2 \frac{|(A \times B)\rho|}{|B\rho|}\, d\rho \\
&= \int_0^1 \log_2 |(A \times B)\rho|\, d\rho - \int_0^1 \log_2 |B\rho|\, d\rho \\
&= U(X, Y) - U(Y). \tag{5.85}
\end{aligned}
$$

This equation is an obvious generalization of Eq. (5.33) for the Hartley measure and is a possibilistic counterpart of Eq. (5.45) for the Shannon entropy. In a similar way, we can derive

$$
U(Y \mid X) = U(X, Y) - U(X). \tag{5.86}
$$

It is known that the U-uncertainty, when expressed by the form (5.82), can be directly generalized to arbitrary basic assignments, which do not necessarily represent possibility measures. Since there are no possibility distributions for basic assignments whose focal elements are not nested, it is obvious that the generalization of the U-uncertainty must be a function that is defined in terms of basic assignments rather than possibility distributions. Let

$$
V: \mathcal{M} \to [0, \infty)
$$

be this function, where $\mathcal{M}$ is the set of all basic assignments. Then,

$$
V(m) = \sum_{A \in \mathscr{F}} m(A)\log_2 |A| \tag{5.87}
$$

measures the *nonspecificity of the body of evidence* $(\mathscr{F}, m)$, where $\mathscr{F} \subseteq \mathcal{P}(X)$.

Function V defined by Eq. (5.87) is known to be the only function that satisfies the following axiomatic requirements:

(V1) Symmetry—V should be invariant with respect to permutations of values of the basic assignment within each group of subsets of X with equal cardinalities.

(V2) Additivity—for any two noninteractive bodies of evidence $(\mathscr{F}_X, m_X)$ and $(\mathscr{F}_Y, m_Y)$, where $\mathscr{F}_X \in \mathcal{P}(X)$ and $\mathscr{F}_Y \in \mathcal{P}(Y)$, V should satisfy the equation

$$
V(m) = V(m_X) + V(m_Y),
$$

where m denotes the joint basic assignment with focal elements in $\mathscr{F}_X \times \mathscr{F}_Y$; the concept of noninteractivity is based on Dempster's rule of combination (as discussed in Sec. 4.2), that is,

$$
m(A \times B) = m_X(A) \cdot m_Y(B)
$$

for all $A \in \mathscr{F}_X$ and all $B \in \mathscr{F}_Y$.

(V3) Subadditivity—for any given joint body of evidence $(\mathscr{F}_X \times \mathscr{F}_Y, m)$, where $\mathscr{F}_x \subset \mathcal{P}(X)$ and $\mathscr{F}_Y \subset \mathcal{P}(Y)$, V should satisfy the inequality

$$V(m) \leq V(m_X) + V(m_Y),$$

where m_X and m_Y are marginal basic assignments of m, that is, projections of m on X and Y, respectively (as defined in Sec. 4.2).

(V4) Branching—let $(\mathscr{F}, m)$, $(\mathscr{F}_1, m_1)$, and $(\mathscr{F}_2, m_2)$ be three bodies of evidence such that

$$\mathscr{F} = \{A, B, C, \ldots\},$$

$$\mathscr{F}_1 = \{A_1, B, C, \ldots\},$$

$$\mathscr{F}_2 = \{A, B_1, C_1, \ldots\},$$

where

$$A_1 \subseteq A, \quad B_1 \subseteq B, \quad C_1 \subseteq C, \ldots,$$

$$|A_1| = |B_1| = |C_1| = \cdots = 1,$$

and

$$m(A) = m_1(A_1) = m_2(A)$$

$$m(B) = m_1(B) = m_2(B_1)$$

$$m(C) = m_1(C) = m_2(C_1)$$

$$\cdots\cdots\cdots\cdots\cdots\cdots$$

Then, V should satisfy the equation

$$V(m) = V(m_1) + V(m_2).$$

(V5) Normalization—when $m(A) = 1$ and $|A| = 2$, then $V(m)$ should be equal to 1.

The requirement of symmetry states that the nonspecificity should depend only on the values of the basic assignment and the cardinalities of the set to which these values are allocated and not on the sets themselves. This requirement is a natural counterpart of the symmetry requirement employed in the characterization of the Shannon entropy. Although no symmetry requirement is employed in the characterization of the U-uncertainty, it is implicitly incorporated in the assumption that the U-uncertainty is defined only on ordered possibility distributions. That is, each unordered possibility distribution is represented by the corresponding ordered distribution, and it is tacitly assumed that both of them have the same degree of nonspecificity.

The requirements of additivity, subadditivity, and normalization are obvious generalizations of the corresponding special requirements employed in the characterization of the U-uncertainty. They are also obvious counterparts of the corresponding requirements through which the Shannon entropy is sometimes characterized.

The branching requirement states the following: given a basic assignment m, if we maintain its values but replace any of the focal elements (e.g., element A) with a singleton (thus obtaining basic assignment m_1) and, separately, replace all the remaining focal elements with singletons (thus obtaining basic assignment m_2), then the sum of the degrees of nonspecificity of the two resulting basic assignments ($V(m_1) + V(m_2)$) should be equal to the degree of nonspecificity of the original basic assignment (i.e., $V(m)$).

The proof that the function V defined by Eq. (5.87) is the only function that satisfies the requirements (V1) through (V5) is rather lengthy and, therefore, it is placed in Appendix A.3.

It follows directly from Eq. (5.87) that function V attains its minimum, $V(m) = 0$, if and only if all focal elements of m are singletons. Hence, $V(m) = 0$ for all probability measures. The maximum of V is attained when $m(X) = 1$; then, $V(m) = \log_2 |X|$. Hence, V has the range

$$0 \le V(m) \le \log_2 |X|. \tag{5.88}$$

Let us now use some examples to compare the three measures of uncertainty that are applicable within the mathematical theory of evidence—the measures of dissonance (E), confusion (C), and nonspecificity (V).

Example 5.10

Let $X = \{x_1, x_2, x_3, x_4\}$. All subsets A of X are listed in Table 5.3, each specified by its characteristic function. Also listed are cardinalities of the subsets. Three basic

TABLE 5.3. COMPARISON OF MEASURES OF DISSONANCE, CONFUSION, AND NONSPECIFICITY FOR THREE BODIES OF EVIDENCE (EXAMPLE 5.10).

$\mathcal{P}(X)$ $X: x_1\ x_2\ x_3\ x_4$	$\lvert A \rvert$	Possibility measure			Probability measure			General evidence measure		
		$m_1(A)$	$Bel_1(A)$	$Pl_1(A)$	$m_2(A)$	$Bel_2(A)$	$Pl_2(A)$	$m_3(A)$	$Bel_3(A)$	$Pl_3(A)$
A: 0 0 0 0	0	0	0	0	0	0	0	0	0	0
0 0 0 1	1	.3	.3	1	.3	.3	.3	0	0	.8
0 0 1 0	1	0	0	.7	.2	.2	.2	0	0	.7
0 0 1 1	2	.2	.5	1	0	.5	.5	.1	.1	1
0 1 0 0	1	0	0	.5	.4	.4	.4	0	0	.6
0 1 0 1	2	0	.3	1	0	.7	.7	.2	.2	1
0 1 1 0	2	0	0	.7	0	.6	.6	.1	.1	1
0 1 1 1	3	.4	.5	1	0	.9	.9	.1	.5	1
1 0 0 0	1	0	0	.1	.1	.1	.1	0	0	.5
1 0 0 1	2	0	.3	1	0	.4	.4	0	0	.9
1 0 1 0	2	0	0	.7	0	.3	.3	0	0	.8
1 0 1 1	3	0	.5	1	0	.6	.6	.3	.4	1
1 1 0 0	2	0	0	.5	0	.5	.5	0	0	.9
1 1 0 1	3	0	.3	1	0	.8	.8	.1	.3	1
1 1 1 0	3	0	.2	.7	0	.7	.7	.1	.2	1
1 1 1 1	4	.1	1	1	0	1	1	0	1	1
$E(m_i)$		0			1.85			0		
$C(m_i)$		1.12			1.85			2.03		
$V(m_i)$		1.03			0			1.35		

assignments on the power set $\mathscr{P}(X)$ are defined in the table and for each of them the degrees of belief and plausibility are calculated. These basic assignments m_1, m_2, m_3 represent a possibility measure, probability measure, and general evidence measure, respectively. For each of these measures, the degrees of dissonance E, confusion C, and nonspecificity V are given at the bottom of the table. Observe that only the underlined entries in the table are employed in the calculation of these values.

We can see that m_1 and m_2 have the same values, but focal elements of m_1 are nested, whereas focal elements of m_2 are singletons. This difference, which exemplifies the distinction between possibility measures and probability measures, results in $E(m_1) = 0$ and $V(m_2) = 0$. Since $Bel_2 = Pl_2$, which is the case for all probability measures, we have $E(m_2) = C(m_2) = 1.85$. Observe also that $E(m_3) = 0$ even though m_3 does not represent a possibility measure. Hence, the evidence captured by m_3 does not involve any conflict (dissonance) even though its degree of confusion is rather high ($C(m_3) = 2.03$).

Let us now show the calculation of the amount of nonspecificity for each of the three measures in the table:

$$V(m_1) = .3 \log_2 1 + .2 \log_2 2 + .4 \log_2 3 + .1 \log_2 4$$

$$= 0 + .2 + .63 + .2 = 1.03,$$

$$V(m_2) = .3 \log_2 1 + .2 \log_2 1 + .4 \log_2 1 + .1 \log_2 1 = 0,$$

$$V(m_3) = .1 \log_2 2 + .2 \log_2 2 + .1 \log_2 2 + .1 \log_2 3$$

$$+ .3 \log_2 3 + .1 \log_2 3 + .1 \log_2 3$$

$$= .4 \log_2 2 + .6 \log_2 3 = .4 + .95 = 1.35.$$

5.7 UNCERTAINTY AND INFORMATION

When our ignorance or uncertainty about some state of affairs is reduced by an act (such as observation, reading, or receiving a message), the act may be viewed as a source of information pertaining to the state of affairs under consideration. The amount of information obtained by the act may then be measured by the difference in uncertainty before and after the act.

A reduction of uncertainty by an act is accomplished only when some options considered possible prior to the act are eliminated after it. This requires a semantic connection between the prospective outcomes of the acts involved (observations, received messages, experimental outcomes) and the entities to which they are applied. When this semantic connection is well established, as in the context of a typical scientific inquiry, the measurement of uncertainty and the associated information is expressed solely in syntactic terms.

In human communication, however, the concept of information is more complex. It almost invariably involves three aspects:

• **syntactic**—regarding the relationship among the signs that are employed in the communication;

- **semantic**—regarding the relationship between the signs and the entities for which they stand, that is, regarding the designation of the meaning of the signs;
- **pragmatic**—regarding the relationship between the signs and their utilities.

It is not our aim in this book to study the semantic and pragmatic aspects of information, which are essential for capturing the rich notion of information in human communication. Instead, we view information strictly in terms of ignorance or uncertainty reduction within a given syntactic and semantic framework, which is assumed to be fixed in each particular application. Let us illustrate, however, that this view is not overly restrictive and can be extended to capture the semantic and pragmatic aspects of information.

Consider, for example, a set X of some signs under consideration (such as states of a system, messages, or experimental outcomes) and a probability distribution $\mathbf{p} = (p(x) \mid x \in X)$ defined on X. Consider, in addition, that for each $x \in X$ there is a set Y_x of meanings of x, which is associated with a probability distribution $\mathbf{q}_x = (q_x(y_x \mid x) \mid y_x \in Y_x)$. Then, we can use the following function $'H$, which is based on Shannon entropy, to express the total uncertainty associated with the two probability distributions:

$$'H(X, Y_x \mid x \in X) = H(X) + \sum_{x \in X} p(x)H(Y_x \mid x). \qquad (5.89)$$

According to this equation, the *total uncertainty* $'H$ is expressed as the sum of the *syntactic uncertainty* $H(X)$ and the *semantic uncertainty*, which is represented by the last term in the equation. Observe that the semantic uncertainty is defined in Eq. (5.89) as the weighted average of the local uncertainties associated with the meanings assigned to each individual sign in X.

The semantic uncertainty is always subjective since the local uncertainties are, in general, different for different individuals (observers, message receivers). For example, if an individual is totally ignorant about the meaning of a sign x, then $q(y_x \mid x) = 1/ \mid Y_x \mid$ for all $y_x \in Y_x$ and

$$H(Y_x \mid x) = \log_2 \mid Y_x \mid.$$

This may happen, for example, when a message is received in a foreign language that is not understood by the receiver or when an individual takes a reading on a meter without knowing what the meter measures. If, on the other hand, the individual has no uncertainty about the meaning of any of the signs in X, then $q(a \mid x) = 1$ and $q(b \mid x) = 0$ for all $b \in Y_x$ such that $b \neq a$; that is, $H(Y_x \mid x) = 0$ for all $x \in X$ so that the semantic part of the total uncertainty disappears. As previously indicated, this can often be arranged by selecting a particular syntactic and semantic framework and keeping it fixed in the course of a specific application.

Our view of information, based strictly upon reduction of ignorance or uncertainty, also has enough flexibility to incorporate pragmatic aspects of information, if desirable. This can be done, for example, by weighting the various signs employed (observations, messages) according to their utilities. Let us illustrate this possibility by the concept of a *weighted Shannon entropy*

$$G(p(x), w(x) \mid x \in X) = -\sum_{x \in X} w(x)p(x)\log_2 p(x), \qquad (5.90)$$

where $p(x)$ are probabilities defined on a finite set X of signs and $w(x)$ are *weights* that characterize utilities of the individual signs $x \in X$; it is only assumed that the weights are nonnegative and finite real numbers, but it is often desirable to normalize them so that they add to 1.

The weighted Shannon entropy defined by Eq. (5.90) is known to be the only function that satisfies five desirable requirements (continuity, symmetry, branching, boundary conditions, and normalization), but we do not intend to pursue this issue here. Observe that when the weights in Eq. (5.90) are all equal and normalized, the pragmatic aspects incorporated in the weighted Shannon entropy disappear and the regular Shannon entropy emerges.

Equations (5.89) and (5.90) are given here only as examples of possible semantic and pragmatic extensions of the basic measures of uncertainty and information. There are certainly other ways of dealing with the semantic and pragmatic aspects of information, but these issues are beyond the scope of this book.

We should note at this point that the concept of information has also been investigated in terms of the theory of computability, independent of the concept of uncertainty. In this approach, the amount of information represented by an object is measured by the length of the shortest possible program written in some standard language (e.g., a program for the standard Turing machine) by which the object is described in the sense that it can be computed. Information of this type, which is defined here in a rather oversimplified manner, is usually referred to as *descriptive information* or *algorithmic information*.

In this book, we are not concerned with descriptive information. Our aim is to expose the many facets of the notion of uncertainty as they emerge within the broad framework of the theory of fuzzy sets and to characterize the relationship between uncertainty and information. We are thus concerned solely with information that is based upon uncertainty; let us call it *uncertainty information*.

Seven types of uncertainty are introduced in previous sections of this chapter. Each of them is expressed by a specific uncertainty measure. A summary of these seven uncertainty measures is given in Fig. 5.6. Each measure is identified by its name, symbol, and the year of its introduction. The four shaded areas in the figure indicate formal frameworks to which the individual measures pertain: crisp sets, fuzzy sets (or possibility measures), probability measures, and evidence measures. There is only one type of uncertainty germane to crisp sets—the Hartley measure—and there is only one type of uncertainty germane to probability theory—the Shannon entropy. For fuzzy sets (and the associated possibility theory), however, two types of uncertainty are applicable: the U-uncertainty, which expresses the nonspecificity of a fuzzy set or of the associated possibility measure, and an appropriate measure of fuzziness. Within the mathematical theory of evidence, we distinguish three germane types of uncertainty, expressed by the measures of dissonance, confusion, and nonspecificity.

For the convenience of the reader, the mathematical formulas by which the seven measures of uncertainty are expressed are given in Table 5.4.

Each type of uncertainty represents a particular type of information. The information is measured in the same units as the underlying uncertainty. When more than one type of uncertainty is applicable, the total uncertainty is a tuple

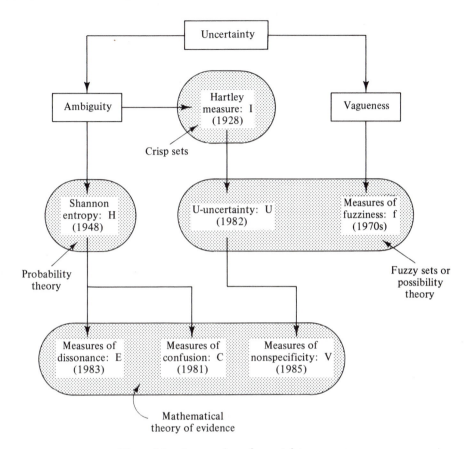

Figure 5.6.   An overview of uncertainty measures.

whose components are uncertainties of the individual types. Information can be obtained by reducing any of the components of the total uncertainty. However, a reduction of uncertainty of one type may often result in an increase of uncertainty of another type. That is, the components of the total uncertainty are interrelated. It is thus important that the types of uncertainty employed as components in the total uncertainty be measured in the same units.

An analogy between uncertainty and energy seems suggestive at this point. Energy, like uncertainty, appears in different forms, such as mechanical, thermal, electric, chemical, or nuclear energy. Different forms of energy require different measurement methods; nevertheless, they all have the same physical dimension and thus can be measured in the same units. This is also similar to uncertainty, whose different measures yield in each application values that are comparable in the sense that they express the same general concept of uncertainty in terms of the same units. Moreover, energy is viewed, in general, as the capacity for doing work. Similarly, uncertainty is viewed as the capacity for obtaining information by a communicative act of some sort, be it communication between variables of

TABLE 5.4. MEASURES OF UNCERTAINTY: A SUMMARY.

Name	Formula	Meaning of symbols
Hartley measure	$I(N) = \log_2 N$	N: cardinality of crisp set
Shannon entropy	$H(\mathbf{p}) = -\sum_{i=1}^{n} p_i \log_2 p_i$	$\mathbf{p}$: probability distribution $\mathbf{p} = (p_1, p_2, \ldots, p_n)$
U-uncertainty	$U(\mathbf{r}) = \sum_{i=1}^{n} (\rho_i - \rho_{i+1}) \log_2 i$	$\mathbf{r}$: possibility distribution $\mathbf{r} = (\rho_1, \rho_2, \ldots, \rho_n)$
Dissonance in evidence	$E(m) = -\sum_{A \in \mathcal{F}} m(A) \log_2 Pl(A)$	m: basic assignment $\mathcal{F}$: set of focal elements
Confusion in evidence	$C(m) = -\sum_{A \in \mathcal{F}} m(A) \log_2 Bel(A)$	m: basic assignment $\mathcal{F}$: set of focal elements
Nonspecificity in evidence	$V(m) = \sum_{A \in \mathcal{F}} m(A) \log_2 |A|$	m: basic assignment $\mathcal{F}$: set of focal elements
Measure of fuzziness	$f_c(A) = |X| - \sum_{x \in x} |\mu_A(x) - c(\mu_A(x))|$	μ_A: membership function or possibility distribution c: fuzzy complement

a system, between persons, or between a person and nature (as in observation or experimentation). We can thus view, at least to some extent, the recent discoveries of different measures of uncertainty as analogous to the discoveries of different forms of energy in the course of the history of science.

5.8 INFORMATION AND COMPLEXITY

Our aim in this section is to establish a connection between the concepts of information and complexity. In particular, we wish to show that the concept of complexity has many faces, while, at the same time, it is associated with some general properties that remain invariant with regard to these faces.

To begin with a broad perspective, let us consult a common dictionary (*Webster's Third International Dictionary*); we find that *complexity* is "the quality or state of being complex," that is,

— having many varied interrelated parts, patterns, or elements and consequently hard to understand fully,
    or
— marked by an involvement of many parts, aspects, details, notions, and necessitating earnest study or examination to understand or cope with.

This commonsense characterization of complexity does not contain any qualification regarding the kind of entities to which it is applicable. As such, it can be

applied to virtually all kinds of entities, material or abstract, natural or manufactured, products of art or science. Regardless of what it is that is actually considered
as being complex or simple, its degree of complexity is, according to the commonsense characterization, associated with the number of recognized parts as
well as with the extent of their interrelationship; in addition, complexity is given
a somewhat subjective connotation since it is related to the ability to understand
or cope with the thing under consideration.

We can see that the commonsense characterization of complexity assumes
an interaction between an object (a part of the world that may have "many varied
interrelated parts . . .") and a human being (or, perhaps, a computer) for whom
it may be difficult "to understand or cope with" the object. This means that the
complexity of an object for a particular human being depends on the way in which
he or she interacts with it (i.e., on his or her interests or capabilities). More
poetically, we may say that *the complexity of an object is in the eyes of the
observer.*

In most cases, there is virtually an unlimited number of ways in which one
can interact with an object. As a consequence, the interaction is almost never
complete. It is based on a limited (and usually rather small) number of attributes
of the object that the observer is capable of distinguishing and that are relevant
to his or her interests. These attributes are not available to the observer directly
but only in terms of their abstract images, which are results of perception or of
some specific measurement procedures. Let these abstract images of attributes
distinguished on various objects be called *variables.*

When a set of variables is established as a result of our interaction with an
object of interest, we say that a *system* is distinguished on the object. The term
system is thus always viewed as an abstraction—or an image—of some aspects
of the object and not as a real thing. In other words, a system is a way of looking
at the world. The important distinction between the notions of object and system
is well characterized by Ashby [1956]:

> At this point we must be clear about how a "system" is to be defined. Our first
> impulse is to point at the pendulum and to say "the system is that thing there." This
> method, however, has a fundamental disadvantage: every material object contains
> no less than an infinity of variables and therefore of possible systems. The real
> pendulum, for instance, has not only length and position, it has also mass, temper-
> ature, electric conductivity, crystalline structure, chemical impurities, some radio-
> activity, velocity, reflecting power, tensile strength, a surface film of moisture, bac-
> terial contamination, an optical absorption, elasticity, shape, specific gravity, and
> so on and on. Any suggestion that we should study "all" the facts is unrealistic and
> actually the attempt is never made. What is necessary is that we should pick out
> and study the facts that are relevant to some main interest that is already given. . . .
> . . . The system now means, not a thing, but a list of variables.

Since we deal with systems distinguished on objects and not with the objects
themselves (in their totality), it is not operationally meaningful to view complexity
as an intrinsic property of objects. This does not mean, however, that we deny
the existence of complexity of objects in the ontological sense. We recognize only

that the notion of object complexity is epistemologically, pragmatically, and methodologically vacuous, in contrast to the notion of systems complexity. This point is well expressed by Ashby in one of his last writings [Ashby, 1973]:

> The word "complex", as it may be applied to systems, has many possible meanings, and I must first make my use of it clear. There is no obvious or pre-eminent meaning, for although all would agree that the brain is complex and a bicycle simple, one has also to remember that to a butcher the brain of a sheep is simple while a bicycle, if studied exhaustively (as the only clue to a crime) may present a very great quantity of significant detail.
>
> Without further justification, I shall follow, in this paper, an interpretation of "complexity" that I have used and found suitable for about ten years. I shall measure the degree of "*complexity*" *by the quantity of information required to describe the vital system.* To the neurophysiologist the brain, as a feltwork of fibers and a soup of enzymes, is certainly complex; and equally the transmission of a detailed description of it would require much time. To a butcher the brain is simple, for he has to distinguish it from only about thirty other "meats", so not more than $\log_2 30$, i.e., about 5 bits, are involved. This method admittedly makes a system's complexity purely *relative to a given observer*, it rejects the attempt to measure an absolute, or intrinsic, complexity; but this acceptance of complexity as something in the eye of the beholder is, in my opinion, the only workable way of measuring complexity.

Complexity is thus associated with systems, that is, some abstractions distinguished on objects that reflect the way in which the objects are interacted with. But, what are these abstractions to which we refer as systems? The answer is not unique; it depends on the conceptual framework within which one operates. Each framework determines the scope of systems that can be described within it and leads to some specific taxonomy of these systems. The notion of complexity has, clearly, different meanings for the different types of systems that are describable within the framework.

Several conceptual frameworks for characterizing a broad scope of systems have been developed. The differences in terminology and mathematical formulation among some of them are considerable. There is enough evidence, however, that these seemingly significant differences are, in fact, minor in substance and can be reconciled. Hence, our restriction to one particular framework does not jeopardize in any significant way the generality of our discussion.

As a basis for our further discussion of complexity of systems, as well as for some applications discussed in Chap. 6, let us briefly describe a *hierarchy of epistemological types of systems,* which stems from a conceptual framework developed by Klir [1985]. This epistemological hierarchy evolved through the process of distilling the notions of systems from the various disciplines, categorizing them, and integrating them in a coherent whole.

Hierarchy of Epistemological Types of Systems

On the lowest epistemological level, referred to as level 0, a system is defined by a set of variables, a set of potential states (values) recognized for each variable, and some operational way of describing the meaning of the variables and their states in terms of the associated real world attributes and their manifestations. The set of variables is always partitioned into two subsets, referred to as *basic* and *supporting* variables. Aggregate states of all supporting variables form a support set within which changes in states of the basic variables occur. The most frequent examples of supporting variables are those representing time, space, or various populations of individuals of the same kind. The term *source systems* is usually used for systems defined at this level to indicate that such systems are, at least potentially, sources of empirical data.

Systems on different higher epistemological levels are distinguished from each other by the level of knowledge regarding the variables of the associated source system. A higher level system entails all knowledge of the corresponding systems on any lower levels and, at the same time, contains some additional knowledge that is not available on the lower levels. Hence, the source system is included in all of the higher-level systems.

When a source system is supplemented by a set of data, that is, by actual states of the basic variables within the defined support set, we view the new system as one associated with epistemological level 1. Systems on this level are called *data systems*. In general, any set of data is defined by a function whose domain and range are, respectively, the support set and the set of overall states of the basic variables. For example, any time function of one or more variables qualifies as a data set.

Higher epistemological levels involve some support-invariant characterization of the constraint among the basic variables involved. In general, the constraint can be utilized for generating states of the basic variables within the support set.

On level 2, each system is represented by one overall support-invariant characterization (time-invariant, space-invariant, population-invariant) of the constraint among the basic variables of the associated source system and, possibly, some additional variables. Each of the additional variables is defined in terms of a basic variable and a specific translation rule in the support set. Since the constraint can be utilized for describing an overall process by which states of the basic variables are generated within the support set, systems at this level are called *generative systems*.

Any adequate characterization of an investigated object by some sort of a support-invariant mathematical relation, such as a differential equation with constant coefficients, Markov chain, or finite-state machine, is an example of a generative system. Each of these forms of mathematical relations characterizes a constraint among the variables involved, which does not change within the relevant support set (time or space set). This constraint characterization makes it possible to generate data sets peculiar to the variables. Observe, for example, that the solution of a set of differential equations for specific initial or boundary conditions represents a data set (a function of time or space).

On epistemological level 3, each system is defined in terms of a set of generative systems (or, sometimes, lower-level systems) viewed as subsystems of an overall system. The subsystems may be coupled (i.e., they may share some variables) or they may interact in some other way. Systems on this level are called *structure systems*.

Structure systems are best depicted by block diagrams. Each block in a block diagram represents a generative system and entries to the block correspond to basic variables involved in the respective generative system. Common entries to two or more blocks indicate variables shared by those blocks, which allow them to interact with each other.

On epistemological level 4 and higher levels, a system defined on some lower level is allowed to change within the support set defined by the associated source system. On level 4, the change is characterized by a support-invariant procedure; such systems are called *metasystems*. On level 5, the characterization is allowed to change according to a support-invariant metaprocedure; such systems are called *meta-metasystems* (or metasystems of second order). In a similar fashion, *metasystems of higher orders* are defined.

Examples of metasystems are time-varying finite state machines, tessellation automata, or differential equations whose coefficients are not constant. Metasystems (of the various orders) are important for capturing systems phenomena that involve change, such as adaptation, self-organization, morphogenesis, autopoiesis, evolution, and so on.

A simplified overview of the epistemological systems hierarchy is given in Fig. 5.7. A more detailed version of the hierarchy is expressed by the diagram in Fig. 5.8. The notation in this diagram is adopted from a recent book by Klir [1985], where individual types of systems as well as the whole epistemological hierarchy are described in great detail. Symbols **S**, **D**, **F** denote source systems, data systems, and generative systems, respectively. When **S** is used as a prefix, it stands for structure systems. For example, **SF** denotes structure systems whose elements are generative systems and **SD** denotes structure systems whose elements are data systems. The symbol S^2 denotes structure systems whose elements are also struc-

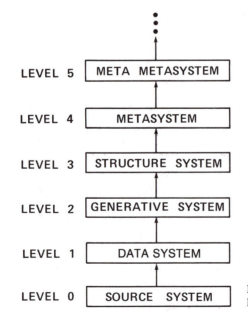

Figure 5.7.   Epistemological systems hierarchy: a simplified overview.

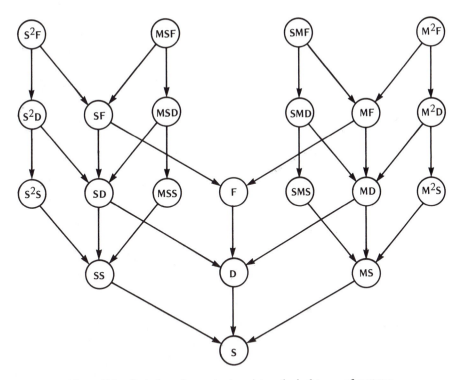

Figure 5.8. Ordering of some basic epistemological types of systems.

ture systems. For example, S^2F denotes structure systems of structure systems whose elements are generative systems. The symbols **M** and M^2 denote metasystems and meta-metasystems, respectively. The combinations **SM** and **MS** denote structure systems whose elements are metasystems and metasystems whose elements are structure systems, respectively. The diagram in Fig. 5.8 describes only a part of the epistemological hierarchy; it can be extended in an obvious way to combinations such as S^3F, S^2MF, **SMSF**, M^2SF, S^2MF, and so forth.

General Principles of Systems Complexity

Systems thus have many different faces, each represented by one of the epistemological systems types. As a consequence, the complexities associated with these types of systems have many faces as well. That is to say, different systems types in the described epistemological hierarchy give the notion of complexity different meanings, each of which requires a special treatment.

Notwithstanding the differences in complexities of the various systems types, *two general principles of systems complexity* can be recognized; they are applicable to any of the systems types and can thus be utilized as guidelines for a comprehensive study of systems complexity.

According to the *first general principle*, the complexity of a system (of any type) should be proportional to the amount of information required to describe the system. Here, the term *information* is used solely in a syntactic sense; no semantic or pragmatic aspects of information are employed. One way of expressing this descriptive complexity, perhaps the simplest one, is to measure it by the number of entities involved in the system (variables, states, components) and the variety of interdependence among the entities. Indeed, everything else being the same, our ability to understand or cope with a system tends to decrease when the number of entities involved or the variety of their interconnections increases. There are, of course, many different ways in which descriptive complexity can be expressed. Each of them, however, must satisfy some general requirements formulated as follows.

Let $\mathscr{X}$ denote the set of all systems of a particular epistemological type, let $\mathscr{P}(\mathscr{X})$ denote the power set of $\mathscr{X}$, and let $W_\mathscr{X}$ denote a *measure of descriptive complexity* within the set $\mathscr{X}$. Then, $W_\mathscr{X}$ is a function

$$W_\mathscr{X} : \mathscr{P}(\mathscr{X}) \to \mathbb{R}$$

that satisfies the following requirements (axioms):

(W1) $W_\mathscr{X}(\varnothing) = 0$.
(W2) If $A \subset B$, then $W_\mathscr{X}(A) \leq W_\mathscr{X}(B)$.
(W3) If A is a homomorphic image of B, then $W_\mathscr{X}(A) \leq W_\mathscr{X}(B)$.
(W4) If A and B are isomorphic, then $W_\mathscr{X}(A) = W_\mathscr{X}(B)$.
(W5) If (1) $A \cap B = \varnothing$, (2) A and B do not interact with each other, and (3) neither A nor B is a homomorphic image of the other, then $W_\mathscr{X}(A \cup B) = W_\mathscr{X}(A) + W_\mathscr{X}(B)$.

Requirements (W1) and (W2) guarantee that the complexity of any system is characterized by a nonnegative number. Requirements (W2) and (W3) deal with fundamental properties of monotonicity: complexity should not increase when the given set of systems is reduced or when less detail is distinguished in the systems. Requirement (W4) is obvious: when some (any) entities in the given systems are relabeled, everything else remaining the same, the complexity should not change. Requirement (W5) describes a desirable property of additivity: if two sets of systems are taken together that in all relevant respects have nothing in common (no common systems, no interaction, no morphic relationship), then the total complexity should be equal to the sum of the two individual complexities.

According to the *second general principle*, systems complexity should be proportional to the amount of information needed to resolve any uncertainty associated with the system involved. Here, again, syntactic information is used, but this information is based on a measure of uncertainty. In general, the more uncertainty is involved in the system, the more complex the system is perceived to be.

This is not to claim that semantic and pragmatic aspects of information are not connected with complexity. As already mentioned, however, these aspects are beyond the scope of this book.

As demonstrated in this chapter in detail, well-justified measures of uncertainty are now available for some formal frameworks within the theory of fuzzy sets. These measures of uncertainty, which are summarized in Fig. 5.6 and Table 5.4, are thus not only applicable as measures of information but are also closely associated with the measurement of complexity of systems.

General Simplification Problem

Systems complexity is primarily studied for the purpose of developing sound methods by which systems that are incomprehensible or unmanageable can be simplified to an acceptable level of complexity. Problems of systems simplification are perhaps the most important of all systems problems.

Given a system of some particular epistemological type, let $\mathscr{X}$ denote the set of all its simplifications that are considered *admissible* in the context of a given problem. Let $\overset{d}{\leq}$ and $\overset{u}{\leq}$ denote the two *preference complexity orderings* on $\mathscr{X}$, which are based on the descriptive and uncertainty information, respectively. These orderings are, in general, only weak orderings (i.e., reflexive and transitive relations on $\mathscr{X}$). The uncertainty ordering may, in fact, stand for several orderings based on different measures of uncertainty. For example, if $\mathscr{X}$ consists of systems formulated in terms of fuzzy sets, then $\overset{u}{\leq}$ may stand for both non-specificity and fuzziness; then

$$x \overset{u}{\leq} y \quad \text{if and only if} \quad x \overset{U}{\leq} y \quad \text{and} \quad x \overset{f}{\leq} y$$

for all $x, y \in \mathscr{X}$, where $\overset{U}{\leq}$ and $\overset{f}{\leq}$ denote, respectively, preference orderings based on the U-uncertainty and some comparable measure of fuzziness. Similarly, for systems characterized in terms of the mathematical theory of evidence,

$$x \overset{u}{\leq} y \quad \text{if and only if} \quad x \overset{E}{\leq} y \quad \text{and} \quad x \overset{C}{\leq} y \quad \text{and} \quad x \overset{V}{\leq} y$$

for all $x, y \in \mathscr{X}$, where $\overset{E}{\leq}, \overset{C}{\leq},$ and $\overset{V}{\leq}$ denote preference orderings based upon the measures of dissonance, confusion, and nonspecificity, respectively. In addition to the basic complexity orderings $\overset{d}{\leq}$ and $\overset{u}{\leq}$, let $\overset{\alpha}{\leq}, \overset{\beta}{\leq}, \ldots$ denote other (optional) orderings on $\mathscr{X}$, which express special preferences specified by the user of the given system. In terms of all the orderings involved, we define a *joint preference ordering* $\overset{*}{\leq}$ by the following formula

$$(\forall x, y \in \mathscr{X})(x \overset{*}{\leq} y \Leftrightarrow x \overset{d}{\leq} y \quad \text{and} \quad x \overset{u}{\leq} y \quad \text{and} \quad x \overset{\alpha}{\leq} y \quad \text{and} \quad x \overset{\beta}{\leq} y \quad \text{and} \ldots).$$

The *solution set* $\mathscr{X}_s$ of the simplification problem consists of those systems in $\mathscr{X}$ that are either equivalent or incomparable in terms of the joint preference ordering $\overset{*}{\leq}$. Formally,

$$\mathscr{X}_s = \{x \in \mathscr{X} \mid (\forall y \in \mathscr{X})(y \overset{*}{\leq} x \Rightarrow x \overset{*}{\leq} y)\}. \tag{5.91}$$

All simplification strategies can be formulated as special cases of this general formulation. They differ from each other in:

— the epistemological type of the system to be simplified;
— the set of simplifications of the given system that are declared as admissible;
— the implied meanings of the two complexity orderings;
— the nature of additional (optional) preference orderings.

It is clear from these considerations that the general formulation subsumes a large class of special simplification problems. Let us look at some important categories of these problems.

One way of simplifying a system of any epistemological type is to exclude some variables from the system or to partition states of some of its variables into equivalence classes. If, for example, the system under consideration is a generative system, the number of variables together with the number of actual overall states of the system (or, perhaps, the product of these two numbers) can be viewed as a simple expression of its descriptive complexity. The uncertainty complexity would be expressed in this case by an appropriate measure of generative uncertainty; it would characterize the average uncertainty associated with the generation of states (or output states) of the system.

An important strategy for making very complex systems manageable is to break them down into appropriate subsystems. One aspect of systems manageability is expressed in terms of the size of computer memory required to store the system; this may in some cases be adopted as the measure of descriptive complexity. Consider, for example, n variables, each of which has k states. When dealing with the overall system of these variables, nk^n memory cells, each of which can store any one of k states, must be made available for storing states of the system. On the other hand, when a structure system consisting of all subsystems with two variables is used, the number of memory cells that are needed for the same purpose is $k^2 n(n - 1)$. This number grows with increasing values of k and n at a considerably lower rate than that of nk^n, as illustrated in Fig. 5.9 for $k = 10$. If the structure system contained only some of the two-variable subsystems, the comparison would be even more favorable. Although for some small values of n and k, structure systems may require more memory space than the corresponding overall system, it is clear that their memory requirements are far less demanding in most cases of practical significance, especially for large values of k and n.

Another aspect of systems manageability is connected with the number of possible systems that must be distinguished in some problems. Consider again n variables, each with k states, and assume that the constraint among the variables is expressed solely by listing the actual overall states. Then, there are 2^{k^n} distinct overall systems of this kind. If the constraint among the same variables can be expressed in terms of a structure system that consists of all two-variable subsystems, the number of all such structure systems is

$$2^{k^2 - 1} n(n - 1).$$

To identify one particular element of a finite set X requires, according to the Hartley measure, $\log_2 |X|$ bits of information, where $|X|$ denotes the number of elements in X. Hence, to identify one overall system requires in our case k^n bits, whereas to identify one structure system requires $(k^2 - 1) + \log_2(n^2 - n)$ bits of information. These information demands are compared for $k = 2$ and $2 \leq n \leq 12$ in Table 5.5. It is obvious that the rapid growth of the difference between these two information demands with increasing n is even more pronounced for larger values of k.

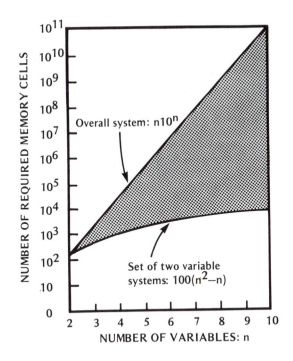

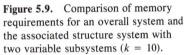

Figure 5.9. Comparison of memory requirements for an overall system and the associated structure system with two variable subsystems ($k = 10$).

The descriptive complexity of a system can thus be reduced by breaking the system into its appropriate subsystems; this is a general principle, which is independent of the adopted measure of descriptive complexity. On the other hand, the uncertainty complexity increases—or, at best, remains the same—when the system is replaced by its subsystems. This means that simplifying systems by breaking them into subsystems is a special case of the general simplification problem formulated by Eq. (5.91). This special, but very important simplification strategy has been extensively investigated since the mid-1970s within the context of *reconstructability analysis*—a set of computer-aided methodological tools for dealing with the relationship between wholes and parts, that is, between overall systems and their various subsystems [Klir, 1985].

Another way, perhaps the most significant one, of dealing with very complex systems is to allow imprecision in the description of properly aggregated data. Here, the imprecision is not of a statistical nature but rather of a more general

TABLE 5.5. INFORMATION DEMANDS IN IDENTIFYING A SINGLE SYSTEM WITH n VARIABLES, EACH WITH TWO STATES.

n	2	3	4	5	6	7	8	9	10	11	12
Overall system	4	8	16	32	64	128	256	512	1024	2048	4096
Structure system	4	5.6	6.6	7.3	7.9	8.4	8.8	9.2	9.5	9.8	10

modality, even though the possibility of imprecise statistical descriptions is included as well. The mathematical apparatus for this new modality is one of the subjects of this book—the theory of fuzzy sets. To express the significance of this theory for dealing with complexity, we can do no better than to quote Lotfi A. Zadeh, the founder of the theory [Zadeh, 1973]:

> Given the deeply entrenched tradition of scientific thinking which equates the understanding of a phenomenon with the ability to analyze it in quantitative terms, one is certain to strike a dissonant note by questioning the growing tendency to analyze the behavior of humanistic systems as if they were mechanistic systems governed by difference, differential, or integral equations.
>
> Essentially, our contention is that the conventional quantitative techniques of system analysis are intrinsically unsuited for dealing with humanistic systems or, for that matter, any system whose complexity is comparable to that of humanistic systems. The basis for this contention rests on what might be called the *principle of incompatibility*. Stated informally, the essence of this principle is that as the complexity of a system increases, our ability to make precise and yet significant statements about its behavior diminishes until a threshold is reached beyond which precision and significance (or relevance) become almost mutually exclusive characteristics.* It is in this sense that precise analyses of the behavior of humanistic systems are not likely to have much relevance to the real-world societal, political, economic, and other types of problems which involve humans either as individuals or in groups.
>
> An alternative approach . . . is based on the premise that the key elements in human thinking are not numbers, but labels of fuzzy sets, that is, classes of objects in which the transition from membership to non-membership is gradual rather than abrupt. Indeed, the pervasiveness of fuzziness in human thought processes suggests that much of the logic behind human reasoning is not the traditional two-valued or even multivalued logic, but a logic with fuzzy truths, fuzzy connectives, and fuzzy rules of inference. In our view, it is this fuzzy, and as yet not well-understood, logic that plays a basic role in what may well be one of the most important facets of human thinking, namely, the ability to *summarize* information—to extract from the collection of masses of data impinging upon the human brain those and only those subcollections which are relevant to the performance of the task at hand.
>
> By its nature, a summary is an approximation to what it summarizes. For many purposes, a very approximate characterization of a collection of data is sufficient because most of the basic tasks performed by humans do not require a high degree of precision in their execution. The human brain takes advantage of this tolerance for imprecision by encoding the "task-relevant" (or "decision-relevant") information into labels of fuzzy sets which bear an approximate relation to the primary data. In this way, the stream of information reaching the brain via the visual, auditory, tactile, and other senses is eventually reduced to the trickle that is needed to perform a specific task with a minimal degree of precision. Thus, the ability to manipulate fuzzy sets and the consequent summarizing capability constitute one of the most important assets of the human mind as well as a fundamental characteristic that distinguishes human intelligence from the type of machine intelligence that is embodied in present-day digital computers.

* A corollary principle may be stated as, "The closer one looks at a real-world problem, the fuzzier becomes its solution."

Viewed in this perspective, the traditional techniques of system analysis are not well suited for dealing with humanistic systems because they fail to come to grips with the reality of the fuzziness of human thinking and behavior. Thus to deal with such systems radically, we need approaches which do not make a fetish of precision, rigor, and mathematical formalism, and which employ instead a methodological framework which is tolerant of imprecision and partial truths.

Observe that the use of imprecision in the sense of fuzzy set theory as a strategy for simplifying complex systems is also a special case of the general formulation of the simplification problem expressed by Eq. (5.91).

Three Ranges of Complexity

In a well-known paper by Warren Weaver [1968], three significant ranges of complexity are distinguished. They differ from each other considerably in the mathematical treatment they require. Weaver calls them organized simplicity, disorganized complexity, and organized complexity.

Organized simplicity is represented by systems that are adequate models of some real world phenomena and yet consist of a very small number of variables (typically two or three), which depend on each other in a highly deterministic fashion. Systems of this sort had been predominant in science prior to the twentieth century. Indeed, the recorded history of the major discoveries in science from the seventeenth to the nineteenth century consists basically of variations on the same theme: a discovery of hidden simplicity in a phenomenon that appears complex. Phenomena of this sort are characterized by small numbers of significant factors and large numbers of negligible factors. This allows the scientist to introduce experimentally acceptable simplifying assumptions, according to which a few significant factors can be isolated from a large number of presumably negligible factors.

Due to their nature, systems with the characteristics of organized simplicity are perfectly suitable for analytic mathematical treatment, usually in terms of the calculus and differential equations. They are best exemplified by systems based upon Newtonian mechanics.

Whereas organized simplicity is characterized by small numbers of variables and high degrees of determinism, *disorganized complexity* possesses characteristics that are exactly opposite to these: it is represented by systems with very large numbers of variables and high degrees of randomness. Interest in systems of this sort began in the late nineteenth century with the investigation of systems representing the motions of gas molecules in a closed space. Such a system would typically consist of, say, 10^{23} molecules. The molecules have tremendous velocities and their paths, affected by incessant impacts, assume the most capricious shapes.

It was obvious that systems with these characteristics could not be studied in terms of the ideas and methods developed for dealing with systems in the category of organized simplicity. A radically new paradigm was needed. It eventually emerged at the beginning of this century in terms of statistical methods.

The purpose of these methods is not to deal directly with the individual variables (e.g., motions of the individual molecules) but to use a small number of calculated average properties. These calculations are based on the assumptions of a large number of components and a high degree of randomness. Disorganized complexity is thus best exemplified by systems based upon principles of statistical mechanics.

While analytical methods are impractical for even a modest number of variables (say five variables) and a small degree of randomness, the relevance and precision of statistical methods increase with a rise in the number of variables and their degree of randomness. These two types of methods are thus highly complementary. They cover the two extremes of the complexity spectrum. Where one of them fails, the other excels. Unfortunately, despite their complementarity, these two methods in fact cover only a tiny fraction of the whole complexity spectrum.

The large range of complexity between the two extremes, which Weaver calls *organized complexity*, is methodologically undeveloped in the sense that neither analytical nor statistical methods are adequate for dealing with systems that fit into it. Instances of systems with the characteristics of organized complexity are abundant, particularly in the life, behavioral, social, and environmental sciences as well as in applied fields such as modern technology or medicine. They are often systems that involve several disciplines.

Systems that exemplify the range of organized complexity are rich in factors that cannot be a priori neglected. That is, if they are neglected without a justification in each particular instance, the systems may lose their relevance as adequate models of the real-world phenomena involved. And, by the same token, they do not involve an adequate number of entities that are sufficiently random to yield meaningful statistical averages. This means that they are not susceptible to either of the two simplification strategies invented by science, and yet simplification is unavoidable in most instances. A simplification is obviously necessary if a problem of concern regarding a highly complex system is computationally intractable or if the desired solution is prohibitively expensive. However, even if the problem can be solved by a computer without any simplification and the computing cost is affordable, the solution must be eventually reduced to a level of complexity that is acceptable to the mind of a user (e.g., decision maker) who is in a position to utilize it. Since neither Newtonian nor statistical simplification strategies are applicable in the range of organized complexity, new avenues to the simplification of systems are needed. The most promising avenue, thus far, seems to rest in skillful use of the various measures of uncertainty and information that emerge from the broad framework of fuzzy set theory.

Computational Complexity

When we formulate the various strategies for simplifying systems, yet another face of complexity emerges—the computational complexity associated with each of the specific simplification problems. In general, *computational complexity* is a characterization of the time or space (memory) requirements for solving a problem by a particular algorithm. Either of these requirements is usually expressed in terms of a single parameter that represents the size of the problem. For the various simplification problems, the problem size may adequately be characterized, for example, by the number of variables in the given system.

Given a particular problem, say one of the simplification problems, let n denote its size. Then, the time requirements of a specific algorithm for solving this problem are expressed by a function

$$f : \mathbb{R} \rightarrow \mathbb{R}$$

such that $f(n)$ is the largest amount of time needed by the algorithm to solve the problem of size n. Function f is usually called a *time complexity function*.

It has been recognized that it is useful to distinguish two classes of algorithms by the rate of growth of their time complexity functions. One class consists of algorithms whose time complexity functions can be expressed in terms of a polynomial

$$f(n) = a_k n^k + a_{k-1} n^{k-1} + \cdots + a_1 n + a_0$$

for some positive integer k. They are usually called *polynomial time algorithms*. The second class of algorithms consists of a formula in which n appears in the exponent—for example, 2^n, 10^n, or 2^{2^n}. They are usually referred to as *exponential time algorithms*.

The distinction between the polynomial and exponential time algorithms is significant, especially when considering large problems (large n). This can be seen by comparing plots of some polynomial and exponential time complexity functions in Fig. 5.10.

Although computational complexity has been predominantly studied in terms of the time it takes to perform a computation, the amount of computer memory required is frequently just as important. This requirement is usually called the space requirement. It is studied in terms of a *space complexity function*, analogous to the time complexity function.

Computational complexity has been extensively investigated since the early 1970s. Many important results are now available in this subject area, but it is beyond the scope of this book to cover them here. A good overview of the main issues regarding computational complexity and results available in the late 1970s was prepared by Garey and Johnson [1979].

Equipped with the rapidly advancing computer technology, we become increasingly more successful in dealing with systems of organized complexity. At the same time, we begin to understand that there are definite limits to our capabilities in this regard. One such limit was determined by Hans Bremermann [1962] by simple considerations based on quantum theory. He expressed it by the following proposition: "No data processing system, whether artificial or living, can process more than 2×10^{47} bits per second per gram of its mass." Bremermann derives his limit from the following considerations based upon quantum physics, in which the phrase "processing of x bits" means the transmission of that many bits over one or several communication channels within the computing system.

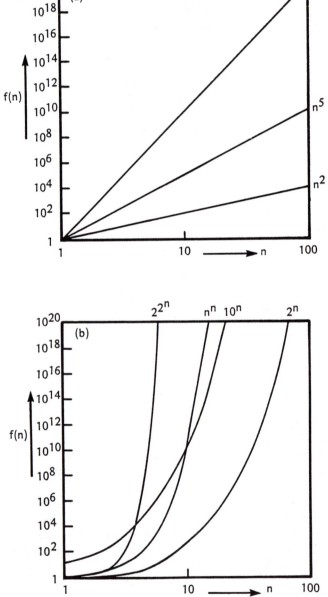

Figure 5.10. Plots of some typical time complexity functions: (a) polynomial; (b) exponential.

It is obvious that information to be acted upon by a machine must be physically encoded in some manner. Assume that it is encoded in terms of energy levels within the interval $[0, E]$ of energy of some sort; E is viewed as the total energy available for this purpose. Assume further that energy levels can be measured with an accuracy of only ΔE. Then the most refined encoding is defined in terms of markers, by which the whole interval is divided into $N = E/\Delta E$ equal subintervals, each associated with the energy amount ΔE. If at each instant no more than one of the levels (represented by the markers) is occupied, then

$$\log_2(N + 1) \quad \text{[bits]}$$

is the maximum amount of information (in the Hartley sense) that is representable by energy E; $N + 1$ is used here to account for the case in which none of the levels is occupied. If, instead of one marker with energy levels in $[0, E]$, K markers $(2 \leq K \leq N)$ are used simultaneously, then

$$K \log_2 \left(1 + \frac{N}{K} \right) \quad \text{[bits]}$$

of information become representable. The optimal utilization of the available amount of energy E is obtained when N markers with levels in the interval $[0, E]$ are used. In this optimal case, N bits of information can be represented.

In order to represent more information by the same amount of energy, it is desirable to reduce ΔE. This is possible only to a certain extent, since the resulting levels must be distinguished by some measurement process, which, regardless of its nature, always has some limited precision. The extreme case is expressed by the Heisenberg principle of uncertainty: energy can be measured to the accuracy of ΔE if the inequality

$$\Delta E \Delta t \geq h$$

is satisfied, where Δt denotes the time duration of the measurement, $h = 6.625 \times 10^{-27}$ ergs/s is Planck's constant, and ΔE is defined as the mean deviation from the expected value of the energy involved. This means that

$$N \leq \frac{E \Delta t}{h} . \tag{5.92}$$

Now, the available energy E can be expressed in terms of the equivalent amount of mass m by Einstein's formula

$$E = mc^2,$$

where $c = 3 \times 10^{10}$ cm/s is the velocity of light in a vacuum. If we take the upper (most optimistic) bound of N in (5.92), we get

$$N = \frac{mc^2 \Delta t}{h} . \tag{5.93}$$

Substituting the numerical values for c and h, we obtain

$$N = 1.36 m \Delta t \times 10^{47}.$$

For a mass of 1 gram ($m = 1$) and time of 1 second ($\Delta t = 1$), we obtain the value

$$N = 1.36 \times 10^{47},$$

which implies the conjecture.

Using the limit of information processing obtained for 1 gram of mass and 1 second of processing time, Bremermann then calculates the total number of bits processed by a hypothetical computer the size of the Earth within a time period equal to the estimated age of the Earth. Since the mass and age of the Earth are estimated to be less than 6×10^{27} grams and 10^{10} years, respectively, and each year contains approximately 3.14×10^7 seconds, this imaginary computer would not be able to process more than 2.56×10^{92} bits or, when rounding up to the nearest power of ten, 10^{93} bits. This last number—10^{93}—is usually referred to as *Bremermann's limit*, and problems that require processing more than 10^{93} bits of information are called *transcomputational problems*.

Bremermann's limit seems at first sight rather discouraging for system problem solving, even though it is quite conservative (more reasonable assumptions would lead to a number smaller than 10^{93}). Indeed, many problems dealing with systems of even modest size exceed it in their information-processing demands. Consider, for example, a system of n variables, each of which can take k different states. The set of all overall states of the variables clearly consists of k^n states. In each particular system, however, the actual overall states are restricted to a subset of this set. There are 2^{k^n} such subsets. Suppose we need to select, identify, distinguish, or classify one system from the set of all possible systems of this sort. Then, under the assumption that the most efficient method of searching is used, in which each bit of information (the answer to a dichotomous question) allows us to cut the remaining choices in half,

$$\log_2 2^{k^n} = k^n$$

bits of information have to be processed. The problem becomes transcomputational when

$$k^n > 10^{93}.$$

That happens, for example, for the following values of k and n:

k	2	3	4	5	6	7	8	9	10
n	308	194	154	133	119	110	102	97	93

The problem of transcomputability arises in various contexts. One of them is pattern recognition. Consider, for example, a $q \times q$ spatial array of the chessboard type, each square of which can have one of k colors. There are clearly k^n color patterns, where $n = q^2$. Suppose we want to determine the best classification (according to certain criteria) of these patterns. This requires a search through all possible classifications of the patterns. In the case of only two classes, the problem becomes isomorphic to the previous one. For two colors, for example, the problem becomes transcomputational when the array is 18×18; for a 10×10 array, the problem becomes transcomputational when 9 colors are used. This pattern recognition problem is directly relevant to physiological studies of the retina, but its complexity is tremendous. The retina contains about a million light-sensitive cells. Even if we consider (for simplicity) that each of the cells has only two states (active and inactive), the attempt to study the retina as a whole would require the processing of

$$2^{1,000,000} \doteq 10^{300,000}$$

bits of information. This is far beyond Bremermann's limit.

Another context in which the same problem occurs is the area of testing large-scale integrated digital circuits. These are tiny electronic chips with considerable complexity and a large number of inputs and outputs. For properly defined electric signals (each usually with two ideal states), the individual outputs should represent some specific logic functions of the logic variables associated with the inputs. To test a particular integrated circuit means to analyze it as a "black box": to determine the actual logic functions it implements solely by manipulating the input variables and observing the output variables. For each output variable, the testing problem is thus basically the same as the problem previously discussed for $k = 2$ (unless a multiple-valued logic is used). It follows that testing of circuits, for example, with 308 inputs and 1 output is a transcomputational problem. However, it is well known that the practical complexity limits of this testing problem are considerably lower. In fact, some currently manufactured large-scale integrated circuits cannot be completely tested. The focus is thus on developing testing methods that can be practically implemented and that guarantee only that the testing be *almost complete*—that well over 90 percent of all possibilities be tested, for instance.

A more detailed characterization of the complexity of this problem, from the practical domain to Bremermann's limit, is expressed by Fig. 5.11. The figure shows the dependence of the time (in years) required to select (identify, classify, distinguish, etc.) one logic function of n variables under the consideration of different information-processing rates in the range from 10 through 10^{100} bits per second. Two significant values of times are also shown in the figure: L indicates the approximate age of the oldest fossil records of life on the Earth and M shows the approximate time since humans first appeared on the Earth.

The testing example is in no way exceptional. Genuine systems problems are notorious for their huge demands on information-processing capabilities. This point is well made by Bremermann in the conclusion of his paper [Bremermann, 1962]:

> The experiences of various groups who work on problem solving, theorem proving and pattern recognition all seem to point in the same direction: These problems are tough. There does not seem to be a royal road or a simple method which at one stroke will solve all our problems. My discussion of ultimate limitations on the speed and amount of data processing may be summarized like this: Problems involving vast numbers of possibilities will not be solved by sheer data processing quantity.

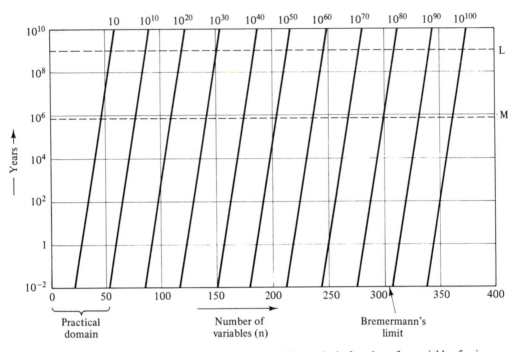

Information-processing rates

Figure 5.11. Time required to select or identify one logic function of n variables for information processing rates of 10, 10^{10}, . . . , 10^{100} bits per second.

We must look for quality, for refinements, for tricks, for every ingenuity that we can think of. Computers faster than those of today will be of great help. We will need them. However, when we are concerned with problems in principle, present day computers are about as fast as they ever will be.

If a problem is transcomputational, it is obvious that it can be dealt with only in some modified form. It is desirable to modify it no more than is necessary to make it manageable. The most natural way of modifying a problem is to soften its requirements. For instance, a requirement of getting the best solution may be replaced with a requirement of getting a good solution, instead of requiring a precise solution we may accept an approximate solution, and so on. Such softening of requirements permits the use of heuristic methods, in which vast numbers of unpromising possibilities are ignored, or of approximate (fuzzy) methods, in which substantial aggregation takes place.

Bremermann's limit allows one to make only the most rudimentary categorization of systems problems by their complexities. It says nothing about the actual, practical computational limits. Nevertheless, it is a useful benchmark for a preliminary evaluation of each problem situation, as emphasized by Ashby [1973]:

One of its most obvious consequences, yet one almost universally neglected today, is that, before the study of a complex system is undertaken, at least a rough estimate of its informational demands should be made. Should the estimate be 2000 bits we have little to worry about, but should it prove to be 10^{300} bits we would know that our whole strategic approach to the system needs re-formulating.

This simple benchmark—10^{93}—must be supplemented, of course, by sharper bounds on problem complexity, derived for specific computer systems. However, it is an indicator of fundamental information limits to our knowledge, as well described by Ashby [1968]:

> The most obvious fact is that we, and our brains, are themselves made of matter, and are thus absolutely subject to the limit. Not only are we subject as individuals, but the whole cooperative organization of World Science is also made of matter, and is therefore subject to it. Thus both the total information that I can use personally, and the information that World Science can use, are limited, on any ordinary scale, to about 10^{80} bits.* Whatever our science will become in the future, all will lie below this ceiling.
>
> We cannot claim any special advantage because of our pre-eminent position in the world of organisms. We have been shaped, and selected to be what we are, by the process of natural selection. As a selection, this process can be measured by an information-measure; it is therefore subject to its limits. In any type of selection, under any planetary conditions, a planetary surface made of matter cannot produce adaptation faster than the rate of the limit. However good we may think we are, 10^{80} measures something that we do not exceed. The science of the future will be built by brains that cannot have had more than 10^{80} bits used in their preparations, and they themselves will advance only by something short of 10^{80}. This is our information universe: what lies beyond is unknowable.

5.9 PRINCIPLES OF UNCERTAINTY AND INFORMATION

The uniqueness of Shannon entropy as the probabilistic measure of information is the crux behind two complementary uncertainty and information principles, which are referred to as principles of minimum and maximum entropy. These principles, which have proven exceedingly successful in many applications, can now be extended to broader principles based upon the uncertainty measures covered in this chapter and summarized in Fig. 5.6. The aim of this section is both to overview the well-established principles of minimum and maximum entropy, which are confined to probability theory, and to introduce some novel uncertainty and information principles that emerge from the broader framework of fuzzy set theory.

The *principle of minimum entropy* is essentially a probabilistic version of

---

* Ashby derives the value 10^{80} from the Bremermann limit for 1 second and 1 gram by considering "centuries of time and tons of computers" (e.g., about 10,000 centuries and 10^{15} tons of mass). It is not important for the argument whether we take 10^{80} or 10^{93}.

the most fundamental *principle of simplification*, which may be expressed as follows: a sound simplification of a system should minimize the loss of relevant uncertainty information with respect to the required reduction of its complexity based upon descriptive information. That is, when we want to reduce the descriptive complexity of a system to some desirable level and there are several ways of doing so, we should select one of those options for which the loss of relevant uncertainty information is minimal or, in other words, one for which the increase in relevant uncertainty is minimal.

The principle of minimum entropy is thus employed as an arbiter that allows us to decide which of several equally complex simplifications (in the descriptive sense) of a probabilistic system to choose in order to preserve as much information of the original system as possible (in the uncertainty sense). When the principle is applied to the general simplification problem expressed by Eq. (5.91), the problem becomes operationally more tractable.

Given a system of some epistemological type, let $\mathscr{X}$ again denote the set of its admissible simplifications. Let $W_{\mathscr{X}}(x)$ denote the descriptive complexity of $x \in \mathscr{X}$ and let $\overset{H}{\leq}$ denote the relevant preference uncertainty ordering expressed in terms of Shannon entropy. Assume that it is required that the descriptive complexity (expressed by function $W_{\mathscr{X}}$) be reduced to some given value K. Then, the solution set $\mathscr{X}_s$ of this simplification problem consists of those systems x in $\mathscr{X}$ for which $W_{\mathscr{X}}(x) = K$ and, at the same time, $x \overset{H}{\leq} y$ for all $y \in \mathscr{X}$. Formally,

$$\mathscr{X}_s = \{x \in \mathscr{X} \mid W_{\mathscr{X}}(x) = K \text{ and } x \overset{H}{\leq} y \text{ for all } y \in \mathscr{X} \text{ such that } W_{\mathscr{X}}(y) = K\}. \quad (5.94)$$

A unique selection from $\mathscr{X}_s$ can then be made on the basis of some additional (optional) preference orderings.

The actual form of the Shannon entropy by which the preference ordering $\overset{H}{\leq}$ is defined depends on the type of the system involved as well as on the simplification strategy. It is usually an appropriate form of conditional Shannon entropy or its generalized form of cross-entropy. Let us illustrate the principle of minimum entropy by a very simple example.

Example 5.11

Consider that a relationship between two variables v_1 and v_2 was established and that it is characterized by the joint probability distribution function $p(\dot{v}_1, \dot{v}_2)$ given in Table 5.6(a). Variable v_1 can assume three states, which represent clusters of low, medium, and high values of the corresponding attribute; variable v_2 has only two states, which represent clusters of low and high values of the associated attribute. Assume that variable v_2 is viewed as dependent on v_1. Then, the relevant uncertainty in this case is expressed by the conditional entropy $H(V_2 \mid V_1)$, where V_1 and V_2 denote the state sets of v_1 and v_2, respectively. Using Eq. (5.46), we have

$$H(V_1, V_2) = -2 \times .2 \log_2.2 - .15 \log_2.15 - .05 \log_2.05$$

$$- .1 \log_2.1 - .3 \log_2.3$$

$$= 2.41,$$

$$H(V_1) = -2 \times .4 \log_2.4 - .2 \log_2.2 = 1.52,$$

$$H(V_2 \mid V_1) = H(V_1, V_2) - H(V_1) = .89.$$

Assume now that we want to simplify the relation by reducing V_1 to two states, low and high. Due to the ordering of V_1 (low < medium < high), there are only two possible simplifications of this sort. In one of them, low and medium merge and form a new cluster designated as low'; in the other, medium and high merge and form a new cluster labeled as high'. These two alternative simplifications are specified in Table 5.6(b) and (c), respectively. Let $^1V_1 = \{\text{low}', \text{high}\}$ and $^2V_1 = \{\text{low}, \text{high}'\}$. Then,

$$H(^1V_1, V_2) = -.35 \log_2 .35 - .25 \log_2 .25 - .1 \log_2 .1 - .3 \log_2 .3$$

$$= 1.88,$$

$$H(^2V_1, V_2) = -2 \times .2 \log_2 .2 - .25 \log_2 .25 - .35 \log_2 .35$$

$$= 1.96,$$

$$H(^1V_1) = H(^2V_1) = -.4 \log_2 .4 - .6 \log_2 .6 = .97,$$

$$H(V_2 \mid {}^1V_1) = H(^1V_1, V_2) - H(^1V_1) = .91,$$

$$H(V_2 \mid {}^2V_1) = H(^2V_1, V_2) - H(^2V_1) = .99.$$

We conclude that the first simplification (Table 5.6(b)) is clearly preferable. It adds only about 2 percent of the uncertainty associated with the original relation (Table 5.6(a)), compared with about 11 percent of this uncertainty added by the second simplification (Table 5.6(c)). In other words, the first simplification preserves more information of the given relation than does the second simplification.

The *principle of maximum entropy* is a probabilistic version of a more general principle for reasoning in which conclusions are not entailed in the given premises. Such reasoning, which is often referred to as *ampliative reasoning*, involves in-

TABLE 5.6.   ILLUSTRATION OF THE MINIMUM ENTROPY PRINCIPLE (EXAMPLE 5.11).

(a)

$\dot{v}_1$	$\dot{v}_2$	$p(\dot{v}_1, \dot{v}_2)$	$p(\dot{v}_1)$
Low	Low	.2	
Low	High	.2	.4
Medium	Low	.15	
Medium	High	.05	.2
High	Low	.1	
High	High	.3	.4

(b)

$\dot{v}_1$	$\dot{v}_2$	$^1p(\dot{v}_1, \dot{v}_2)$	$^1p(\dot{v}_1)$
Low'	Low	.35	
Low'	High	.25	.6
High	Low	.1	
High	High	.3	.4

(c)

$\dot{v}_1$	$\dot{v}_2$	$^2p(\dot{v}_1, \dot{v}_2)$	$^2p(\dot{v}_1)$
Low	Low	.2	
Low	High	.2	.4
High'	Low	.25	
High'	High	.35	.6

ferences whose content is beyond the evidence on hand. The general principle of ampliative reasoning may be expressed as follows: in any ampliative inference, *use all but no more information than is available*. This principle had in fact been recognized by the ancient Chinese philosopher Lao Tsu as early as the sixth century B.C. and expressed in his book *Tao Te Ching** by the following two simple statements of remarkable clarity:

> Knowing ignorance is strength.
> Ignoring knowledge is sickness.

When applied within the framework of probability theory, this principle is made operational by employing Shannon entropy as the unique measure of information.

To make sure that our ignorance is fully recognized when we attempt to estimate a probability distribution from insufficient evidence, we must ensure that from among all probability distributions that conform to the evidence, the chosen one has maximal uncertainty (i.e., minimal information). Hence, if we have no evidence pertaining to the estimated probability distribution and are thus totally ignorant in the matter involved, the principle requires that we choose the uniform distribution (one with equal probabilities on the universal set), for which the Shannon entropy reaches its absolute maximum.

In the case of total ignorance, the maximum entropy principle is equivalent to an old principle of probability theory, presumably stated by Bernoulli in 1713, which is usually called a *principle of insufficient reason* or *principle of indifference*. According to this principle, if there is no reason (no evidence) to believe that some of the considered alternatives are more likely than others, then the best strategy is to assign them equal probabilities.

We recall, however, that the general principle of ampliative reasoning, which is exemplified in probability theory by the principle of maximum entropy, also requires that our inferences be based on all relevant information contained in the available evidence. To ensure that all information pertaining to the estimated probability distribution is fully utilized, we must eliminate all probability distributions that conflict with the available evidence. Among the remaining probability distributions, we must then fully recognize our ignorance and, consequently, choose the one with maximum entropy. These two steps can be accomplished by solving an optimization problem in which we search for the maximum of Shannon entropy within the domain of relevant probability distributions restricted by constraints expressing our evidence. This is the essence of the maximum entropy principle.

This preliminary discussion should be helpful in understanding the following *mathematical formulation of the maximum entropy principle*, in which we assume that we deal with probability distributions defined on a finite set with n elements.

Determine a probability distribution $\mathbf{p} = (p_1, p_2, \ldots, p_n)$ that maximizes the function

$$H(p_1, p_2, \ldots, p_n) = -\sum_{i=1}^{n} p_i \log_2 p_i$$

* Vintage Books (Random House), New York, Chapter 71, 1972.

subject to constraints $w_1, w_2, \ldots$, which represent the available evidence relevant to the matter of concern, as well as the general constraints

$$p_i > 0 \quad \text{for all } i \in \mathbb{N}_n$$

and

$$\sum_{i=1}^{n} p_i = 1,$$

which must be satisfied by all probability distributions.

Let ${}^n\mathcal{P}_w$ denote the subset of all probability distributions of length n that satisfy the constraints of this optimization problem. It is well established that the solution to the problem is unique if and only if the constraints $w_1, w_2, \ldots$ are consistent and the set ${}^n\mathcal{P}_w$ is convex, that is, if $\mathbf{p}, \mathbf{q} \in {}^n\mathcal{P}_w$, then also $a\mathbf{p} + (1 - a)\mathbf{q} \in {}^n\mathcal{P}_w$ for all $a \in [0, 1]$. If the constraints are inconsistent, the problem has no solution; if ${}^n\mathcal{P}_w$ is not convex, it may have more than one solution.

The most frequent types of constraints employed by the maximum entropy principle are mean values of one or more random variables or various marginal probability distributions of an unknown joint distribution. In these cases and in most other cases of empirical evidence, the constraints represent a convex set of probability distributions and, consequently, the uniqueness of the solution to the optimization problem is guaranteed (assuming consistency in evidence).

Solving the various maximum entropy optimization problems requires the knowledge of optimization methods. Since this nontrivial subject is not of our interest in this book, we illustrate the optimization problem only by a simple example in which we employ the classical Lagrange multiplier optimization method. For more information on this issue, see Note 5.22.

Consider a random variable x whose possible values are $x_1, x_2, \ldots, x_n$. Let $x_i \in \mathbb{R}^+$ for all $i \in \mathbb{N}_n$ and let p_i denote the unknown probability of x_i ($i \in \mathbb{N}_n$). Assume that we want to estimate the probability distribution $\mathbf{p} = (p_1, p_2, \ldots, p_n)$ provided that we know the mean (expected) value $E(x)$ of the variable. The use of the maximum entropy principle for this purpose leads to the following formulation:*

Maximize the function

$$H(p_1, p_2, \ldots, p_n) = -\sum_{i=1}^{n} p_i \ln p_i$$

subject to the constraints

$$E(x) = \sum_{i=1}^{n} p_i x_i \tag{5.95}$$

and

$$p_i \geq 0 \quad (i \in \mathbb{N}_n), \qquad \sum_{i=1}^{n} p_i = 1. \tag{5.96}$$

---

* For the sake of simplicity, we use the natural logarithm in the definition of Shannon entropy in this formulation.

Equation (5.95) represents the available information; Eq. (5.96) represents the standard constraints imposed upon **p** by probability theory.

First, we form the Lagrange function

$$L = -\sum_{i=1}^{n} p_i \ln p_i - \alpha \left(\sum_{i=1}^{n} p_i - 1 \right) - \beta \left(\sum_{i=1}^{n} p_i x_i - E(x) \right),$$

where α and β are the Lagrange multipliers that correspond to the two constraints. Second, we form the partial derivatives of L with respect to p_i ($i \in \mathbb{N}_n$), α, and β and set them equal to zero; this results in the equations

$$\frac{\partial L}{\partial p_i} = -\ln p_i - 1 - \alpha - \beta x_i = 0 \qquad \text{for each } i \in \mathbb{N}_n,$$

$$\frac{\partial L}{\partial \alpha} = 1 - \sum_{i=1}^{n} p_i = 0,$$

$$\frac{\partial L}{\partial \beta} = E(x) - \sum_{i=1}^{n} p_i x_i = 0.$$

The last two equations are exactly the same as the constraints in this optimization problem. The first n equations can be rewritten as

$$p_1 = e^{-1-\alpha-\beta x_1} = e^{-(1+\alpha)} e^{-\beta x_1}$$

$$p_2 = e^{-1-\alpha-\beta x_2} = e^{-(1+\alpha)} e^{-\beta x_2}$$

$$\vdots \qquad\qquad \vdots$$

$$p_n = e^{-1-\alpha-\beta x_n} = e^{-(1+\alpha)} e^{-\beta x_n}$$

When we divide each of these equations by the sum of all of them, we obtain

$$p_i = \frac{e^{-\beta x_i}}{\sum_{k=1}^{n} e^{-\beta x_k}} \tag{5.97}$$

for each $i \in \mathbb{N}_n$. In order to determine the value of β, we multiply the ith equation in (5.97) by x_i and add all the resulting equations, thus obtaining

$$E(x) = \frac{\sum_{i=1}^{n} x_i e^{-\beta x_i}}{\sum_{i=1}^{n} e^{-\beta x_i}}$$

and

$$\sum_{i=1}^{n} x_i e^{-\beta x_i} - E(x) \sum_{i=1}^{n} e^{-\beta x_i} = 0.$$

Multipying this equation by $e^{\beta E(x)}$ results in

$$\sum_{i=1}^{n} [x_i - E(x)]e^{-\beta[x_i - E(x)]} = 0. \tag{5.98}$$

This equation must now be solved (numerically) for β and the solution substituted for β in Eq. (5.97), which results in the estimated maximum entropy probabilities p_i ($i \in \mathbb{N}_n$).

Example 5.12

Consider first an "honest" (unbiased) die. Here, $x_i = i$ for $i \in \mathbb{N}_6$ and $E(x) = 3.5$. Equation (5.98) has the form

$$-2.5e^{2.5\beta} - 1.5e^{1.5\beta} - .5e^{0.5\beta} + .5e^{-0.5\beta} + 1.5e^{-1.5\beta} + 2.5e^{-2.5\beta} = 0.$$

The solution is clearly $\beta = 0$; when it is substituted for β in Eq. (5.97), we obtain $p_i = \frac{1}{6}$ for all $i \in \mathbb{N}_6$.

Consider now a biased die for which it is known only that $E(x) = 4.5$. Here, Eq. (5.98) assumes a different form:

$$-3.5e^{3.5\beta} - 2.5e^{2.5\beta} - 1.5e^{1.5\beta} - .5e^{0.5\beta} + .5e^{-0.5\beta} + 1.5e^{-1.5\beta} = 0.$$

When solving this equation (by an appropriate numerical method), we obtain $\beta = -0.37105$. Substitution of this value in Eq. (5.97) yields the maximum entropy probability distribution

$$p_1 = \frac{1.45}{26.66} = .05 \qquad \text{.045} \quad \text{.053}$$

$$p_2 = \frac{2.10}{26.66} = .08 \qquad \text{.087}$$

$$p_3 = \frac{3.04}{26.66} = .11 \qquad \text{.12}$$

$$p_4 = \frac{4.41}{26.66} = .17 \qquad \text{.2}$$

$$p_5 = \frac{6.39}{26.66} = .24 \qquad \text{.3}$$

$$p_6 = \frac{9.27}{26.66} = .35 \qquad \text{.497}$$

A generalization of the principle of maximum entropy is the *principle of minimum cross-entropy*. It can be stated as follows: given a prior probability distribution $^\circ p$ and some relevant new evidence, determine a new probability distribution p that minimizes the cross-entropy

$$H(p(x), {}^\circ p(x) \mid x \in X) = \sum_{x \in X} p(x) \log_2 \frac{p(x)}{{}^\circ p(x)} \tag{5.99}$$

subject to the constraints $w_1, w_2, \ldots$, which represent the new evidence, as well as to the standard constraints of probability theory.

Since we now recognize several types of uncertainty and have well-justified measures for them, several alternative principles can be formulated in addition to the classical principles of minimum and maximum entropy. Particularly important uncertainty principles are likely to be based upon the measure of nonspecificity. In analogy to the minimum and maximum entropy principles, we may envision corresponding principles of minimum and maximum nonspecificity. When properly developed, these principles will have a broad applicability. While the entropy principles are restricted to probability theory, the nonspecificity principles are potentially applicable to possibility theory, fuzzy sets, and evidence theory.

Like the minimum entropy principle, the corresponding *principle of minimum nonspecificity* is basically another version of the fundamental simplification principle. Thus, for example, in simplifying systems formalized in terms of possibility theory, the principle of minimum nonspecificity (minimum U-uncertainty in this case) can be employed to decide which of several equally complex simplifications of a given system we should select in order to preserve as much information (specificity) of the original system as possible. Let us illustrate this principle by an example analogous to Example 5.11.

Example 5.13

Consider a relationship between two variables v_1 and v_2 (similar to those introduced in Example 5.11), which is characterized by the joint possibility distribution function $r(\dot{v}_1, \dot{v}_2)$ given in Table 5.7(a). Assuming, as in Example 5.11, that variable v_2 is dependent on v_1, we use Eq. (5.86) to calculate the relevant conditional nonspecificity as follows:

$$U(V_1, V_2) = .2 \log_2 3 + .2 \log_2 4$$
$$+ .4 \log_2 5 + .2 \log_2 6 = 2.16,$$
$$U(V_1) = .2 \log_2 2 + .8 \log_2 3 = 1.47,$$
$$U(V_2 \mid V_1) = U(V_1, V_2) - U(V_1) = .69.$$

Two simplifications in which the state set V_1 is reduced to two states are given in Table 5.7(b) and (c). In order to decide which of them is preferable, we calculate the relevant conditional nonspecificities:

$$U(^1V_1, V_2) = .4 \log_2 3 + .6 \log_2 4 = 1.83,$$
$$U(^2V_1, V_2) = .2 \log_2 2 + .6 \log_2 3 + .2 \log_2 4 = 1.55,$$
$$U(^1V_1) = \log_2 2 = 1, \quad U(^2V_1) = .2 \log_2 1 + .8 \log_2 2 = .8,$$
$$U(V_2 \mid {}^1V_1) = U(^1V_1, V_2) - U(^1V_1) = .83,$$
$$U(V_2 \mid {}^2V_1) = U(^2V_1, V_2) - U(^2V_1) = .75.$$

The second simplification (Table 5.7(c)) is clearly preferable because it preserves more of the specificity of the given relation. It increases the nonspecificity of the given system by about 9 percent, whereas the other (competing) simplification increases it by 20 percent.

Let us now discuss the prospective *principle of maximum nonspecificity*. It should be similar in spirit to the principle of maximum entropy, as both of them are special instances of the general principle of ampliative reasoning. The difference between the principles of maximum entropy and maximum nonspecificity is in the types of uncertainties with which they deal and in their applicability. While the maximum entropy principle is applicable only within probability theory, the principle of maximum nonspecificity is applicable to all the formal frameworks introduced in this book except probability theory. These two principles thus complement each other perfectly.

The following is a *mathematical formulation of the principle of maximum nonspecificity* in terms of the general function of nonspecificity V.

Determine a basic assignment $m(A)$, $A \in \mathcal{P}(X)$, that maximizes the function

$$V(m) = \sum_{A \in \mathcal{P}(X)} m(A) \log_2 |A|$$

subject to constraints $w_1, w_2, \ldots$, which represent the available evidence relevant to the matter of concern, as well as to the general constraints

$$m(\varnothing) = 0, \qquad m(A) \geq 0 \qquad \text{for all } A \in \mathcal{P}(X),$$

and

$$\sum_{A \in \mathcal{P}(X)} m(A) = 1,$$

which must be satisfied by all basic assignments in the evidence theory. Due to the additivity of basic assignments in various calculations such as the determi-

TABLE 5.7. ILLUSTRATION OF THE PRINCIPLE OF MINIMUM NONSPECIFICITY IN POSSIBILITY THEORY (EXAMPLE 5.13).

(a)

$\dot{v}_1$	$\dot{v}_2$	$r(\dot{v}_1, \dot{v}_2)$	$r(\dot{v}_1)$
Low	Low	.8	
Low	High	.2	.8
Medium	Low	1	
Medium	High	1	1
High	Low	.6	
High	High	1	1

(b)

$\dot{v}_1$	$\dot{v}_2$	$^1r(\dot{v}_1, \dot{v}_2)$	$^1r(\dot{v}_1)$
Low'	Low	1	
Low'	High	1	1
High	Low	.6	
High	High	1	1

(c)

$\dot{v}_1$	$\dot{v}_2$	$^2r(\dot{v}_1, \dot{v}_2)$	$^2r(\dot{v}_1)$
Low	Low	.8	
Low	High	.2	.8
High'	Low	1	
High'	High	1	1

nation of total beliefs or marginal basic assignments, typical constraints in this optimization problem are likely to have the form

$$\sum_{A \in \mathscr{A}} m(A) = a,$$

where $\mathscr{A}$ is some family of subsets of the universal set and a is a given number in the unit interval $[0, 1]$. This means that the principle of maximum nonspecificity is likely to have the form of a linear programming problem. Let us illustrate the principle by a very simple example.

Example 5.14

Consider a universal set X, three nonempty subsets of which are of our interest: A, B and $A \cap B$. Assume that the only evidence we have is expressed in terms of the total belief focusing on A and the total belief focusing on B. Our aim is to estimate the basic assignment values for X, A, B, and $A \cap B$. The use of the principle of maximum nonspecificity leads in this case to the following optimization problem.

Determine values $m(X)$, $m(A)$, $m(B)$, and $m(A \cap B)$ for which the function

$$m(X)\log_2 |X| + m(A)\log_2 |A| + m(B)\log_2 |B| + m(A \cap B)\log_2 |A \cap B|$$

reaches its maximum subject to the constraints

$$m(A) + m(A \cap B) = a,$$

$$m(B) + m(A \cap B) = b,$$

$$m(X) + m(A) + m(B) + m(A \cap B) = 1,$$

$$m(X), m(A), m(B), m(A \cap B) \geq 0,$$

where $a, b \in [0, 1]$ are given numbers.

The constraints are represented in this case by three linear algebraic equations of four unknowns and, in addition, by the requirement that the unknowns be non-negative real numbers. The equations are consistent and independent. Hence, they involve one degree of freedom. Selecting, for example, $m(A \cap B)$ as the free variable, we readily obtain

$$m(A) = a - m(A \cap B),$$

$$m(B) = b - m(A \cap B), \tag{a}$$

$$m(X) = 1 - a - b + m(A \cap B).$$

Since all the unknowns must be nonnegative, the first two equations set the upper bound of $m(A \cap B)$, whereas the third equation specifies its lower bound; the bounds are

$$\max(0, a + b - 1) \leq m(A \cap B) \leq \min(a, b). \tag{b}$$

Using Eqs. (a), the objective function can now be expressed solely in terms of the free variable $m(A \cap B)$. After a simple rearrangement of terms, we obtain

$$m(A \cap B) [\log_2 |X| - \log_2 |A| - \log_2 |B| + \log_2 |A \cap B|]$$

$$+ (1 - a - b)\log_2 |X| + a \log_2 |A| + b \log_2 |B|.$$

Clearly, only the first term in this expression can influence its value, so that we may rewrite the expression as

$$m(A \cap B)\log_2 K_1 + K_2, \tag{c}$$

where

$$K_1 = \frac{|X| \cdot |A \cap B|}{|A| \cdot |B|}$$

and

$$K_2 = (1 - a - b)\log_2 |X| + a \log_2 |A| + b \log_2 |B|$$

are constant coefficients. The solution to the optimization problem depends only on the value of K_1. Since A, B, and $A \cap B$ are assumed to be nonempty subsets of X, $K_1 > 0$. If $K_1 < 1$, then $\log_2 K_1 < 0$ and we must minimize $m(A \cap B)$ to obtain the maximum of (c); hence, $m(A \cap B) = \max(0, a + b - 1)$ due to (b). If $K_1 > 1$, then $\log_2 K_1 > 0$, and we must maximize $m(A \cap B)$; hence, $m(A \cap B) = \min(a, b)$ by (b). When $K = 1$, $\log_2 K_1 = 0$, and (c) is independent of $m(A \cap B)$; this implies that the solution is not unique or, more precisely, that any value of $m(A \cap B)$ in the range (b) is a solution to the optimization problem. The complete solution can thus be expressed by the following equations:

$$m(A \cap B) = \begin{cases} \max(0, a + b - 1) & \text{when } K_1 < 1 \\ [\max(0, a + b - 1), \min(a, b)] & \text{when } K_1 = 1 \\ \min(a, b) & \text{when } K_1 > 1, \end{cases}$$

$$m(A) = a - m(A \cap B),$$

$$m(B) = b - m(A \cap B),$$

$$m(X) = 1 - a - b + m(A \cap B).$$

As demonstrated previously in this chapter and summarized in Fig. 5.6, measures of several types of uncertainty are now available within some mathematical frameworks. In particular, nonspecificity and fuzziness are both applicable to possibility theory as well as to fuzzy sets; dissonance, confusion, and nonspecificity are applicable to evidence theory. In these cases, any comprehensive principle of uncertainty must incorporate all types of uncertainty that are applicable. That is, it must be formulated as an optimization problem with several objective functions, each representing one of the applicable types of uncertainty.

The various minimum uncertainty principles are basically instances of the general simplification problem formulated in Sec. 5.8. Maximum uncertainty principles, which are instances of the general principle of ampliative reasoning, are formulated similarly, but the preference orderings based on uncertainty measures are inverted. As an illustration, let us describe a comprehensive principle of maximum uncertainty in the theory of evidence, which is an extension of the principle of maximum nonspecificity discussed previously.

Given a universal set X and some incomplete evidence regarding a particular question involving subsets of X (e.g., relevant marginal basic assignments), let $\mathscr{E}$ denote the set of all basic assignments on $\mathscr{P}(X)$ that conform to the evidence and let $\overset{E}{\leq}, \overset{C}{\leq}, \overset{V}{\leq}$ denote the preference orderings on $\mathscr{E}$ based upon the measures of dissonance, confusion, and nonspecificity, respectively (Table 5.4). Let $\overset{*}{\leq}$ denote a joint preference ordering defined for each pair $x, y \in \mathscr{E}$ by

$$x \overset{*}{\leq} y \Leftrightarrow x \overset{E}{\leq} y \quad \text{and} \quad x \overset{C}{\leq} y \quad \text{and} \quad x \overset{V}{\leq} y.$$

Then, the maximum uncertainty principle requires us to determine the set

$$\mathscr{E}_s = \{x \in \mathscr{E} \mid (\forall y \in \mathscr{E}) \, (y \overset{*}{\geq} x \Rightarrow x \overset{*}{\geq} y)\}$$

of feasible solutions. This solution set consists of all basic assignments on $\mathcal{P}(X)$ that conform to the available evidence and are either equivalent or incomparable in terms of the joint preference ordering.

We must admit that comprehensive uncertainty principles, such as the one just formulated, have not been adequately developed as yet and are currently a subject of active research. Therefore, we do not pursue this subject any further here.

NOTES

5.1. The *measure of fuzziness* defined by Eq. (5.1) was introduced and investigated by DeLuca and Termini [1972, 1974, 1977]. This measure, which is usually called an *entropy of fuzzy sets*, was also studied by Emptoz [1981] and Trillas and Riera [1978]. The term *entropy* was apparently chosen due to the similarity of the product terms in function (5.1) with the product terms in Shannon entropy (5.37). This choice is unfortunate, since the two functions measure fundamentally different types of uncertainties.

5.2. The *concept of entropy* arose in physics in the context of the theory of heat in the middle of the nineteenth century [Grad, 1961, Fast, 1962, Carnap, 1977]. The term was coined by Clausius in 1854 for a measure of the disorganization of a physical system. Its mathematical form emerged from the theory of ideal gases developed by Boltzmann [1896]. The name *entropy* was adopted by Shannon [1948] for his measure of uncertainty, presumably due to the similarity between the two mathematical forms. As shown in Sec. 5.3, however, Boltzmann entropy and Shannon entropy have fundamentally different meanings.

5.3. The *index of fuzziness*, expressed in a general form by Eq. (5.4), was proposed by Kaufmann [1975] in presumably the first book on fuzzy set theory.

5.4. The fact that *measures of fuzziness* expressed by Eqs. (5.1) and (5.4) are only special cases of a larger class of measures of fuzziness expressed by Eq. (5.6) was demonstrated by Knopfmacher [1975] and Loo [1977].

5.5. The idea that the degree of fuzziness of a fuzzy set can be most naturally expressed in terms of the lack of distinction between the set and its complement was proposed by Yager [1979b, 1980a]. A general formulation based upon this idea, which is applicable to all possible fuzzy complements, was developed by Higashi and Klir [1982]; they proved that every measure of fuzziness of this type can be expressed in terms of a metric distance that is based on appropriate aggregation of the absolute values of the individual differences between membership grades of the fuzzy set of concern and its complement for all elements of the universal set.

5.6. The discussion of the *Hartley measure* in Sec. 5.3 is primarily based on the original paper by Hartley [1928], but the axiomatic treatment is adopted from Rényi [1970].

5.7. Literature on *information theory* based on the Shannon entropy is abundant. The original paper by Shannon [1948] is also available in a book form [Shannon and Weaver, 1964]. Among the many books providing general coverage of the theory, particularly notable are books by Ash [1965], Billingsley [1965], Csiszár and Körner [1981], Feinstein [1958], Goldman [1953], Guiasu [1977], Jelinek [1968], Jones [1979], Khinchin [1957], Reza [1961], and Yaglom and Yaglom [1983]. The role of information theory in science is well described in books by Brillouin [1956, 1964], Denbigh and Denbigh [1985], and Watanabe [1969]. Other books focus on more specific areas such as economics [Georgescu-Roegen, 1971, Theil, 1967], engineering [Bell, 1953, Reza, 1961], biology [Brooks and Wiley, 1986, Gatlin, 1972], psychology [Quastler, 1955, Garner, 1962], or geography [Webber, 1979].

5.8. Various subsets of the axioms for a probabilistic measure of uncertainty that are presented in Sec. 5.3 were shown to be sufficient for proving the *uniqueness of Shannon entropy* by Aczél, Forte, and Ng [1974], Feinstein [1958], Forte [1975], Khinchin [1957], Rényi [1960, 1970], and others. The uniqueness proof presented in Appendix A.1 is adopted from a book by Ash [1965]. Excellent overviews of the various axiomatic treatments of Shannon entropy can be found in books by Aczél and Daróczy [1975] and Mathai and Rathie [1975]. Both of these books are heavily based on the use of functional equations. In fact, functional equations have played an important role not only in information theory but also in characterizing classes of operations on fuzzy sets. An excellent and comprehensive monograph on functional equations was prepared by Aczél [1966].

5.9. Several classes of functions that subsume the Shannon entropy as a special case have been proposed and studied. They include:

1. *Rényi's entropies* (also called entropies of degree α), which are defined for all real numbers $\alpha \neq 1$ by the function

$$H_\alpha(p_1, p_2, \ldots, p_n) = \frac{1}{1 - \alpha} \log_2 \sum_{i=1}^{n} p_i^\alpha. \qquad (5.100)$$

It is well known that the limit of H for $\alpha \to 1$ is the Shannon entropy. For $\alpha = 0$, we obtain

$$H(p_1, p_2, \ldots, p_n) = \log_2 \sum_{i=1}^{n} p_i^o. \qquad (5.101)$$

This function represents one of the probabilistic interpretations of the Hartley information as a measure that is insensitive to actual values of the given probabilities and distinguishes only between zero and nonzero probabilities. As the name suggests, Rényi entropies were proposed and investigated by Rényi [1960, 1970].

2. *Entropies of order* β, introduced by Daróczy (1970), which have the form

$$H_\beta(p_1, p_2, \ldots, p_n) = \frac{1}{2^{1-\beta} - 1} \left(\sum_{i=1}^{n} p_i^\beta - 1 \right) \qquad (5.102)$$

for all $\beta \neq 1$. As in the case of Rényi's entropies, the limit of H_β for $\beta \rightarrow 1$ results in the Shannon entropy.

3. *R-norm entropies*, which are defined for all $R \neq 1$ by the function

$$H_R(p_1, p_2, \ldots, p_n) = \frac{R}{R-1} \left[1 - \left(\sum_{i=1}^{n} p_i^R \right)^{1/R} \right]. \quad (5.103)$$

As in the other two classes of functions, the limit of H_R for $R \rightarrow 1$ is the Shannon entropy. This class of functions was proposed by Boekee and Van der Lubbe [1980] and was further investigated by Van der Lubbe [1984].

Formulas converting entropies from one class to the other classes are well known. Conversion formulas between H_α and H_β were derived by Aczél and Daróczy [1975, p. 185]; formulas for converting H_α and H_β into H_R were derived by Van der Lubbe [1980, p. 144].

Except for the Shannon entropy, these classes of functions are not adequate measures of uncertainty, since each of them violates some essential requirement for such a measure. For example, when α, β, $R > 0$, Rényi's entropies violate subadditivity, entropies of order β violate additivity, and R-norm entropies violate both subadditivity and additivity. The significance of these functions in the context of information theory is thus primarily theoretical as they help us better to understand Shannon entropy as a limiting case in these classes of functions. Strong arguments supporting this claim can be found in papers by Aczél, Forte, and Ng [1974] and Forte [1975].

5.10. Using Theorems 5.2 through 5.5 as a base, numerous other theorems regarding the relationship among the information transmission and basic, conditional, and joint Shannon entropies can be derived by simple algebraic manipulations and by mathematical induction to obtain generalizations. Conant [1981b] offers some useful ideas in this regard. A good summary of practical theorems for Shannon entropy was prepared by Ashby [1969]; see also Conant [1981a].

5.11. An excellent examination of the difference between *Shannon* and *Boltzmann entropies* is made by Reza [1961]. This issue is also discussed by Ash [1965], Guiasu [1977], and Jones [1979].

5.12. The *measure of dissonance* given by Eq. (5.68) was derived by Yager [1983a]. He also showed that the measure becomes equivalent to the Shannon entropy within probability theory (Theorem 5.6). The *measure of confusion* was introduced by Höhle [1981, 1982]. Both of these measures were further investigated by Dubois and Prade [1987].

5.13. For further information on the concept of *antichain*, which is referred to in Sec. 5.5, we recommend the book *Combinatorial Theory* by M. Aigner (Springer-Verlag, New York, 1979).

5.14. The *U-uncertainty* was proposed by Higashi and Klir [1983a]. They also formulated the axiomatic requirements for a *possibilistic measure of uncertainty* given in Sec. 5.6 (except the branching axiom) and proved that the *U-uncertainty* satisfies them. The possibilistic branching requirement was formulated by Klir and Mariano [1987] and employed as an axiom for proving the uniqueness of the *U-uncertainty* as a possibilistic measure of uncertainty and information. Their proof is given in Appendix A.2. As shown by Ramer and Lander [1987], the branching requirement is essential for obtaining a unique measure of possibilistic uncertainty.

5.15. Dubois and Prade [1985b] suggested the *measure of nonspecificity* given by Eq. (5.87) as a generalization of the *U*-uncertainty for evidence theory. The uniqueness of this measure was proven by Ramer [1987]; the proof is described in Appendix A.3. The measure was also investigated and compared with other measures of uncertainty by Dubois and Prade [1987]. Among other results, they showed that the measures of nonspecificity *V* and measures of dissonance *E* are monotonic with respect to appropriate ordering among bodies of evidence; the measures of confusion are not monotonic in this sense. They also showed that *V* is the only subadditive measure of the three measures of uncertainty.

5.16. An alternative *measure of nonspecificity* was proposed by Yager [1982c, 1983a]. It is characterized by a function *a* defined by the formula

$$a(m) = 1 - \sum_{A \in \mathcal{F}} \frac{m(A)}{|A|} \qquad (5.104)$$

where *m* is a given basic assignment and $\mathcal{F}$ is the set of associated focal elements. This measure possesses only some properties required for a measure of nonspecificity. In particular, it does not satisfy the additivity, subadditivity, and branching requirements. In addition, it does not coincide with the well-established Hartley measure for crisp possibility distributions.

5.17. Syntactic, semantic, and pragmatic aspects involving signs in all their forms and manifestations are studied by three areas of *semiotics*: syntactics (or syntax), semantics, and pragmatics. Semiotics (or a general theory of signs) was introduced by Charles Morris [1938, 1946, 1964]. Syntactic, semantic, and pragmatic aspects of information in human communication are well discussed by Cherry [1957]. Dretske [1981] demonstrates that the concepts of semantic information, meaning, and knowledge can be developed in terms of the Shannon entropy.

5.18. The concept of weighted Shannon entropy was introduced and investigated by Guiasu [1971, 1977]. The notion of algorithmic information was proposed by Kolmogorov [1965]; a good overview was prepared by Chaitin [1977].

5.19. Some of the systems frameworks referred to in Sec. 5.8 were developed by Mesarovic and Takahara [1975], Wymore [1969, 1976], and Zeigler [1976, 1984]. Islam [1974] rigorously compared two of the frameworks, seemingly totally different, and established that they were almost isomorphic and could be made totally isomorphic under a minor modification.

5.20. Fundamental issues of descriptive complexity are well discussed by Löfgren [1977]. Further references on this topic can be found in a bibliography prepared by Cornacchio [1977]. Descriptive complexity is closely connected with the concept of algorithmic (or descriptive) information (Note 5.18).

5.21. The *principle of maximum entropy* has been justified by at least three distinct arguments:

1. The maximum entropy probability distribution is the only *unbiased distribution*, that is, the distribution that takes into account all available information but no additional (unsupported) information (bias). This follows directly from the facts that (1) all available information (but nothing else) is required to form the constraints of the optimization problem, and (2) the chosen probability distribution is required to be the one that represents the maximum uncertainty (entropy) within the constrained set of probability distributions. Indeed, any reduction of

uncertainty is an equal gain of information. Hence, a reduction of uncertainty from its maximum value, which would occur when any distribution other than the one with maximum entropy were chosen, would mean that some information from outside the available evidence was implicitly added.

This argument of justifying the maximum entropy principle is covered in the literature quite extensively. Perhaps its best and most thorough presentation is given in a paper by Jaynes [1979], which also contains an excellent historical survey of related developments in probability theory, and in a book by Christensen [1981a].

2. It was shown by Jaynes [1968], strictly on combinatorial grounds, that the maximum probability distribution is the *most likely distribution* in any given situation.

3. It was demonstrated by Shore and Johnson [1980] that the principle of maximum entropy can be deductively derived from the following *consistency axioms* for inductive (or ampliative) reasoning:

- *Uniqueness*. The result should be unique.

- *Invariance*. The choice of coordinate system (permutation of variables) should not matter.

- *System independence*. It should not matter whether one accounts for independent systems separately in terms of marginal probabilities or together in terms of joint probabilities.

- *Subset independence*. It should not matter whether one treats an independent subset of system states in terms of separate conditional probabilities or in terms of full system probabilities.

The rationale for choosing these axioms is expressed by Shore and Johnson as follows: any acceptable method of inference must be such that different ways of using it to take the same information into account lead to consistent results. Using the axioms, they derive the following proposition: given some information in terms of constraints regarding the probabilities to be estimated, there is only one probability distribution satisfying the constraints that can be chosen by a method that satisfies the consistency axioms; this unique distribution can be attained by maximizing the Shannon entropy (or any other function that has exactly the same maxima as the entropy) subject to the given constraints.

Alternative derivations of the principle of maximum entropy were demonstrated by Smith [1974] and Avgers [1983]. An excellent historical survey of the principle with an extensive bibliography was prepared by Kapur [1983]. Other notable publications regarding the principle include: Good [1963], Tzannes and Noonan [1973], Jaynes [1982, 1983], Christensen [1980–1986], and Smith and Grandy [1985]. Williams [1980] examines the principle of minimum cross-entropy and shows that the well-known Bayesian rule of conditionalization is a special case of this principle.

5.22. For a general background on optimization methods relevant to the maximum entropy principle, particularly the Lagrange multiplier optimization method, we recommend the text *Introduction to Nonlinear Optimization* by D. A. Wismer and R. Chattergy (North-Holland, New York, 1978) or the excellent monograph *Constrained Optimization and Lagrange Multiplier Methods* by D. P. Bertsekas (Academic Press, New York, 1982). For optimization with multiple objective functions, we recommend the text *Multiple Criteria Decision Making* by M. Zeleny (McGraw-Hill, New York, 1982), or the survey book *Multiple Attribute Decision Making* by C.-L. Hwang and K. Yoon (Springer-Verlag, New York, 1981).

5.23. When logarithm base 2 in the various measures of uncertainty is replaced with the natural logarithm or the base 10 logarithm, the measurement unit of uncertainty is usually called *nat* or *hartley*, respectively.

EXERCISES

5.1. Using Eq. (5.14) and the standard fuzzy complement, calculate the degree of fuzziness for some of the fuzzy binary relations defined in Table 3.4 (p. 104). Also calculate the corresponding normalized degree of fuzziness by Eq. (5.15).

5.2. Repeat Exercise 5.1 for the following alternative definitions of the measure of fuzziness:
  (a) Equation (5.14) and some nonstandard fuzzy complements;
  (b) the measure of fuzziness defined by Eq. (5.1);
  (c) the measure of fuzziness defined by Eq. (5.2).

5.3. Show that the measures of fuzziness defined by both Eqs. (5.1) and (5.14) express fuzziness in bits.

5.4. Show that the maximum of the measure of fuzziness defined by Eq. (5.1) is $|X|$.

5.5. Define some particular measures of fuzziness in the class defined by Eq. (5.6) in addition to those given in Sec. 5.2 as examples.

5.6. Using the measure of fuzziness defined by Eq. (5.18) for $w = 1$ and the standard fuzzy complement, calculate the degree of fuzziness of the fuzzy set defined on the interval [0, 10] of real numbers by the following membership function:
  (a) $\mu_A(x) = x/10$;
  (b) $\mu_B(x) = 2^{-x}$;
  (c) $\mu_C(x) = x/5$ for $x \le 5$ and $\mu_C(x) = 2 - .2x$ for $x \ge 5$.

5.7. Using Eq. (5.14) and the standard fuzzy complement, calculate the degree of fuzziness for each relation involved in the procedure for obtaining transitive closure in Example 3.14 (p. 81).

5.8. For all α-cuts of some of the fuzzy binary relations given in Table 3.4 (p. 104), calculate all applicable types of Hartley information, that is, the simple information of each of the two sets involved, joint information, both forms of the conditional information, and information transmission.

5.9. Use some additional branching schemes in Example 5.4 to calculate the value of Shannon entropy for the given probability distribution, for example:
  (a) $p_A = p_1 + p_4$, $p_B = p_2 + p_3$;
  (b) $p_A = p_1 + p_2 + p_4$, $p_B = p_3$;
  (c) scheme IV in which p_1 is exchanged with p_4 and p_2 is exchanged with p_3.

5.10. Assume that the probabilities p_1, p_2, p_3, p_4 in Example 5.4 are different from each other. How many different alternatives are there for each of the four types of branching schemes in Fig. 5.4 and what is the total number of all possible branching schemes in this example?

5.11. Consider two joint probability distributions on $X \times Y$ ($X = \{a, b, c\}$, $Y = \mathbb{N}_5$) defined by the matrices

$$
\begin{array}{c}
 \\ a \\ b \\ c
\end{array}
\begin{array}{c}
\begin{array}{ccccc} 1 & 2 & 3 & 4 & 5 \end{array} \\
\left[\begin{array}{ccccc}
.1 & 0 & 0 & .05 & 0 \\
0 & .3 & .1 & 0 & .2 \\
.05 & 0 & 0 & .1 & .1
\end{array} \right]
\end{array}
\quad \text{and} \quad
\begin{array}{c}
 \\ a \\ b \\ c
\end{array}
\begin{array}{c}
\begin{array}{ccccc} 1 & 2 & 3 & 4 & 5 \end{array} \\
\left[\begin{array}{ccccc}
.25 & .05 & .05 & .05 & 0 \\
0 & 0 & 0 & 0 & .05 \\
.1 & .15 & .1 & .1 & .1
\end{array} \right]
\end{array}.
$$

(a) Calculate $H(X)$, $H(Y)$, $H(X, Y)$, $H(X \mid Y)$, $H(Y \mid X)$, and $T(X, Y)$ for both of the distributions.

(b) Verify that Theorems 5.3 through 5.5 hold for these distributions.

(c) Make some general statements comparing the distributions.

5.12. Derive the generalized form (5.47) for Theorem 5.3.

5.13. Derive the generalized form (5.49) for Theorem 5.4.

5.14. The so-called Q-factor, which is defined by the equation

$$H(X, Y, Z) = H(X, Y) + H(X, Z) + H(Y, Z) - H(X) - H(Y) - H(Z) + Q(X, Y, Z)$$

is often used in classical information theory. Express $Q(X, Y, Z)$ solely in terms of the various information transmissions.

5.15. For each of the probability density functions $q(x)$ shown in Fig. 5.12, calculate the Boltzmann entropy and demonstrate that it is negative, zero, or positive depending on the values of a, b, c. *Note*: remember that the condition

$$\int_{-\infty}^{\infty} q(x) \, dx = 1$$

must be satisfied.

5.16. Calculate the following two cross-entropies of the probability distributions ^1p and ^2p on states $(\dot{v}_1, \dot{v}_2)$ given in Table 5.6:

(a) cross-entropy in which ^1p and ^2p correspond to p and p' in formula (5.63), respectively;

(b) cross-entropy in which the correspondence specified in (a) is inverted.

5.17. Consider the following basic assignments defined on $X = \{a, b, c\}$:

$$m_1(\{c\}) = .2, \qquad m_1(\{b, c\}) = .3, \qquad m_1(X) = .5;$$

$$m_2(\{a\}) = .5, \qquad m_2(\{b\}) = .2, \qquad m_2(\{c\}) = .3;$$

$$m_3(\{c\}) = .2, \qquad m_3(\{a, b\}) = .3, \qquad m_3(X) = .5.$$

Determine the amount of dissonance, confusion, and nonspecificity for each of these basic assignments. Do any of the basic assignments represent probability or possibility measures?

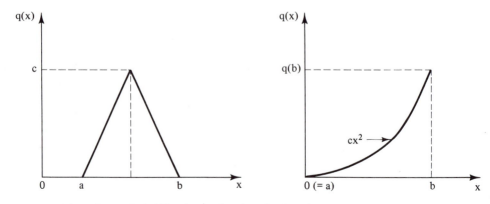

Figure 5.12. Probability density functions for Exercise 5.15.

TABLE 5.8. JOINT BASIC ASSIGNMENTS FOR EXERCISE 5.20.

R_i	1a	1b	1c	2a	2b	2c	3a	3b	3c	$m_1(R_i)$	$m_2(R_i)$	$m_3(R_i)$	$m_4(R_i)$	$m_5(R_i)$	$m_6(R_i)$
R_1	1	0	0	0	0	0	0	0	0	.2	.25	0	0	0	0
R_2	0	1	0	0	0	0	0	0	0	.05	.15	0	0	0	0
R_3	0	0	1	0	0	0	0	0	0	.1	.1	0	0	0	0
R_4	0	0	0	1	0	0	0	0	0	.05	.2	0	0	0	0
R_5	0	0	0	0	1	0	0	0	0	.25	.12	0	0	0	0
R_6	0	0	0	0	0	1	0	0	0	.04	.08	0	0	0	0
R_7	0	0	0	0	0	0	1	0	0	.25	.05	.3	0	.02	.05
R_8	0	0	0	0	0	0	0	1	0	.03	.03	0	0	0	0
R_9	0	0	0	0	0	0	0	0	1	.03	.02	0	0	0	0
R_{10}	0	0	0	1	0	0	1	0	0	0	0	.1	.02	.03	.03
R_{11}	0	0	0	0	0	0	1	0	1	0	0	0	0	.06	.10
R_{12}	0	0	1	0	0	1	0	0	0	0	0	0	0	0	0
R_{13}	1	0	0	1	0	0	0	0	0	0	0	0	.03	.025	.07
R_{14}	1	0	0	1	0	0	1	0	0	0	0	0	.05	.025	.05
R_{15}	0	0	0	0	0	0	1	1	1	0	0	0	0	.12	.12
R_{16}	1	1	1	0	0	0	0	0	0	0	0	0	0	0	0
R_{17}	0	0	0	0	1	1	0	1	1	0	0	0	.06	0	0
R_{18}	0	0	0	1	0	1	1	0	1	0	0	.2	.04	.09	.09
R_{19}	1	0	1	1	0	1	0	0	0	0	0	0	.06	.075	.02
R_{20}	0	1	1	0	1	1	0	0	0	0	0	0	.09	0	0
R_{21}	1	0	1	1	0	1	1	0	1	0	0	.1	.1	0	0
R_{22}	0	0	0	1	1	1	1	1	1	0	0	0	.08	.18	.07
R_{23}	1	1	1	1	1	1	0	0	0	0	0	0	.12	.15	.10
R_{24}	0	1	1	0	1	1	0	1	1	0	0	0	.15	0	0
R_{25}	1	1	1	0	0	0	1	1	1	0	0	0	0	.075	.17
R_{26}	1	1	1	1	1	1	1	1	1	0	0	.3	.2	.15	.13

5.18. Prove Theorems 5.7 through 5.9.

5.19. Prove Theorem 5.11 by utilizing its analogy with Theorem 5.10.

5.20. Six joint basic assignments on $X \times Y$, where $X = \{1, 2, 3\}$ and $Y = \{a, b, c\}$ are given in Table 5.8. Focal elements are specified in the table by their characteristic functions. Determine:

    (a) which of the basic assignments represent probability or possibility measures;

    (b) which of the basic assignments represent noninteractive marginal bodies of evidence;

    (c) measures of dissonance, confusion, and nonspecificity for all the given joint basic assignments as well as the associated marginal basic assignments (projections) and check the additivity properties of the three measures for basic assignments that represent noninteractive marginals;

    (d) joint basic assignments reconstructed from their marginals on the assumption of noninteraction and their measures of dissonance, confusion, and nonspecificity.

5.21. Prove that the U-uncertainty is subadditive.

5.22. Consider the situation depicted in Fig. 5.13, where $X = \{a, b, c, d\}$ is a set of some signs and Y_x is a set of meanings of x ($x \in X$). Each of these sets is associated with a probability distribution given in Fig. 5.13. Using Eq. (5.89), calculate the total Shannon entropy which incorporates both syntactic and semantic aspects of the situation.

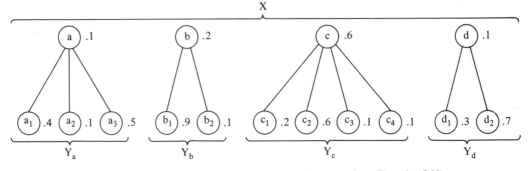

Figure 5.13. Syntactic and semantic uncertainty (Exercise 5.22).

5.23. Assume that the signs $x \in X$ in Fig. 5.13 are associated with weights $w(a) = .4$, $w(b) = .2$, $w(c) = .1$, and $w(d) = .3$, which characterize their utilities. Using Eq. (5.90), calculate the weighted Shannon entropy in this case and compare it with the regular Shannon entropy for the given probability distribution $(p(x) \mid x \in X)$.

5.24. Consider some problems in your area of interest and check for each of them whether or not it is transcomputational in the sense of the Bremermann limit.

5.25. Calculate some entries in Table 5.5.

5.26. Repeat Example 5.11 for the following probability distributions (given in the same order as the probability distribution in Table 5.6):

(a) $(.4, 0, .1, 0, .2, .3)$;
(b) $(.1, .2, .1, .2, .2, .2)$;
(c) $(.1, .1, .6, .1, 0, .1)$.

5.27. Repeat Example 5.13 for the following possibility distributions (given in the same order as the possibility distribution in Table 5.7):

(a) $(.6, .7, .8, .9, 1, 1)$;
(b) $(.9, .9, 1, 1, .8, .8)$;
(c) $(.1, .5, 1, .6, 1, .7)$.

5.28. Consider a random variable x that takes values in the set $\mathbb{N}_{10}$. It is known that the mean (expected) value $E(x)$ of the variable is 2.7. Estimate, using the maximum entropy principle, the probability distribution associated with the variable.

5.29. Let $p_1 = (.7, .3)$ and $p_2 = (.4, .6)$ be marginal probability distributions of two-valued random variables. Estimate, using the maximum entropy principle, their joint probability distribution.

5.30. Consider a universal set X, four subsets of which are of our interest: $A \cap B$, $A \cap C$, $B \cap C$, and $A \cap B \cap C$. The only evidence we have is expressed by the equations

$$m(A \cap B) + m(A \cap B \cap C) = .2,$$

$$m(A \cap C) + m(A \cap B \cap C) = .5,$$

$$m(B \cap C) + m(A \cap B \cap C) = .1.$$

Estimate, using the maximum nonspecificity principle, the values of $m(A \cap B)$, $m(A \cap C)$, $m(B \cap C)$, $m(A \cap B \cap C)$, and $m(X)$.

5.31. In analogy with the total Shannon entropy, which combines syntactic and semantic aspects of relevant situations, propose a similar total measure of nonspecificity.

6

APPLICATIONS

6.1 GENERAL DISCUSSION

We see our problems today differently than we did in the past. We are no longer merely facing the world trying to win its secrets or battling to shape it to meet our desires. We are now facing ourselves and our own knowledge and trying to cope with the complexity we have created. Therefore, we no longer formulate the problems in our scientific inquiries or in our engineering attempts simply in terms of missing facts or inadequate measuring devices or materials. Problems are seen to be of a different order; that is, they concern questions about knowledge and information itself and can therefore be formulated independently of any particular area of knowledge. Indeed, questions in fields as diverse as psychology and engineering, economics and medicine, sociology and genetics, artificial intelligence and meteorology, or decision making and architecture begin to take on a similar form when seen in light of their common problems of handling information, uncertainty, and complexity.

New problems call for new solutions. We want to know how to simplify complexity, how to utilize information, and how to deal with uncertainty, not in any particular context but in the form of general principles and representations that hold true in any given context. For this reason, fuzzy set theory and the other mathematical frameworks introduced in this book that deal with information and uncertainty are applicable in any field in which issues of complexity arise. It is, in fact, difficult to name a field in which complexity does not arise, either as the actual focus of the study or as a by-product or attribute of the content of the inquiry. This is not to say, however, that the mathematics of information and uncertainty will be all things to all people. This field is still relatively young, and more research is needed before a clear determination can be made of exactly what these mathematical frameworks can and cannot offer. Most early attempts to discover the usefulness of these frameworks have consisted in applying them to already existing solutions, for instance, by "fuzzifying" algorithms in order to achieve better performance. In many cases, this better performance has, in fact, been achieved. The aim of the mathematics of information and uncertainty, however, is not merely to do the same thing in a different way. It is, instead, to approach problems from a higher level of question about the possibilities and limits of knowledge itself. The success of this approach in application to real-world problems, while quite promising thus far, will be fully realized only in time.

The examples presented in this chapter are intended to give a flavor of the types of application research currently taking place in the field of fuzzy sets, uncertainty, and information. The successful applications of these mathematical models and tools is extremely diverse and widespread, and the sample included here is by no means exhaustive. Each section includes a brief overview of the application area followed by some specific illustrative examples drawn from the literature. Key references in each area are incorporated into the text and additional, more specific references are given in the notes at the end of the chapter. The number of books and papers presenting these applications is quite large and the references that are included are not intended to stand as full surveys of these fields; their purpose is to aid the reader in pursuing a more extensive study in selected areas.

6.2 NATURAL, LIFE, AND SOCIAL SCIENCES

The various natural, life, and social sciences have been active areas for application of the mathematics of uncertainty and information, both for the purpose of understanding human systems and for making natural systems understandable to human thought. The applications of fuzzy set theory include explorations within psychology and cognitive science of concept formation and manipulation, memory and learning, as well as studies in the fields of sociology, economics, ecology, meteorology, biology, and others.

The applications of information theory are even more extensive; they cover virtually the whole spectrum of science. They are, however, almost exclusively based on the classical Shannon or Hartley measures of information. The principles of maximum and minimum entropy have proven to be the most important for applications; their use is best exemplified by the work of Ronald Christensen [1981b, 1986]. The alternative principles of uncertainty, such as the maximum nonspecificity principle suggested in Sec. 5.9, are too new to have been demonstrated in specific applications thus far. There is little doubt, however, that their impact on a broad spectrum of applications in science will be profound.

In this section, we illustrate the vastly diverse applications in the sciences by two simple applications of fuzzy set theory: one in the field of meteorology and the other in the area of interpersonal communication.

Meteorology

The science of meteorology deals with a vastly complex system which, in truth, encompasses the entire planet. For this reason, meteorological descriptions as well as forecasts have always relied on the kind of robust summary offered by vague linguistic terms such as *hot weather*, *drought*, or *low pressure*. Applications of fuzzy set theory to meteorology, therefore, constitute attempts to deal with the complexity of the study by taking advantage of the representation of vagueness offered in this mathematical formalism. Three different applications of fuzzy set theory within the field of meteorology are presented by Cao and Chen [1983]. We briefly examine the first of these, which makes use of a fuzzy clustering technique. (The area of fuzzy clustering is covered briefly in Sec. 6.8.)

This particular application focuses on an understanding of climatological changes based on the examination of statistical records of weather patterns collected over a period of time. For simplicity, only the two climatological stages of drought periods and wet periods are considered in this study. The data used were taken at a single monitoring station during the years 1886 through 1979 in Shanghai, China, and consisted of records of annual precipitation levels. Each of the n original measurements of annual precipitation p_i ($i \in \mathbb{N}_n$) is preprocessed by a 10-year running mean calculated by

$$p_i = \frac{p_{i-4} + \cdots + p_{i-1} + p_i + p_{i+1} + \cdots + p_{i+5}}{10}.$$

After this preprocessing, a total of 85 records were obtained from the original data.

The interest of the study is to examine the points and features of alternations between wet and drought periods. Obviously, each period is characterized by annual precipitation levels that are similar to one another, and turning points or boundaries between periods are characterized by the appearance of dissimilarity between the values of annual precipitations. The exact point of change between any two of these periods may be difficult to determine crisply. In order to accommodate the inherent fuzziness between boundaries of drought and wet periods, a fuzzy similarity relation is constructed from the annual precipitation values in order to cluster together into periods those years whose precipitation levels are "similar" to one another. This similarity relation is created by first calculating a similarity index s'_{ij} between precipitations p_i and p_j ($i, j \in \mathbb{N}_{85}$). These indices are given by

$$s'_{ij} = 1 - |p_i - p_j|.$$

The indices are then normalized to the interval [0, 1] by the formula

$$s_{ij} = \frac{s'_{ij} - s'_{min}}{s'_{max} - s'_{min}},$$

where s'_{min} and s'_{max} are the maximum and minimum similarity indices, respectively. The resulting normalized similarity indices can be viewed as constituting the membership grades of a fuzzy relation R such that

$$\mu_R(p_i, p_j) = s_{ij}.$$

The relation R is reflexive and symmetric but is generally not transitive. Therefore, the transitive closure of R must be calculated; this can be done with the algorithm given in Sec. 3.3. The resulting relation R_T is the desired similarity relation. By taking α-cuts of this relation, where $\alpha \in [0, 1]$, we arrive at crisp clusters of precipitation levels, which are similar to each other to the degree α. Figure 6.1 illustrates the matrix representation of this type of clustering that was performed on the data collected by Cao and Chen for the similarity level $\alpha = .95$. An ijth entry of 1 in this matrix indicates the presence of similarity between record i and record j of a degree at least equal to α. As can be seen from the figure, five different climatological stages or clusters are apparent. The periods of drought are found in records 1–17, 31–52, and 68–85, whereas records 18–30 and 53–67 constitute the wet periods. The changes between periods are clearly displayed in the matrix representation of the α-cut. Such an application forms a useful tool for the investigation of climatological change and forecasting by mathematically representing the fuzziness or imprecision of boundaries between the stages of interest.

Interpersonal Communication

The process of interpersonal communication consists of a vast array of different types of simultaneously communicated signals (words, voice tone, body posture, clothing, etc.), many of which conflict with each other. It is therefore difficult to determine the precise intention and meaning of the communication, both because of distortion from environmental noise and because of ambivalence on the part of the sender. Nevertheless, the receiver must respond appropriately in the face of this fuzzy or vague information. We outline here an approach suggested by Yager [1980h], which models this process and the vagueness associated with it through the use of fuzzy set theory.

Suppose that X constitutes the universal set of all possible signals x that may be communicated by the sender. Because of the distorting factors mentioned above, a clear, unique signal may not be available. Instead, the message received is a fuzzy subset M of X, in which $\mu_M(x)$ denotes the degree of certainty of the receipt of the specific signal x. In order to determine whether an appropriate response can be chosen based on the message received or whether some error was involved in the communication, an assessment of the quality of the transmission must be made. Let the maximum value of membership which any $x \in X$ attains in the set M correspond to the *strength* of the transmission. If the set M has no unique maximum, then the message is called *ambiguous*. If the support of M is large, then M is considered to be *general*. The clarity of the message can be measured by the distance between the maximum membership grade attained in M and the next largest grade of any signal x in M. When the message received is strong, unambiguous, and clear, then the signal attaining the maximum mem-

```
    1111111111111111110000000000000011111111111111111111100000000000000010111111111111111
    1111111111111111110000000000000011111111111111111111100000000000000010111111111111111
    1111111111111111110000000000000011111111111111111111100000000000000010111111111111111
    1111111111111111110000000000000011111111111111111111100000000000000010111111111111111
    1111111111111111110000000000000011111111111111111111100000000000000010111111111111111
    1111111111111111110000000000000011111111111111111111100000000000000010111111111111111
    1111111111111111110000000000000011111111111111111111100000000000000010111111111111111
    1111111111111111110000000000000011111111111111111111100000000000000010111111111111111
 10 1111111111111111110000000000000011111111111111111111100000000000000010111111111111111 10
    1111111111111111110000000000000011111111111111111111100000000000000010111111111111111
    1111111111111111110000000000000011111111111111111111100000000000000010111111111111111
    1111111111111111110000000000000011111111111111111111100000000000000010111111111111111
    1111111111111111110000000000000011111111111111111111100000000000000010111111111111111
    1111111111111111110000000000000011111111111111111111100000000000000010111111111111111
    1111111111111111110000000000000011111111111111111111100000000000000010111111111111111
    1111111111111111110000000000000011111111111111111111100000000000000010111111111111111
    0000000000000000111111111111100000000000000000000001111110110111101000000000000000000
    0000000000000000111111111111100000000000000000000001111110110111101000000000000000000
 20 0000000000000000111111111111100000000000000000000001111110110111101000000000000000000 20
    0000000000000000111111111111100000000000000000000001111110110111101000000000000000000
    0000000000000000111111111111100000000000000000000001111110110111101000000000000000000
    0000000000000000111111111111100000000000000000000001111110110111101000000000000000000
    0000000000000000111111111111100000000000000000000001111110110111101000000000000000000
    0000000000000000111111111111100000000000000000000001111110110111101000000000000000000
    0000000000000000111111111111100000000000000000000001111110110111101000000000000000000
    0000000000000000111111111111100000000000000000000001111110110111101000000000000000000
    0000000000000000111111111111100000000000000000000001111110110111101000000000000000000
    0000000000000000111111111111100000000000000000000001111110110111101000000000000000000
 30 0000000000000000111111111111100000000000000000000001111110110111101000000000000000000 30
    1111111111111111110000000000000011111111111111111111100000000000000010111111111111111
    1111111111111111110000000000000011111111111111111111100000000000000010111111111111111
    1111111111111111110000000000000011111111111111111111100000000000000010111111111111111
    1111111111111111110000000000000011111111111111111111100000000000000010111111111111111
    1111111111111111110000000000000011111111111111111111100000000000000010111111111111111
    1111111111111111110000000000000011111111111111111111100000000000000010111111111111111
    1111111111111111110000000000000011111111111111111111100000000000000010111111111111111
    1111111111111111110000000000000011111111111111111111100000000000000010111111111111111
 40 1111111111111111110000000000000011111111111111111111100000000000000010111111111111111 40
    1111111111111111110000000000000011111111111111111111100000000000000010111111111111111
    1111111111111111110000000000000011111111111111111111100000000000000010111111111111111
    1111111111111111110000000000000011111111111111111111100000000000000010111111111111111
    1111111111111111110000000000000011111111111111111111100000000000000010111111111111111
    1111111111111111110000000000000011111111111111111111100000000000000010111111111111111
    1111111111111111110000000000000011111111111111111111100000000000000010111111111111111
    1111111111111111110000000000000011111111111111111111100000000000000010111111111111111
    1111111111111111110000000000000011111111111111111111100000000000000010111111111111111
 50 1111111111111111110000000000000011111111111111111111100000000000000010111111111111111 50
    1111111111111111110000000000000011111111111111111111100000000000000010111111111111111
    1111111111111111110000000000000011111111111111111111100000000000000010111111111111111
    0000000000000000111111111111100000000000000000000001111110110111101111111111111111111
    0000000000000000111111111111100000000000000000000001111110110111101111111111111111111
    0000000000000000111111111111100000000000000000000001111110110111101111111111111111111
    0000000000000000111111111111100000000000000000000001111110110111101111111111111111111
    0000000000000000111111111111100000000000000000000001111110110111101111111111111111111
    0000000000000000111111111111100000000000000000000001111110110111101111111111111111111
    0000000000000000111111111111100000000000000000000001111110110111101111111111111111111
 60 0000000000000000000000000000000000000000000000000000010010000000000000000000000000000 60
    0000000000000000111111111111100000000000000000000001111110110111101000000000000000000
    0000000000000000111111111111100000000000000000000001111110110111101000000000000000000
    0000000000000000000000000000000000000000000000000000010010000000000000000000000000000
    0000000000000000111111111111100000000000000000000001111110110111101000000000000000000
    0000000000000000111111111111100000000000000000000001111110110111101000000000000000000
    0000000000000000111111111111100000000000000000000001111110110111101000000000000000000
    0000000000000000111111111111100000000000000000000001111110110111101000000000000000000
    1111111111111111110000000000000011111111111111111111100000000000000010111111111111111
    0000000000000000111111111111100000000000000000000001111110110111101000000000000000000
 70 1111111111111111110000000000000011111111111111111111100000000000000010111111111111111 70
    1111111111111111110000000000000011111111111111111111100000000000000010111111111111111
    1111111111111111110000000000000011111111111111111111100000000000000010111111111111111
    1111111111111111110000000000000011111111111111111111100000000000000010111111111111111
    1111111111111111110000000000000011111111111111111111100000000000000010111111111111111
    1111111111111111110000000000000011111111111111111111100000000000000010111111111111111
    1111111111111111110000000000000011111111111111111111100000000000000010111111111111111
    1111111111111111110000000000000011111111111111111111100000000000000010111111111111111
    1111111111111111110000000000000011111111111111111111100000000000000010111111111111111
 80 1111111111111111110000000000000011111111111111111111100000000000000010111111111111111 80
    1111111111111111110000000000000011111111111111111111100000000000000010111111111111111
    1111111111111111110000000000000011111111111111111111100000000000000010111111111111111
    1111111111111111110000000000000011111111111111111111100000000000000010111111111111111
    1111111111111111110000000000000011111111111111111111100000000000000010111111111111111
    1111111111111111110000000000000011111111111111111111100000000000000010111111111111111
```

Figure 6.1.   Fuzzy clustering of climatological periods: α-cut of S for $\alpha = .95$.

bership grade in M can easily be selected as the most obvious intended communication. Difficulty occurs, however, when the message is weak, ambiguous, or unclear. In this case, the receiver must determine whether the problem in the communication lies in some environmental distortion (in which case a repetition of the signal may be requested) or in the sender of the message (in which case a response must be made, which is, as far as possible, appropriate).

Usually, the receiver of the communication possesses some background information in the form of probabilities or possibilities of the signals which can be expected. If $p(x_1), p(x_2), \ldots, p(x_n)$ represent the probabilities associated with each of the signals $x_1, x_2, \ldots, x_n \in X$, then the probability of the fuzzy event of the receipt of message M is given by

$$p(M) = \sum_{x \in X} \mu_M(x) p(x).$$

The receiver can use this information to assess the consistency of the received message with his or her expectations. If the probability of the received message is high, then it can be assumed that little distortion was introduced by the environment. On the other hand, if the message is very clear and unambiguous, then an appropriate response can be made even if the probability of the signal was low.

Instead of the expectation or background information being given in probabilistic form, this information may be given in the form of a possibility distribution $\mathbf{r}$ on X. In this case, $r(x) \in [0, 1]$ indicates the receiver's belief in the possibility of signal x being sent. The total possibility of the fuzzy message M is calculated as

$$r(M) = \max_{x \in X}[\min(\mu_M(x), r(x))].$$

As in the case of probabilistic expectation, if the received message conflicts with the expected possibility of communication, then the receiver may attempt clarification by requesting a repetition of the transmission. Before this new transmission is sent, the receiver will probably have already modified his or her expectations based on the previous message. If $\mathbf{r}_0$ indicates the initial possibilistic expectations of the receiver and $\mathbf{r}_1$ is the modified expectations subsequent to the receipt of message M, then

$$r_1(x) = \min[r_0^\alpha(x), \mu_M(x)]$$

for each $x \in X$, where α indicates the degree to which past messages are considered relevant in the modification of expectations. Our procedure for signal detection now consists of the following: a test of the consistency of M against the expectations and a test of the message M for strength and clarity. If both of these values are high, the signal attaining the maximum value in M can be comfortably assumed to be the intended signal. If both tests yield low values, the expectations are modified and a repetition is requested. If only one of these tests yields a satisfactory value, then either a new signal is requested or a response is made despite the presence of doubt.

An additional complication is introduced when we consider that the receiver may also introduce distortion in the message because of inconsistency with the expectations. Let

$$s(M, r) = \max_{x \in X}[\min(\mu_M(x), r(x))] \qquad (6.1)$$

correspond to the consistency of the received message with the possibilistic expectations. Then, let M' denote the message which the receiver actually hears, where

$$\mu_{M'}(x) = \mu_M^s(x) \qquad (6.2)$$

for each $x \in X$. The less consistent M is with the expectations, the less M' resembles M. Since the receiver will be modifying his or her expectations based on the message thought to have been received, the new possibilistic expectation structure is given by

$$r_1(x) = \min[r_0^{1-s}(x), \mu_{M'}(x)] \qquad (6.3)$$

for each $x \in X$.

Finally, once a determination has been made of the signal $x \in X$ that was sent, an appropriate response must be chosen. Let Y be the universal set of all responses and let $R \in Y \times X$ be a fuzzy binary relation in which $\mu_R(y, x)$ indicates the degree of appropriateness of response y given signal x. A fuzzy response set $A \in Y$ can be generated by composing the appropriateness relation R with the fuzzy message M,

$$A = R \circ M$$

or

$$\mu_A(y) = \max_{x \in X}[\min(\mu_R(y, x), \mu_M(x))] \qquad (6.4)$$

for each $y \in Y$. The membership grade of each possible message y in fuzzy set A thus corresponds to the degree to which it is an appropriate response to the message M. A more interesting case occurs when the elements $y \in Y$ are not actual messages but instead indicate characteristics or attributes that the appropriate message should possess. This allows for creativity in formulating the actual response. The following example illustrates the use of this model of interpersonal communication.

Suppose that a young man has just proposed marriage to a young woman and is now eagerly awaiting her response. Let us assume that her answer will be chosen from the set X of the following responses:

$x_1 =$ simple yes

$x_2 =$ simple no

$x_3 =$ request for time to think it over

$x_4 =$ request for young man to ask permission of the young woman's parents

x_5 = derisive laughter

x_6 = joyful tears

Assume also that the young man has expectations of her response represented by the possibility distribution

$$\mathbf{r}_0 = (.9, .1, .7, .3, .1, .6).$$

We can see from this distribution that the young man expects a positive answer. Suppose, however, that the following message M_1 is received:

$$M_1 = .1/x_1 + .8/x_2 + .4/x_3 + .1/x_5.$$

This message, although relatively strong, unambiguous, and clear, is rather inconsistent with the young man's expectations. As measured by Eq. (6.1), the consistency is

$$s(M_1, \mathbf{r}_0) = \max[.1, .1, .4, .1] = .4.$$

Because the message is contrary to the young man's expectations, let us assume that he introduces some distortion, as specified by Eq. (6.2), such that the message he hears is

$$M_1' = .4/x_1 + .9/x_2 + .7/x_3 + .4/x_5.$$

Based on this message, he modifies his expectations according to Eq. (6.3) such that

$$r_1(x) = \min[r_0^6(x), \mu_{M_1'}(x)]$$

for each $x \in X$, or

$$\mathbf{r}_1 = .4/x_1 + .25/x_2 + .7/x_3 + .25/x_5.$$

The young man has thus greatly diminished his expectation of a simple yes, somewhat increased his expectation of a simple no and of derisive laughter, and has given up all hope of the possibility of joyful tears. Suppose now that, in disbelief, he asks the young woman to repeat her answer and receives the following message:

$$M_2 = .9/x_2 + .4/x_5.$$

This message is stronger, clearer, and less general than the first answer. Its consistency with the young man's new expectations is

$$s(M_2, \mathbf{r}_1) = .25.$$

Thus, the message is highly contrary even to the revised expectations of the young man, so let us suppose that he distorts the message such that he hears

$$M_2' = .97/x_2 + .8/x_5.$$

His surprise has thus diminished the clarity of the message heard and has led him to exaggerate the degree to which he believes that the young woman has responded with derisive laughter. Let us now suppose that the response which the young man makes will have characteristics chosen from the following set Y:

y_1 = happiness y_2 = pain y_3 = surprise

y_4 = anger y_5 = patience y_6 = impatience

y_7 = affection

Let the fuzzy relation $R \in Y \times X$ represent the degree to which the young man plans to respond to a given signal x with a response having the attribute y. This relation is given by the following matrix:

$$
\begin{array}{c}
\\
y_1 \\
y_2 \\
y_3 \\
y_4 \\
y_5 \\
y_6 \\
y_7
\end{array}
\begin{array}{cccccc}
x_1 & x_2 & x_3 & x_4 & x_5 & x_6 \\
\left[\begin{array}{cccccc}
.9 & 0 & .2 & 0 & 0 & 1 \\
0 & .9 & .1 & .2 & 1 & 0 \\
.1 & .9 & .2 & .9 & 1 & .3 \\
0 & .5 & 0 & .6 & .7 & 0 \\
.1 & 0 & .9 & 0 & 0 & .5 \\
0 & .3 & .2 & .3 & .4 & 0 \\
.9 & 0 & .9 & .3 & 0 & 1
\end{array}\right]
\end{array}
$$

Using Eq. (6.4), we can now calculate the response which the young man will make to the message M_2':

$$A = R \circ M_2' = .9/y_2 + .9/y_3 + .7/y_4 + .4/y_6.$$

The young man's response, therefore, will have the characteristics of a great deal of pain and surprise, a large degree of anger, and some impatience.

6.3 ENGINEERING

Applications of information theory in engineering have been extensive. In fact, both the classical measures of information, the Hartley and Shannon measures, were conceived in the context of engineering applications, particularly in the area of communication engineering.

Although the use of fuzzy set theory in engineering has been less extensive than the use of information theory, a broad spectrum of applications has occurred in the literature, ranging from civil engineering and architecture to automatic control and robotics. In this section, we focus on the area of fuzzy control, which is by far the most developed area of application of fuzzy set theory in engineering.

The purpose of automatic control is to achieve or maintain some desired state of a process or system by monitoring certain state variables and by taking appropriate control actions. Many industrial processes make use of automatic control in lieu of or in addition to a human operator. Aircraft autopilots and traffic-light control systems are further examples of common applications of automatic control. The design of an effective automatic controller, however, usually requires a precise mathematical model of the process or system involved. For many complex processes, the construction of such a model is made difficult or impossible due, for instance, to a nonlinear, time-varying nature or due to poor quality of the available measurements. In these cases, a human operator may offer superior control to that of a conventional automatic controller, apparently through the use of an imprecise and robust protocol. As an alternative to a precise model of the system, an imprecise description of the process or of the manner of control can often be articulated by the expert human operator; this linguistic description generally makes use of vague or fuzzy concepts. For instance, such heuristic control rules or "rules of thumb" may take the form

IF the temperature is very high

AND the pressure is slightly low

THEN the heat change should be slightly negative,

where temperature and pressure are the observed state variables of the process and heat change is the action to be taken by the controller. The vague terms *very high*, *slightly low*, and *slightly negative* can be conveniently represented by fuzzy sets defined on the universe of discourse of temperature values, pressure values, and heat change values, respectively. This type of linguistic rule has formed the basis for the design of several different prototype fuzzy controllers, many for industrial processes, including a cement kiln, heat-exchange process, and steam engine, as well as others. In addition, fuzzy controllers have been created for the control of automobile speed, traffic junctions, robots, and aircraft autopilots.

The operation of a fuzzy controller consists of four stages. First, a measurement must be taken of the state variables to be monitored during the process. Next, these values must be translated by a component known as a *condition interface* into fuzzy linguistic terms; these terms are specified by the membership functions of the fuzzy sets, which are defined on the appropriate universe of discourse. These linguistic terms are then used in the evaluation of the fuzzy control rules. The result of application of the rules is itself a fuzzy set defined on the universe of possible control actions. Finally, based on this fuzzy set, the component called an *action interface* must produce the crisp control action, which is then executed. The design of a fuzzy controller, therefore, requires specification of all the fuzzy sets and their membership functions defined for each input and output variable, compilation of an appropriate and complete set of heuristic control rules that operate on these fuzzy sets, and a determination of the method of choosing a crisp output action based on the fuzzy result generated by the control rules. Figure 6.2 depicts the five major components of a fuzzy controller.

Fuzzy control rules are evaluated by means of the compositional rule of inference. This rule is essentially a fuzzy generalization of the traditional modus ponens rule of logical inference. Given statements of the form "If A then B," in which A and B are fuzzy sets defined on universal sets X of inputs and Y of outputs, respectively, then the rule base composed of these statements can be fully captured by a fuzzy relation R defined on $X \times Y$. Each statement S is itself represented by a relation specified by

$$\mu_S(x, y) = \min(\mu_A(x), \mu_B(y)),$$

where $x \in X$ and $y \in Y$. The relations S are then aggregated to form the rule base

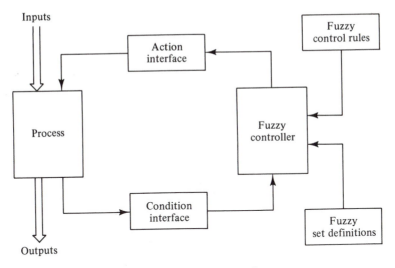

Figure 6.2. Fuzzy controller.

relation R. This is accomplished by specifying the grade of membership of (x, y) in R to be the maximum of its membership grades in any of the fuzzy relations S. This corresponds to concatenating the control statements with "else." Once the relation R is defined, then an actual input fuzzy set A' can be used to compute the resulting output fuzzy set B', which is induced on Y by

$$B' = A' \circ R.$$

The max-min composition is used here such that

$$\mu_{B'}(y) = \max_{x \in X}[\min(\mu_{A'}(x), \mu_R(x, y))].$$

This inference rule can be extended for any number of fuzzy inputs and outputs.

One method of assessing the effectiveness of control involves examining the shape of the membership functions for the fuzzy control actions calculated. A single peak indicates one dominant control rule, whereas the appearance of two distinct peaks indicates the presence of contradictory rules. A very low maximum membership grade indicates that some control rules are missing.

Fuzzy Traffic Control

Automatic control of traffic signals based on a stochastic model has long been in use in many cities. Simulations of an alternative approach that makes use of fuzzy heuristic control rules have shown significantly improved delay times over the conventional methods of control. These fuzzy control rules mimic the protocol used by a human operator to decide the time intervals of opposing green lights. For instance, if there are *few* cars waiting for an east-west green light, the operator may delay changing the north-south light to red until the north-south traffic is *thinner* or until the east-west light seems to have been red *too long*. The terms *few*, *thinner*, and *too long* constitute labels of fuzzy sets defined on the universal set of the number of cars (for *few* and *thinner*) and time (for *too long*). In addition, the appropriateness of these rules may change given different traffic conditions. For example, the fact that *some* cars are waiting at a red light during rush hour may call for a different response than if *some* cars are waiting at three in the morning. Optimum control, therefore, would allow for flexibility in the set of heuristics employed.

A *fuzzy logic controller* (FLC) was designed and tested by Mamdani and Pappis [1977] for a single intersection of two one-way streets. The fuzzy control rules employed have the form

$$
\begin{aligned}
\text{IF} \quad & T = medium \\
\text{AND} \quad & A = mt(medium) \\
\text{AND} \quad & Q = lt(small) \\
\text{THEN} \quad & E = medium
\end{aligned}
$$

ELSE

$$
\begin{aligned}
\text{IF} \quad & T = long \\
\text{AND} \quad & A = mt(many) \\
\text{AND} \quad & Q = lt(medium) \\
\text{THEN} \quad & E = long
\end{aligned}
$$

The input variables T, A, and Q represent the time since the last change, the number of arrivals through the green light, and the number of cars waiting in a queue at the opposing red light, respectively. The output variable E represents the extension of time before the light is to be changed. The symbols mt and lt stand for the terms *more than* and *less than*, respectively. Each term such as *more than many* or *less than small* is modeled as a fuzzy set on the appropriate universe of discourse. If the membership grade of variable x in the fuzzy set A (where A represents *many*, or *small*, or *few*, etc.) is given by $\mu_A(x)$, then the membership grade of x in the set *more than A* is given by

$$
\mu_{mt(A)}(x) = \begin{cases} 0 & \text{for } x \le x_0 \\ 1 - \mu_A(x) & \text{for } x > x_0, \end{cases}
$$

where x_0 is the element attaining the maximum membership grade in A. Likewise, the membership grade of x in the fuzzy set *less than A* is given by

$$
\mu_{lt(A)}(x) = \begin{cases} 0 & \text{for } x \ge x_0 \\ 1 - \mu_A(x) & \text{for } x < x_0. \end{cases}
$$

Every 10 s, a different set of five control rules is evaluated by the controller for times equal to 1 through 10 s in order to determine the time of continued extension of the current state of the light. Detection pads are assumed to be placed at a sufficient distance from the light to allow projection of arrivals for the full 10 s. This controller allows a maximum time of 57 s for a green light. In order to select

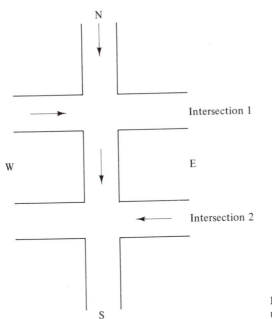

Figure 6.3. Two traffic intersections under fuzzy control.

a crisp control action, a decision table is used, which assigns to each crisp extension time the degree of confidence with which this action could be taken given the actual conditions. The controller chooses the extension time that maximizes the confidence level. Simulations of this FLC and a standard *vehicle actuated controller* (VC) showed a consistent reduction in delay times by the FLC as compared to those of the VC, although some trial and error was needed in order to create a set of satisfactory fuzzy control rules.

This concept of an FLC traffic controller was extended for use in two consecutive east-west intersections along a north-south arterial road by Nakatsuyama, Nagahashi, and Nishizuka [1984]. Figure 6.3 shows the configuration of these traffic junctions. Simulations were run to compare the performance of the FLC and the VC for these two consecutive intersections. The results of this simulation are given in Table 6.1. Except at the heaviest of rates of traffic flow (1440 vehicles per hour), the FLC produced impressively shorter average delay times than the VC.

This study further expanded the FLC concept to that of a *fuzzy logic phase controller* (FPC). A phase controller at traffic intersection 2 performs light changes at times offset from those changes performed by the FLC at intersection 1, thus coordinating the two systems. In other words, if the light at intersection 1 changes from red to green at time t, then the light at intersection 2 will change to green at time $t + n$, where n is the offset computed by fuzzy control rules based on traffic conditions. Likewise, if the light at intersection 1 changes from green to amber at time s, then intersection 2 will change to amber at time $s + m$. Thus, the FPC uses fuzzy control rules to compute both the offset n, extending the red signal, and the offset m, extending the green signal. The FPC gave shorter average

TABLE 6.1. COMPARISON OF FUZZY LOGIC CONTROLLER (FLC) AND VEHICLE
ACTUATED CONTROLLER (VC) AT TWO CONSECUTIVE INTERSECTIONS

N-S traffic rate inter. 1 veh./h	E-W traffic rate inter. 1 veh./h	N-S traffic rate inter. 2 veh./h	E-W traffic rate inter. 2 veh./h	average delay s/veh VC	FLC
360	360	360	360	9.5	7.3
720	360	540	360	10.0	7.8
1080	360	720	360	10.8	8.4
1440	360	900	360	11.9	9.1
1800	360	1080	360	12.6	9.2
2160	360	1260	360	13.7	10.9
2520	360	1440	360	16.4	15.2
720	720	720	720	11.9	9.1
1080	720	900	720	14.0	9.9
1440	720	1080	720	16.4	11.6
1800	720	1260	720	19.0	14.5
2160	720	1440	720	24.5	20.8
1080	1080	1080	1080	17.4	13.3
1440	1080	1260	1080	22.4	17.2
1880	1080	1440	1080	30.2	23.7
1440	1440	1440	1440	29.2	30.6

delay times in all cases except those in which the north-south traffic was the same at intersections 1 and 2.

The FLC and FPC have different ranges of superior performance based on the amount of traffic flow. Therefore, this study proposes an *interchange controller* (IC), which examines the current traffic rates and determines the control of intersection 2 by an FLC or FPC appropriately. This decision is made at a fixed time after a signal change is executed by the FLC at intersection 1; it is based on inputs A_1 and A_2 of arrivals in the north-south and the east-west directions, respectively. These control rules have the form

$$IF \quad A_1 = \textit{not small}$$

$$AND \quad A_2 = \textit{medium}$$

$$THEN \quad C = \textit{FPC}$$

$$ELSE$$

$$IF \quad A_1 = \textit{not large}$$

$$AND \quad A_2 = \textit{any}$$

$$THEN \quad C = \textit{FLC}$$

and so on. Again, *not large*, *medium*, and *any* are fuzzy sets defined on the universal set of arrivals at the intersections. Simulations were run to compare the performance of the FLC alone, the FPC alone, and the IC that alternates the use of the FLC and FPC at intersection 2. For sufficiently large east-west traffic volume at intersection 2, the IC offered shorter delay times than either the FLC or the FPC alone. Simulations showed, in fact, that the cooperation of the FLC and FPC offered by the IC performs quite well when the rapid traffic-flow changes common at rush hours occur.

Fuzzy Aircraft Control

The final approach and landing of an aircraft involve maneuvering and maintaining flight in an appropriate course to the airport and then along the optimum "glide path" trajectory to touchdown on the runway. This path is usually provided to the pilot by an *instrument landing system* (ILS), which transmits two radio signals to the aircraft as a navigational aid. These orthogonal radio beams are known as the *localizer* and *glide slope* and are transmitted from the ends of the runway in order to provide the approaching aircraft with the correct trajectory for landing. The pilot executing such a landing must monitor cockpit instruments that display the position of the aircraft relative to the desired flight path and make appropriate corrections to the controls. These corrections are often accomplished by successive approximation and can be described linguistically as heuristic "rules of thumb" by expert pilots. Larkin [1985] designed and implemented a fuzzy autopilot controller, which demonstrated good results when tested on a flight simulator. This autopilot controls final approach and landing of the aircraft until just prior to touchdown. Three state inputs are monitored by the controller: rate of aircraft descent (in feet per minute), deviation from desired speed (in miles per hour), and deviation from glide-slope trajectory (in degrees). Outputs or control actions are engine-speed change (in rotations per minute) and elevator-angle change (in degrees). All inputs and outputs are described with the use of five fuzzy sets: *negative big* (NB), *negative medium* (NM), *insignificant change* (IC), *positive medium* (PM), and *positive big* (PB). Rules were elicited from an experienced pilot and have the form

IF rate of descent = PM, airspeed = NB, and glide-slope = PB

THEN rpm change = PM and elevator angle change = IC.

Two different methods are employed for selecting crisp control actions from the fuzzy directions provided by the control rules. One of these calculates the center of gravity g of the fuzzy set C indicating the control action as follows:

$$g = \sum_{x \in X} \frac{x \cdot \mu_C(x)}{\sum_{x \in X} \mu_C(x)},$$

where X is the appropriate universal set of output actions, each represented by a control change magnitude. The value g is then the crisp control action executed. The second method calculates the mean of all the maxima of the fuzzy set C. In other words, if the set possesses a single maximum, this value is the control action taken. If several local maxima appear in the set, then the control action to be executed is the mean value of these maxima.

The major challenges in the design of fuzzy controllers involve the definition of appropriate fuzzy sets at the correct quantization level and the specification of the control rules themselves. Although no formal procedure for rule selection exists, the possible sources for the control rules include the experience and actions of an expert operator, the knowledge of the control engineer, a fuzzy model of the operator's control actions, or a fuzzy model of the process itself.

Prototype applications of fuzzy controllers have, on the whole, achieved positive results with performance as good, or in some cases, better than conventional controllers. Proposals for future advancements in the area of fuzzy control include the addition of learning capabilities and the use of more complex rule structures and decision-making processes similar to those found in expert systems.

6.4 MEDICINE

Imprecision and uncertainty play a large role in the field of medicine. The field has, for this reason, become one of the most fruitful and active areas of application for the theory of fuzzy sets and the theory of evidence. Overviews of these applications are presented by Adlassnig [1982] and Gupta, Martin-Clouaire, and Nikiforuk [1984]. Within this field, it is the uncertainty found in the process of diagnosis of disease that has most frequently been the focus of these applications. With the increased volume of information available to physicians from new medical technologies, the process of classifying different sets of symptoms under a single name and determining the appropriate therapeutic actions becomes increasingly difficult. A single disease may manifest itself quite differently in different patients and at different disease stages. Further, a single symptom may be indicative of several different diseases, and the presence of several diseases in a single patient may disrupt the expected symptom pattern of any one of them. The best and most useful descriptions of disease entities often use linguistic terms that are irreducibly vague. Although medical knowledge concerning the symptom-disease relationship constitutes one source of imprecision and uncertainty in the diagnostic process, the knowledge concerning the state of the patient constitutes another. The physician generally gathers knowledge about the patient from the past history, physical examination, laboratory test results, and other investigative procedures such as X-ray and ultrasonics. The knowledge provided by each of these sources carries with it varying degrees of uncertainty. The past history offered by the patient may be subjective, exaggerated, underestimated, or incomplete. Mistakes may be made in the physical examination and symptoms may be overlooked. The measurements provided by laboratory tests are often of limited precision and the exact borderline between normal and pathological is often unclear. X-rays and other similar procedures require a correct interpretation of the results. Thus, the state and symptoms of the patient can be known by the physician with only a limited degree of precision. In the face of the uncertainty concerning the observed symptoms of the patient as well as the uncertainty concerning the relation of the symptoms to a disease entity, it is nevertheless crucial that the physician determine the diagnostic label that will entail the appropriate therapeutic regimen. The desire to better understand and teach this difficult and important technique of medical diagnosis has prompted attempts to model this process, most recently, with the use of fuzzy sets. These models vary in the degree to which they attempt to deal with different complicating aspects of medical diagnosis such as the relative importance of symptoms, the varied symptom patterns of different disease stages, relations between diseases themselves, and the stages of hypothesis formation, preliminary diagnosis, and final diagnosis within the diagnostic process itself. These models also form the basis for computerized medical expert systems, which are usually designed to aid the physician in the diagnosis of some specified category of diseases.

The fuzzy set framework has been utilized in several different approaches to modeling the diagnostic process. In the approach formulated by Sanchez [1979b], the physician's medical knowledge is represented as a fuzzy relation between symptoms and diseases. Thus, given the fuzzy set A of the symptoms observed in the patient and the fuzzy relation R representing the medical knowledge that relates the symptoms in set S to the diseases in set D, then the fuzzy set B of the possible diseases of the patient can be inferred by means of the compositional rule of inference

$$B = A \circ R \tag{6.5}$$

or

$$\mu_B(d) = \max_{s \in S}[\min(\mu_A(s), \mu_R(s, d))]$$

for each $d \in D$. This max-min composition corresponds to the fuzzy conditional statement "If A then B by R." The membership grades of observed symptoms in fuzzy set A may represent the degree of certainty of the presence of the symptom or its severity. The membership grades in fuzzy set B denote the degree of certainty with which we can attach each possible diagnostic label to the patient. The fuzzy relation R of medical knowledge should constitute the greatest relation such that given the fuzzy relation Q on the set P of patients and S of symptoms and the fuzzy relation T on the set P of patients and D of diseases, then

$$T = Q \circ R. \tag{6.6}$$

Thus, relations Q and T may represent, respectively, the symptoms that were present and diagnoses that were consequently made for a number of known cases. Figure 6.4 summarizes the meanings and uses of fuzzy relations Q, T, and R and fuzzy sets A and B. By solving the fuzzy relation equation (6.6) for R, the accumulated medical experience can be used to specify the relation between symptoms and diseases that was evidenced in the previous diagnoses. The method of solving Eq. (6.6) differs from the method presented in Sec. 3.8 (which is applicable only for solving Eq. (6.6) for relation Q); this method is discussed by Sanchez [1977]. The maximal solution to Eq. (6.6) must be chosen for R in order to avoid arriving at a relation that is more specific than our information warrants. However, this can lead to cases in which R shows more symptom-disease association than exists in reality. Therefore, it may be necessary to interpret the results of applying

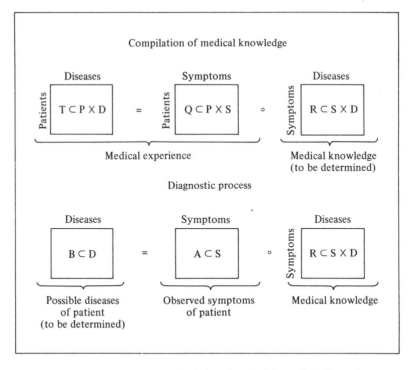

Figure 6.4. Fuzzy sets and relations involved in medical diagnosis.

relation R to a specific set of symptoms as a diagnostic hypothesis rather than as a confirmed diagnosis. Adlassnig and Kolarz [1982] and Adlassnig [1986] elaborate on this relational model in the design of CADIAG-2, a computerized system for diagnosis assistance. We illustrate their approach with a somewhat simplified version of part of this design.

The model proposes two types of relations to exist between symptoms and diseases: an occurrence relation and a confirmability relation. The first provides knowledge about the tendency or frequency of appearance of a symptom when the specific disease is present; it corresponds to the question, "How often does symptom s occur with disease d?" The second relation describes the discriminating power of the symptom to confirm the presence of the disease; it corresponds to the question, "How strongly does symptom s confirm disease d?" The distinction between occurrence and confirmability is useful because a symptom may be quite likely to occur with a given disease but may also commonly occur with several other diseases, therefore limiting its power as a discriminating factor among them. Another symptom, on the other hand, may be relatively rare with a given disease, but its presence may nevertheless constitute almost certain confirmation of the presence of the disease.

For this example, let S denote the crisp universal set of all symptoms, D be the crisp universal set of all diseases, and P be the crisp universal set of all patients. Let us define a fuzzy relation R_S on the set $P \times S$ in which membership grades

$\mu_{RS}(p, s)$ (where $p \in P$, $s \in S$) indicate the degree to which the symptom s is present in patient p. For instance, if s represents the symptom of increased potassium level and the normal test result range is roughly 3.5 to 5.2, then a test result of 5.2 for patient p could lead to a membership grade $\mu_{RS}(p, s) = .5$. Let us further define a fuzzy relation R_O on the universal set $S \times D$, where $\mu_{RO}(s, d)$ ($s \in S$, $d \in D$) indicates the frequency of occurrence of symptom s with disease d. Let R_C also be a fuzzy relation on the same universal set, where $\mu_{RC}(s, d)$ corresponds to the degree to which symptom s confirms the presence of disease d.

In this example, we will determine the fuzzy occurrence and confirmability relations from expert medical documentation. Since this documentation usually takes the form of statements such as "Symptom s seldom occurs in disease d" or "Symptom s always indicates disease d," we assign membership grades of 1, .75, .5, .25, and 0 in fuzzy sets R_O and R_C for the linguistic terms *always*, *often*, *unspecific*, *seldom*, and *never*, respectively. We use a concentration operation to model the linguistic modifier *very* such that

$$\mu_{very\ A}(x) = \mu_A^2(x).$$

Assume that the following medical documentation exists concerning the relations of symptoms s_1, s_2, and s_3 to diseases d_1 and d_2:

- Symptom s_1 occurs very seldom in patients with disease d_1.
- Symptom s_1 often occurs in patients with disease d_2 but seldom confirms the presence of disease d_2.
- Symptom s_2 always occurs with disease d_1 and always confirms the presence of disease d_1; s_2 never occurs with disease d_2 and (obviously) its presence never confirms disease d_2.
- Symptom s_3 very often occurs with disease d_2 and often confirms the presence of d_2.
- Symptom s_3 seldom occurs in patients with disease d_1.

All missing relational pairs of symptoms and diseases are assumed to be *unspecified* and are given a membership grade of .5. From our medical documentation we construct the following matrices of relations R_O, $R_C \in S \times D$:*

$$\mathbf{R}_O = \begin{array}{c} s_1 \\ s_2 \\ s_3 \end{array} \begin{array}{cc} d_1 & d_2 \\ \left[\begin{array}{cc} .06 & .75 \\ 1 & 0 \\ .25 & .56 \end{array}\right] \end{array}$$

$$\mathbf{R}_C = \begin{array}{c} s_1 \\ s_2 \\ s_3 \end{array} \begin{array}{cc} d_1 & d_2 \\ \left[\begin{array}{cc} .5 & .25 \\ 1 & 0 \\ .5 & .75 \end{array}\right] \end{array}$$

* For the sake of simplicity, we denote the matrix representation of any relation R by $\mathbf{R}$.

Now assume that we are given a fuzzy relation R_S specifying the degree of presence of symptoms s_1, s_2, and s_3 for three patients p_1, p_2, and p_3 as follows:

$$\mathbf{R}_S = \begin{array}{c} \\ p_1 \\ p_2 \\ p_3 \end{array} \begin{array}{ccc} s_1 & s_2 & s_3 \\ \left[\begin{array}{ccc} .4 & .8 & .7 \\ .6 & .9 & 0 \\ .9 & 0 & 1 \end{array}\right] \end{array}$$

Using relations R_S, R_O, and R_C, we can now calculate four different indication relations defined on the set $P \times D$ of patients and diseases. The first is the occurrence indication R_1 defined as

$$R_1 = R_S \circ R_O.$$

For our example, R_1 is given by the following matrix:

$$\mathbf{R}_1 = \begin{array}{c} \\ p_1 \\ p_2 \\ p_3 \end{array} \begin{array}{cc} d_1 & d_2 \\ \left[\begin{array}{cc} .8 & .56 \\ .9 & .6 \\ .25 & .75 \end{array}\right] \end{array}$$

The confirmability indication relation R_2 is calculated by

$$R_2 = R_S \circ R_C;$$

this results in

$$\mathbf{R}_2 = \begin{array}{c} \\ p_1 \\ p_2 \\ p_3 \end{array} \begin{array}{cc} d_1 & d_2 \\ \left[\begin{array}{cc} .8 & .7 \\ .9 & .25 \\ .5 & .75 \end{array}\right] \end{array}$$

The nonoccurrence indication R_3 is defined as

$$R_3 = R_S \circ (1 - R_O)$$

and specified here by

$$\mathbf{R}_3 = \begin{array}{c} \\ p_1 \\ p_2 \\ p_3 \end{array} \begin{array}{cc} d_1 & d_2 \\ \left[\begin{array}{cc} .7 & .8 \\ .6 & .9 \\ .9 & .44 \end{array}\right] \end{array}$$

Finally, the nonsymptom indication R_4 is given by

$$R_4 = (1 - R_S) \circ R_O$$

and equals

$$\mathbf{R}_4 = \begin{array}{c} \\ p_1 \\ p_2 \\ p_3 \end{array} \begin{array}{cc} d_1 & d_2 \\ \left[\begin{array}{cc} .25 & .6 \\ .25 & .56 \\ 1 & .1 \end{array}\right] \end{array}$$

From these four indication relations we may draw different types of diagnostic conclusions. For instance, we may make a confirmed diagnosis of a disease d for patient p if $\mu_{R_2}(p, d) = 1$. Although this is not the case for any of our three patients, R_2 does seem to indicate, for instance, that disease d_1 is strongly confirmed for patient p_2. We may make an excluded diagnosis for a disease d in patient p if $\mu_{R_3}(p, d) = 1$ or if $\mu_{R_4}(p, d) = 1$. In our example, we may exclude disease d_1 as a possible diagnosis for patient p_3. Finally, we may include in our set of diagnostic hypotheses for patient p any disease d such that the inequality

$$.5 < \max[\mu_{R_1}(p, d), \mu_{R_2}(p, d)]$$

is satisfied. In our example, both diseases d_1 and d_2 are suitable diagnostic hypotheses for patients p_1 and p_2, whereas the only acceptable diagnostic hypothesis for patient p_3 is disease d_2. Our three types of diagnostic results, therefore, seem to point to the presence of disease d_2 in patient p_3 and disease d_1 in patient p_2, whereas the symptoms of patient p_1 do not strongly resemble the symptom pattern of either disease d_1 or d_2 alone.

The actual CADIAG-2 system incorporates relations not only between symptoms and diseases but also between diseases themselves, between symptoms themselves, and between combinations of symptoms and diseases. Partial testing of the system on patients with rheumatological diseases produced an accuracy of 94.5 percent in achieving a correct diagnosis.

An alternative model of medical diagnosis proposed by Smets [1981b] combines the use of fuzzy sets and belief functions to handle two different sources of uncertainty in the diagnostic process. In this model, diagnostic entities or disease labels are fuzzy; each diagnostic label describes a fuzzy set A defined on the universal set X of patients that represents the class of patients with the disease. Complete evidence concerning the state of the patient in question would allow the physician to assign a membership grade for the patient in each of the possible fuzzy sets corresponding to the diagnostic labels. However, because the physician may be ignorant of the existence or importance of certain key pieces of information or because he or she may be unable to obtain the information either at all or to a sufficient degree of precision, only incomplete and uncertain information regarding the patient's state is available. Let X denote all the available information concerning the patients, Y be all the unavailable information, and $X(p) = x$, $Y(p) = y$ be the available and unavailable information, respectively, that specifically concerns patient p. Then the membership grade of patient p in any diagnostic class A depends on the values of x and y such that

$$\mu_A(p) = f(x, y),$$

where f is some appropriate function from $X \times Y$ to $[0, 1]$. Although the value of x is given, the physician has only a degree of belief concerning the value of y. Therefore, the physician can, at best, express only a degree of belief, $Bel(A)$, indicating the expectation concerning the patient's grade of membership in each fuzzy diagnostic category A. A generalization of the probabilistic Bayes' theorem of inference to the theory of evidence can then be used to calculate a degree of belief that the patient belongs to any particular diagnostic category given the degree of belief relative to the observation of symptoms indicative of that specific pathology. Through the use of both fuzzy sets and degrees of belief, this model deals with two basic sources of uncertainty in the diagnostic process: the vagueness inherent in the diagnostic labels themselves and the limited accuracy and completeness of information or evidence concerning the status of the patient.

Another alternative approach to modeling the medical diagnostic process utilizes fuzzy cluster analysis. This type of technique is used by Fordon and Bezdek [1979], and Esogbue and Elder [1979, 1980, 1983]. The technique of clustering examines the elements of some universal set and groups them according to similarity. Thus, elements grouped in one cluster are similar to each other and dissimilar to the members of other clusters. Since the boundaries of these clusters are not precisely defined, each cluster is, in effect, a fuzzy set in which the grade of membership of any element indicates the similarity of that element to other members of the cluster. Any particular element can, of course, belong with varying degrees to several different clusters. (References for fuzzy clustering analysis are provided in Sec. 6.8.)

Models of medical diagnosis that use cluster analysis usually perform a clustering algorithm on the set of patients by examining the similarity of the presence and severity of symptom patterns exhibited by each. (The severity of the symptoms present can be designated with degrees of membership in fuzzy sets representing each symptom category.) Often the similarity measure is computed between the symptoms of the patient in question and the symptoms of a patient possessing the prototypical symptom pattern for each possible disease. The patient to be diagnosed is then clustered to varying degrees with the prototypical patients whose symptoms are most similar. The most likely diagnostic candidates are those disease clusters in which the patient's degree of membership is the greatest.

Several different methods of fuzzy clustering exist. One group of common methods uses some form of distance measure to determine the similarity between observed attributes (symptoms) and those present in the existing diagnostic clusters. We use a simplified adaptation of the method employed by Esogbue and Elder [1979, 1980] to illustrate this technique.

Let us assume that we are given a patient x who displays the symptoms s_1, s_2, s_3, and s_4 at the levels of severity given by the following vector:

$$\begin{array}{cccc} s_1 & s_2 & s_3 & s_4 \\ \mathbf{x} = [\ .1 & .7 & .4 & .6\ ] \end{array}$$

Let $\mu_x(s_i) \in [0, 1]$ denote the grade of membership in the fuzzy set characterizing patient x and defined on the set $S = \{s_1, s_2, s_3, s_4\}$, which indicates the severity level of the symptom s_i for the patient. We must determine a diagnosis for this patient among three possible diseases d_1, d_2, and d_3. Each of these diseases is described by a matrix giving the upper and lower bounds of the normal range of severity of each of the four symptoms that can be expected in a patient with the disease. The diseases d_1, d_2, and d_3 are described in this way by the matrices

$$\mathbf{d}_1 = \begin{array}{c} \\ \text{lower} \\ \text{upper} \end{array} \begin{array}{cccc} s_1 & s_2 & s_3 & s_4 \\ \begin{bmatrix} 0 & .6 & .5 & 0 \\ .2 & 1 & .7 & 0 \end{bmatrix} \end{array}$$

$$\mathbf{d}_2 = \begin{array}{c} \\ \text{lower} \\ \text{upper} \end{array} \begin{array}{cccc} s_1 & s_2 & s_3 & s_4 \\ \begin{bmatrix} 0 & .9 & .3 & .2 \\ 0 & 1 & 1 & .4 \end{bmatrix} \end{array}$$

$$\mathbf{d}_3 = \begin{array}{c} \\ \text{lower} \\ \text{upper} \end{array} \begin{array}{cccc} s_1 & s_2 & s_3 & s_4 \\ \begin{bmatrix} 0 & 0 & .7 & 0 \\ .3 & 0 & .9 & 0 \end{bmatrix} \end{array}$$

Let $\mu_{d_jl}(s_i) \in [0, 1]$ denote the lower bound of the symptom i for disease j, and let $\mu_{d_ju}(s_i) \in [0, 1]$ denote the upper bound of the fuzzy symptom i for disease j. We further define a fuzzy relation W on the set of symptoms and diseases that specifies the pertinence or importance of symptom s in the diagnosis of disease d. The relation W of these weights of relevance is given by

$$\mathbf{W} = \begin{array}{c} \\ s_1 \\ s_2 \\ s_3 \\ s_4 \end{array} \begin{array}{ccc} d_1 & d_2 & d_3 \\ \begin{bmatrix} .4 & .8 & 1 \\ .5 & .6 & .3 \\ .7 & .1 & .9 \\ .9 & .6 & .3 \end{bmatrix} \end{array}$$

Let $\mu_W(s_i, d_j)$ denote the weight of symptom s_i for disease d_j. In order to diagnose the patient x, we use a clustering technique to determine to which diagnostic cluster (as specified by matrices d_1, d_2, and d_3) the patient is most similar. This clustering is performed by computing a similarity measure between the patient's symptoms and those typical of each disease d_j. To compute this similarity, we use a distance measure based on the Minkowski distance that is appropriately modified; it is given by the formula

$$Dp(d_j, x) = [\sum_{i \in A_j} | \mu_W(s_i, d_j)(\mu_{d_jl}(s_i) - \mu_x(s_i)) |^p$$
$$+ \sum_{i \in B_j} | \mu_W(s_i, d_j)(\mu_{d_ju}(s_i) - \mu_x(s_i) |^p]^{1/p} \qquad (6.7)$$

where

$$A_j = \{i \mid \mu_x(s_i) < \mu_{d_jl}(s_i), 1 \le i \le m\}$$
$$B_j = \{i \mid \mu_x(s_i) > \mu_{d_ju}(s_i), 1 \le i \le m\}$$

where m equals the total number of symptoms. For this example, we give a value of 2 to the parameter p of the distance measure, thus creating a modified Euclidean metric. We use Eq. (6.7) with $p = 2$ to calculate the similarity between patient

x and diseases d_1, d_2, and d_3 as follows:

$$D_2(d_1, x) = [\,|\,(.7)(.5 - .4)\,|^2 + |\,(.9)(0 - .6)\,|^2]^{1/2}$$

$$= .54$$

$$D_2(d_2, x) = [\,|\,(.6)(.9 - .7)\,|^2 + |\,(.8)(0 - .1)\,|^2$$

$$+ |\,(.6)(.4 - .6)\,|^2]^{1/2}$$

$$= .19$$

$$D_2(d_3, x) = [\,|\,(.9)(.7 - .4)\,|^2 + |\,(.3)(0 - .7)\,|^2$$

$$+ |\,(.3)(0 - .6)\,|^2]^{1/2}$$

$$= .39$$

The most likely disease candidate is the one for which the similarity measure attains the minimum value; in this case, the patient's symptoms are most similar to those typical of disease d_2.

6.5 MANAGEMENT AND DECISION MAKING

The subject of the field of decision making is, as the name suggests, the study both of how decisions are actually made and how they can be made better or more successfully. Much of the focus in this field has been in the area of management, in which the decision-making process is of key importance for functions such as inventory control, investments, personnel actions, new-product development, allocation of resources, and many others. Decision making itself, however, is broadly defined to include any choice or selection of alternatives and is therefore of importance in many fields in both the "soft" social sciences and the "hard" natural and engineering sciences.

Applications of fuzzy sets within the field of decision making have, for the most part, consisted of extensions or "fuzzifications" of the classical theories of decision making. While decision making under conditions of risk and uncertainty have been modeled by probabilistic decision theories and by game theories, fuzzy decision theories attempt to deal with the vagueness or fuzziness inherent in subjective or imprecise determinations of preferences, constraints, and goals. In this section, we briefly introduce some of the simple applications of fuzzy sets in some selected areas of decision making.

Classical decision making generally deals with a set of alternatives comprising the decision space, a set of states of nature comprising the state space, a relation indicating the state or outcome to be expected from each alternative action, and, finally, a utility or objective function, which orders these outcomes according to their desirability. A decision is said to be made under conditions of *certainty* when the outcome for each action can be determined and ordered precisely. In this case, the alternative that leads to the outcome yielding the highest utility is chosen. A decision is made under conditions of *risk*, on the other hand, when the only available knowledge concerning the outcome states is their probability distributions. Again, this information can be used to optimize the utility function. When knowledge of the probabilities of the outcome states is unknown, decisions must be made under conditions of *uncertainty*. In this case, fuzzy decision theories may be used to accommodate this vagueness.

There exist several major approaches within the theories of classical crisp decision making. Decisions may, for instance, be considered to occur in a single stage or in multiple stages. The decision maker may be a single person, or a collection of multiple decision makers may be involved. The decision may involve the simple optimization of a utility function, an optimization under constraints, or an optimization given multiple criteria. The first of these three types of decision problems corresponds to statistical decision theory, the second to mathematical (linear or nonlinear) programming, and the last to theories of multicriteria decision making.

Fuzziness can be introduced at several points in the existing models of decision making. For instance, under a simple optimization of utility, any uncertainty concerning the states of the system can be handled by modeling the states as well as the utility assigned to each state with fuzzy sets. Bellman and Zadeh [1970] suggest a fuzzy model of decisions that must accommodate certain constraints C and goals G. Here, both constraints and goals are treated as fuzzy sets characterized by membership functions

$$\mu_C : X \rightarrow [0, 1]$$

and

$$\mu_G : X \rightarrow [0, 1],$$

where X is the universal set of alternative actions. The expected outcomes that result from these actions can, in many cases, be assumed to remain deterministic or probabilistic, thus restricting the introduction of fuzziness only to the goals and constraints themselves. This fuzziness allows the decision maker to frame the goals and constraints in vague, linguistic terms, which may more accurately reflect the actual state of knowledge or preference concerning these. The membership function of the fuzzy goal in this case serves much the same purpose as a utility or objective function that orders the outcomes according to preferability. Unlike the classical theory of decision making under constraints, however, in which constraints are defined on the set X of alternatives and goals are modeled as performance functions from X to another space, the symmetry between the goals and constraints under this fuzzy model allows them to be treated in exactly the same manner. This model can be extended to allow goals and constraints to be defined on different universal sets, for instance, the set X of possible actions and the set Y of possible effects or outcomes. In this case, the fuzzy constraints may be defined on the set X and the fuzzy goals on the set Y such that

$$\mu_C : X \rightarrow [0, 1]$$

and

$$\mu_G : Y \rightarrow [0, 1].$$

A function f can then be defined as a mapping from the set of actions X to the set of outcomes Y,

$$f : X \rightarrow Y,$$

such that a fuzzy goal G defined on set Y induces a corresponding fuzzy goal G' on set X. Thus,

$$\mu_{G'}(x) = \mu_G(f(x)).$$

A fuzzy decision D may then be defined as the choice that satisfies both the goals G and the constraints C. If we interpret this as a logical "and," we can model it with the intersection of the fuzzy sets G and C,

$$D = G \cap C,$$

which can easily be extended for any number of goals and constraints. If the classical fuzzy set intersection is used, the fuzzy decision D is then specified by the membership function

$$\mu_D(x) = \min[\mu_G(x), \mu_C(x)],$$

where $x \in X$. This definition of the intersection does not allow, however, for any interdependence, interaction, or trade-off between the goals and constraints under consideration. For many decision applications, this lack of compensation may not be appropriate; the full compensation or trade-off offered by the union operation that corresponds to the logical "or" (the max operator) may be inappropriate as well. Therefore, an alternative fuzzy set intersection or aggregation operation may be used to reflect a situation in which some degree of positive compensation exists among the goals and constraints.

This fuzzy model can be further extended to accommodate the relative importance of the various goals and constraints by the use of weighting coefficients. In this case, the fuzzy decision D can be arrived at by a convex combination of the n weighted goals and m weighted constraints such that

$$\mu_D(x) = \sum_{i=1}^{n} u_i \mu_{G_i}(x) + \sum_{j=1}^{m} v_j \mu_{C_j}(x)$$

where u_i and v_j are weights attached to each fuzzy goal G_i ($i \in \mathbb{N}_n$) and each fuzzy constraint C_j ($j \in \mathbb{N}_m$), respectively, such that

$$\sum_{i=1}^{n} u_i + \sum_{j=1}^{m} v_j = 1.$$

Once a fuzzy decision has been arrived at, it may be necessary to choose the "best" single crisp alternative from this fuzzy set. This may be accomplished in a straightforward manner by choosing the alternative $x \in X$ that attains the maximum membership grade in D. Since this method ignores information concerning any of the other alternatives, it may not be desirable in all situations. Methods that calculate the mean or center of gravity of the fuzzy set D may therefore be used instead. These concepts have been effectively utilized to extend conventional crisp mathematical programming into methods of fuzzy mathematical programming.

The fuzzy model of decision making proposed by Bellman and Zadeh [1970] can be illustrated by a simple example. Suppose we must choose one of four different possible jobs a, b, c, and d, the salaries of which are given by the function f such that

$$f(a) = 30,000$$
$$f(b) = 25,000$$
$$f(c) = 20,000$$
$$f(d) = 15,000$$

Our goal is to choose the job that will give us a high salary given the constraints that the job is interesting and within close driving distance. This first constraint of interest value is represented by the fuzzy set C_1 defined on our universal set of alternative jobs as follows:

$$C_1 = .4/a + .6/b + .8/c + .6/d.$$

Our second constraint concerning the driving distance to each job is defined by the fuzzy set C_2 such that

$$C_2 = .1/a + .9/b + .7/c + 1/d.$$

The fuzzy goal G of a high salary is defined on the universal set X of salaries by the membership function

$$\mu_G(x) = \begin{cases} 1 & \text{for } x > 40,000 \\ -.00125 \left(\dfrac{x}{1000} - 40\right)^2 + 1 & \text{for } 13,000 \le x \le 40,000 \\ 0 & \text{for } x < 13,000, \end{cases}$$

where $x \in X$, and the corresponding goal G' induced on the set of alternative jobs by the function f is given by

$$G' = .875/a + .7/b + .5/c + .2/d.$$

Taking the standard fuzzy set intersection of these three fuzzy sets, we obtain the fuzzy decision D, where

$$D = G' \cap C_1 \cap C_2 = .1/a + .6/b + .5/c + .2/d.$$

Finally, we take the maximum of this set to obtain alternative b as the choice that seems best to satisfy our goal and constraints. Note that no real distinction exists here between a goal and a constraint; that is, the two concepts are symmetric.

When decisions which are made by more than one person are modeled, two differences from the case of a single decision maker can be considered: first, the goals of the individual decision makers may differ such that each places a different ordering on the alternatives; second, the individual decision makers may have access to different information upon which to base their decision. Theories known as *n*-person game theories deal with both of these considerations, team theories of decision making deal only with the second, and group-decision theories deal only with the first.

A fuzzy model of group decision was proposed by Blin [1974] and Blin and Whinston [1973]. Here, each member of a group of *n* individual decision makers is assumed to have a reflexive, antisymmetric, and transitive preference ordering P_k, $k \in \mathbb{N}_n$, which totally or partially orders a set X of alternatives. A "social choice" function must then be found which, given the individual preference orderings, produces the most acceptable overall group preference ordering. Basically, this model allows for the individual decision makers to possess different aims and values while still assuming that the overall purpose is to reach a common, acceptable decision. In order to deal with the multiplicity of opinion evidenced in the group, the social preference S may be defined as a fuzzy binary relation with membership grade function

$$\mu_S : X \times X \to [0, 1],$$

which assigns the membership grade $\mu_S(x_i, x_j)$ indicating the degree of group preference of alternative x_i over alternative x_j. The expression of this group preference requires some appropriate means of aggregating the individual preferences. One simple method computes the relative popularity of alternative x_i over x_j by dividing the number of persons preferring x_i to x_j, denoted by $N(x_i, x_j)$, by the total number of decision makers, *n*. This scheme corresponds to the simple majority vote. Thus,

$$\mu_S(x_i, x_j) = \frac{N(x_i, x_j)}{n} . \tag{6.8}$$

Other methods of aggregating the individual preferences may be used to accommodate different degrees of influence exercised by the individuals in the group. For instance, a dictatorial situation can be modeled by the group preference relation S for which

$$\mu_S(x_i, x_j) = \begin{cases} 1 & \text{if } x_i \overset{k}{>} x_j \text{ for some individual } k \\ 0 & \text{otherwise,} \end{cases}$$

where $\overset{k}{>}$ represents the preference ordering of the one individual k who exercises complete control over the group decision.

Once the fuzzy relationship S has been defined, the final nonfuzzy group preference can be determined by converting S into its resolution form

$$S = \bigcup_\alpha \alpha S_\alpha,$$

which is the union of the crisp relations S_α comprising the α-cuts of the fuzzy relation S, $\alpha \in \Lambda_S$ (the level set of S), each scaled by α. Each value α essentially represents the level of agreement between the individuals concerning the particular crisp ordering S_α. One procedure that maximizes the final agreement level consists of intersecting the classes of crisp total orderings that are compatible with the pairs in the α-cuts S_α for increasingly smaller values of α until a single crisp total ordering is achieved. In this process, any pairs (x_i, x_j) that lead to an intransitivity are removed. The largest value α for which the unique compatible ordering on $X \times X$ is found represents the maximized agreement level of the group and the crisp ordering itself represents the group decision. This procedure is illustrated in the following example.

Assume that each individual of a group of eight decision makers has a total preference ordering $\mathbf{P}_i$ ($i \in \mathbb{N}_8$) on a set of alternatives $X = \{w, x, y, z\}$ as follows:

$$\mathbf{P}_1 = (w, x, y, z)$$

$$\mathbf{P}_2 = \mathbf{P}_5 = (z, y, x, w)$$

$$\mathbf{P}_3 = \mathbf{P}_7 = (x, w, y, z)$$

$$\mathbf{P}_4 = \mathbf{P}_8 = (w, z, x, y)$$

$$\mathbf{P}_6 = (z, w, x, y)$$

Using the membership function given in Eq. (6.8) for the fuzzy group preference ordering relation S (where $n = 8$), we arrive at the following fuzzy social preference relation:

$$S = \begin{array}{c} \\ w \\ x \\ y \\ z \end{array} \begin{array}{cccc} w & x & y & z \\ \left[\begin{array}{cccc} 0 & .5 & .75 & .625 \\ .5 & 0 & .75 & .375 \\ .25 & .25 & 0 & .375 \\ .375 & .625 & .625 & 0 \end{array} \right] \end{array}$$

The α-cuts of this fuzzy relation S are:

$$S_1 = \varnothing$$

$$S_{.75} = \{(w, y), (x, y)\}$$

$$S_{.625} = \{(w, z), (z, x), (z, y), (w, y), (x, y)\}$$

$$S_{.5} = \{(x, w), (w, x), (w, z), (z, x), (z, y), (w, y), (x, y)\}$$

$$S_{.375} = \{(z, w), (x, z), (y, z), (x, w), (w, x), (w, z),$$
$$(z, x), (z, y), (w, y), (x, y)\}$$

$$S_{.25} = \{(y, w), (y, x), (z, w), (x, z), (y, z), (x, w), (w, x),$$
$$(w, z), (z, x), (z, y), (w, y), (x, y)\}$$

We can now apply the procedure to arrive at the unique crisp ordering which constitutes the group choice. All total orderings on $X \times X$ are, of course, compatible with the empty set of S_1. The total orderings $O_{.75}$ that are compatible with the pairs in the crisp relation $S_{.75}$ are

$$O_{.75} = \{(z, w, x, y), (w, x, y, z), (w, z, x, y), (w, x, z, y),$$
$$(z, x, w, y), (x, w, y, z), (x, z, w, y), (x, w, z, y)\}.$$

Thus,

$$O_1 \cap O_{.75} = O_{.75}.$$

The orderings compatible with $S_{.625}$ are

$$O_{.625} = \{(w, z, x, y), (w, z, y, x)\}$$

and

$$O_1 \cap O_{.75} \cap O_{.625} = \{(w, z, x, y)\}.$$

Thus, the value .625 represents the group level of agreement concerning the social choice denoted by the total ordering (w, z, x, y).

6.6 COMPUTER SCIENCE

Applications of the mathematics of uncertainty and information within the field of computer science have been quite extensive, particularly in those endeavors concerned with the storage and manipulation of knowledge in a manner compatible with human thinking. This includes the construction of database and information storage and retrieval systems as well as the design of computerized expert systems. In this section, we give a brief overview and example of the application of fuzzy set theory in these two areas.

Database

The motivation for the application of fuzzy set theory to the design of databases and information storage and retrieval systems lies in the need to handle information that is less than ideal in the sense of being incomplete, indeterministic, contradictory, vague, imprecise, and so on. The database that can accommodate imprecise information can store and manipulate not only precise facts but also subjective expert opinions, judgments, and values that can be specified in linguistic terms. This type of information can be quite useful when the database is to be used as a decision aid in areas such as medical diagnosis, employment, investment, and geological exploration, where "soft," subjective, and imprecise data are not only common but quite valuable. In addition, it is also desirable to relieve the user of the constraint of having to formulate queries to the database in precise terms. Vague queries such as, "Which employment candidates are highly educated and moderately experienced?" or "What industries are forecasted to experience significant growth by a substantial number of experts?" often capture the relevant concerns of database users more accurately and easily than precise queries. It is important, however, that the database system incorporating imprecision be able to propagate appropriately the level of uncertainty associated with the data to the level of uncertainty associated with answers or conclusions based on the data. Precise answers should not be generated from imprecise data. An overview of some of the problems and possibilities in database technology and a discussion of the uses of imprecision in database design can be found in a paper by Gaines [1981].

Buckles and Petry [1982a, 1982b, 1983, 1984] present a model for a fuzzy relational database that contains as a special case the classical crisp model of a relational database. The model of a classical relational database consists of a series of *n*-dimensional relations conceptualized as tables. The columns of these tables correspond to fields or attributes and are usually called *domains*. Each domain is defined on an appropriate domain base (or universal) set. The rows are elements of the relation; they correspond to records or entries and are called *tuples*. Access to the database is accomplished through a relational algebra. This algebra consists of the procedural application of operations containing four basic elements: an operation name, the names of relations and the names of domains to be operated on, and an optional conditional expression. For instance, if our database contains a ternary relation STUDENT with domains NAME, ADDRESS, and MAJOR, we can obtain the names and addresses of all students whose major is computer science by constructing a new relation with domains NAME and ADDRESS as a projection of the original relation. The algebraic operation performing this task would be

Project(STUDENT:NAME,ADDRESS) where

MAJOR = ''computer science''

The algebra also contains other relational operations such as **complement**, **union**, **intersection**, or **join**, which perform the corresponding task on the relation and domains specified in order to produce the desired information.

The fuzzy relational database proposed by Buckles and Petry differs from this crisp model in two ways: first, elements of the tuples contained in the relations may be subsets of the domain universal set and, second, a similarity relation is defined on each domain universal set. The first qualification allows the elements of tuples to consist either of singletons of the domain universal sets (as in the conventional relational database model) or of crisp subsets of the domain universal sets as in the relation MARKETS with the domains AREA, SIZE, and POTENTIAL represented by the following table:

RELATION: MARKETS		
AREA	SIZE	POTENTIAL
East	*Large*	*Good*
Midwest	*{Large, Medium}*	*{Moderate, Good}*
South	*Small*	*{Good, Excellent}*

Domain values that are not singletons may indicate, for instance, the merging of the opinions or judgments of several experts.

The second qualification is based on the assumption that in the classical database model, a crisp equivalence relation is defined on each domain universal set which groups together elements which are strictly equivalent. This identity is utilized, for example, when redundant tuples are to be eliminated or ignored. Most often, the equivalence classes generated by this relation are simply the singletons of the universal set. In the fuzzy database model, this equivalence relation is generalized to a fuzzy similarity relation. This introduction of fuzziness provides an interesting element of flexibility, since the value or meaning structures of different individual database users may be reflected by modifying the domain similarity relations appropriately. The fuzzy relational algebra used to access this fuzzy database consists of the same four components as the conventional relational algebra and, in addition, allows for the specification of a threshold level defining the minimum acceptable degree of similarity between elements in some specified domain. In the special case of the conventional database, all threshold levels are implicitly assumed to be equal to 1, thus requiring strict equivalence for the merging or elimination of tuples. In the fuzzy database, tuples may be merged if they are considered sufficiently similar.

As an example of the use of this fuzzy database model and its associated fuzzy relational algebra, suppose our database contains the opinions of a group of experts on three policy options X, Y, and Z. Two relations are contained within the database: EXPERT, which has domains NAME and FIELD and which associates the name and field of each expert, and ASSESSMENT, which has domains OPTION, NAME, and OPINION and associates the name of each expert with their expressed opinions on the policy options. These two relations are specified in Table 6.2. In addition, the following similarity relation is defined for the domain OPINION on the domain universal set {*highly favorable* (HF), *favorable* (F), *slightly favorable* (SF), *slightly negative* (SN), *negative* (N), and *highly negative* (HN)}:

$$
\begin{array}{c@{\quad}c}
 & \begin{array}{cccccc} \text{HF} & \text{F} & \text{SF} & \text{SN} & \text{N} & \text{HN} \end{array} \\
\begin{array}{c} \text{HF} \\ \text{F} \\ \text{SF} \\ \text{SN} \\ \text{N} \\ \text{HN} \end{array} &
\left[\begin{array}{cccccc}
1 & .8 & .6 & .2 & 0 & 0 \\
.8 & 1 & .8 & .6 & .2 & 0 \\
.6 & .8 & 1 & .8 & .6 & .2 \\
.2 & .6 & .8 & 1 & .8 & .6 \\
0 & .2 & .6 & .8 & 1 & .8 \\
0 & 0 & .2 & .6 & .8 & 1
\end{array} \right]
\end{array}
$$

Crisp equivalence relations in which equivalence classes are singletons are assumed to be defined on domains NAME, FIELD, and OPTION.

Suppose now that our query to this fuzzy database consists of the following question:

> Which sociologists are in *considerable agreement* with Kass concerning policy option Y?

The first step is to retrieve the opinion of Kass concerning option Y. This is accomplished with the relational algebraic operation

(Project(Select ASSESSMENT where NAME = Kass and

OPTION = Y) over OPINION) giving R_1.

TABLE 6.2. EXAMPLES OF RELATIONS IN RELATIONAL DATABASE

Relation: EXPERT	
NAME	FIELD
Cohen	*Sociologist*
Fadem	*Economist*
Fee	*Attorney*
Feldman	*Economist*
Kass	*Physician*
Osborn	*Sociologist*
Schreiber	*Sociologist*
Specterman	*Sociologist*

Relation: ASSESSMENT		
OPTION	NAME	OPINION
X	Osborn	*Favorable*
X	Fee	*Negative*
X	Fadem	*Slightly favorable*
X	Feldman	*Highly favorable*
Y	Cohen	*Slightly negative*
Y	Osborn	*Slightly favorable*
Y	Fee	*Highly favorable*
Y	Schreiber	*Favorable*
Y	Kass	*Favorable*
Y	Fadem	*Negative*
Y	Specterman	*Highly favorable*
Y	Feldman	*Slightly negative*
Z	Osborn	*Negative*
Z	Kass	*Slightly negative*
Z	Fee	*Slightly favorable*

The resulting temporary relation R_1 on domain OPINION is given by

RELATION: R_1
OPINION
Favorable

The next step involves the selection of all sociologists from the table of experts. This is accomplished by the operation

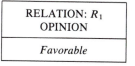

$$(\textbf{Project}(\textbf{Select}\ \text{EXPERT where FIELD} = Sociologist)$$

$$\text{over NAME}) \text{ giving } R_2.$$

Here R_2 is a temporary relation on domain NAME listing only sociologists. It is equal to

RELATION: R_2
NAME
Osborn
Schreiber
Cohen
Specterman

Next, temporary relation R_3 must be constructed on domains NAME and OPIN-

ION, which lists the opinions of the sociologists in R_2 about option Y. The algebraic expression accomplishing this is

(Project(Select(Join R_2 and ASSESSMENT over NAME)

where OPTION $=$ Y) over NAME, OPINION) giving R_3.

The relation R_3 is given by this table:

RELATION: R_3	
NAME	OPINION
Osborn	Slightly favorable
Schreiber	Favorable
Cohen	Slightly negative
Specterman	Highly favorable

Finally, we perform a join of relations R_1 (giving the opinion of Kass) and R_3 (giving the opinion of the sociologists) that specifies a threshold similarity level of .75 on the domain OPINION, which is chosen for this example to represent the condition of *considerable agreement*. The algebraic expression for this task is

(Join R_3 and R_1 over OPINION) with

THRES(OPINION) $\geq$ .75, and THRES(NAME) $\geq$ 0.

The specification of a zero similarity threshold level for NAME is necessary to allow the merging of names into sets as shown in the results given by

NAME	OPINION
{Osborn, Schreiber, Specterman}	{Slightly favorable, favorable, highly favorable}

Note that the resulting response is less precise and contains less information than a response of

NAME	OPINION
Osborn	Slightly favorable
Schreiber	Favorable
Specterman	Highly favorable

In this way, the uncertainty contained in the specification of *considerable agreement* and in the similarity defined over the possible opinions is propagated to the response given. Buckles and Petry [1983] present an interesting investigation into the use of probabilistic and fuzzy information measures to determine the precision of data representation and of response generation of their fuzzy database.

Expert Systems

A computerized expert system, as the name suggests, models the reasoning process of a human expert within a specific domain of knowledge in order to make the experience, understanding, and problem-solving capabilities of the expert available to the nonexpert for purposes of consultation, diagnosis, learning, decision support, or research. Usually an expert system is distinguished from a sophisticated lookup table (which merely maps questions to answers) by the attempt to include in the expert system some sense of an understanding of the meaning and relevance of questions and information and an ability to draw nontrivial inferences from data. Computerized rule-driven control systems (which we discuss in Sec. 6.3) are sometimes considered to constitute a subset of expert systems, which emulate the reasoning of an expert human operator.

Many expert systems are designed to interact directly with the user in a dialogue format; the user generally provides the parameters of the problem of concern and the expert system provides relevant advice, judgments, or information. Often the system can also provide the user with an explanation of the reasoning process employed to arrive at the conclusions. Some specific domains for which expert systems have been designed or proposed include the areas of medical diagnosis and treatment, chemistry, technical troubleshooting, geological exploration, quality control, damage assessment, management forecast, and investment advice. Most expert systems consist of a domain-specific *knowledge base* and problem-solving or reasoning algorithms known as an *inference engine*. Some systems consist of an inference engine *shell* to which various knowledge bases can be added, thus allowing the creation of different "experts." In addition, some means of *knowledge acquisition* must be incorporated into the system. The methods of eliciting human expert knowledge (which may be implicit even to the human expert) and translating it into a form suitable for use by a computerized expert system is known as *knowledge engineering*; it constitutes one of the major challenges in the field of expert systems. Knowledge acquisition may also take place through a learning capability; this highly desirable feature allows the expert system to collect knowledge or modify rules based on feedback during operation.

The facts, relations, judgments, opinions, and "rules of thumb" contained within the expert knowledge base usually manifest varying degrees of imprecision and uncertainty. It is nevertheless desirable for an expert system to be able, like the human expert, to draw nontrivial inferences from imprecise data and vague heuristics. Thus, the management of uncertainty in the design of expert systems is of key importance for the successful modeling of the reasoning process, and the utility of fuzzy set theory and the theory of evidence for this purpose has been and continues to be extensively studied.

The inference engine of an expert system operates on a series of statements known as *production rules*, which connect antecedents with consequences, premises with conclusions, or conditions with actions. They most commonly have the form *IF A*, *THEN B*. These rules offer the advantage of a high degree of modularity; they are easily added, modified, or deleted from the rule base. There exist two major approaches to evaluating these production rules. The first is data driven and corresponds to the logical inference method of modus ponens. In this case, the available data is supplied to the expert system, which then uses it to evaluate the production rules and to draw all possible conclusions. An alternative method of evaluation is goal driven; it corresponds to the modus tollens form of logical inference. Here, the expert system searches for the data specified in the *IF* clauses of rules that will lead to the objective; these data are found either in the knowledge base, in the *THEN* clauses of other production rules, or by querying the user. Since the data driven method proceeds from *IF* clauses to *THEN* clauses in a chain through the production rules, it is commonly called *forward chaining*; because the goal driven method proceeds from the *THEN* clauses (objectives) to the *IF* clauses in its search for the required data, it is commonly called *backward chaining*. Backward chaining has the advantage of speed, since only those rules leading to the objective will be evaluated. Furthermore, if certain data are difficult to provide and only potentially necessary, then the backward chaining method is clearly superior.

The rules employed by human experts are often of an imprecise or heuristic nature; the use of bivalent truth values or the requirement of exact satisfaction of *IF* clause conditions for rule evaluation often seems unnatural in the context of human reasoning. Therefore, a great deal of research has taken place on the applications of approximate reasoning and fuzzy logic in the inference process. Whalen and Schott [1985b] describe one use of fuzzy production rules in an interactive expert system designed to aid in the selection of appropriate techniques of forecasting sales of commercial products. This system, called fINDex, operates in a goal-driven manner by asking the user an increasingly specific series of questions concerning constraints on the forecast technique to be used. These constraints consist of the type and quality of outputs required from the forecast technique and the resources such as data, time, and money that are available for implementation of the forecast. The system does not perform the forecast itself but employs the constraints input by the user to produce a small list of possible forecast techniques. This list is a fuzzy set, where membership grades of the techniques indicate their possibilities under the constraints as evaluated by the fuzzy production rules. These fuzzy production rules have the form illustrated by the following examples:

IF the long-term historical data available are at least fair, THEN regression analysis is possible.

IF medium-term accuracy is required at a level of quality at most good to very good, THEN regression analysis is possible.

The final possibility of any one technique is determined to be the minimum possibility specified by any of the production rules.

An interactive session with fINDex begins with the system querying the user about some very basic constraints such as the time and money available for the forecast. The user may answer the questions as precisely or as vaguely as desired or may decline to answer at all; this allows an accurate reflection of the precision of the user's current knowledge concerning the constraints. At any point, the user may request that the system take the answers provided thus far and evaluate the possibilities of the various techniques. In response to this request, fINDex performs a forward chaining on the production rules based on the available data in order to determine the possibility of each technique. The user may either accept this list and end the session or request a further winnowing of the techniques. In order to accomplish this latter task, fINDex must first compile a list of questions ordered by importance, which seek from the user the data necessary for further determination of the possibilities of the techniques. The most important data concern highly relevant constraints on those techniques that are currently highly possible. The relevance of a constraint such as time or money for a particular technique is based on the degree to which the technique is possible only for a narrow range of constraint values. Thus, time is a more relevant constraint for a lengthy technique than for a short one. The importance of each constraint for each technique is determined by a version of backward chaining on the production rules. The ordered list of *askable* questions about these constraints is then generated, where *askability* is high if the corresponding constraint is important (*strict* for some *highly possible* techniques), if the system possesses only vague knowledge about it, and if the question has not been asked recently. The session then continues in the question-answer mode until the user requests processing of the data.

The fINDex system manipulates vague linguistic terms such as *highly possible*, *strict*, *at least fair*, and *very good* with the use of fuzzy sets. In addition, fuzzy sets defining linguistic terms *days*, *weeks*, *months*, and *one* (*figure*) through *six* (*figures*) (for money) are defined. The use of these linguistic terms allows the compilation of fuzzy production rules to be made directly from published documentation concerning forecasting techniques. They also allow natural language responses to be made by the user at various levels of precision. The following illustrates a sample session with fINDex:

fINDex: Please give information, if possible, on the following condition. If you cannot supply information on the condition, just press return. Time available?

USER: ⟨return⟩ (no response; interpreted as unknown)

fINDex: Do you want to respond to more questions (type q) or do you want to get a reaction to the information provided already (type r)?

USER: q

fINDex: Please give information, if possible, on the following condition. If you cannot supply information on the condition, just press return. Knowledge of structural relations available?

USER: Very good

fINDex: Long term accuracy required?

USER: Poor

fINDex: More questions or a reaction?

USER: q

fINDex: Recent historical data available?

USER: Fair

fINDex: More questions or a reaction?

USER: r

fINDex: Possibility   Technique

Possibility	Technique
1	Delphi method
1	Input-output analysis
1	Panel consensus
1	Visionary forecast
1	Historical analogy
1	Trend projection
1	Regression analysis
1	Econometric analysis
.35	Life-cycle analysis
.26	Moving averages method
.26	Exponential smoothing method
.26	Box-Jenkins method
.26	X-11
.26	Intention-to-buy and surveys
.26	Economic input-output model
.26	Diffusion indexes
.26	Leading indicators
0	Market research

Do you want to stop now (type s) or continue (type c)?

USER: c

fINDEX: Delphi-coordinator available?

USER: ⟨return⟩ (no response)

fINDex: More questions or a reaction?

USER: q

fINDex: Panel of experts available?

USER: ⟨return⟩ (no response)

fINDex: More questions or a reaction?

USER: q

fINDex: Short-term accuracy required?

USER: Good to very good

fINDex: More questions or a reaction?

USER: r

```
fINDex: Possibility    Technique*
           1              Delphi method
           1              Trend projection
           1              Regression analysis
           1              Econometric analysis
           1              Input-output analysis
          .26             Exponential smoothing method
          .26             Box-Jenkins method
          .26             X-11
          .26             Economic input-output model
          .01             Moving averages method
          .01             Intention-to-buy and surveys
          .01             Diffusion indexes
          .01             Leading indicators
```

Do you want to s̲top now (type s) or c̲ontinue (type c)?

USER: c

fINDex: Turning point identification required?

USER: Very good

fINDex: More q̲uestions or a r̲eaction?

USER: r

```
fINDex: Possibility    Technique
           1              Regression analysis
           1              Econometric analysis
          .26             X-11
```

Do you want to s̲top now (type s) or c̲ontinue (type c)?

USER: c

fINDex: Funds available?

USER: Lower three (three indicates a three-figure amount)

fINDex: More q̲uestions or a r̲eaction?

USER: r

```
fINDex: Possibility    Technique
           1              Regression analysis
          .26             X-11
          .18             Econometric analysis
```

Do you want to s̲top (type s) or c̲ontinue (type c)?

USER: s

One of the major concerns in the design of both information storage and retrieval systems and expert systems is efficiency; these systems must be able to perform quickly enough to make interaction with human users or with a controlled process feasible, despite large amounts of stored data and large numbers of in-

---

* Only techniques with nonzero possibility are listed from this point on.

ference rules. The implementation of fuzzy databases and expert systems utilizing fuzzy logics may soon be made far more efficient with the introduction of integrated circuits designed specifically to implement fuzzy logic. Scattered reports indicate that prototypes for these chips have already been produced and tested. A particular VLSI implementation of fuzzy logic is described in a paper by Togai and Watanabe [1986].

6.7 SYSTEMS SCIENCE

It has recently been argued that one of the main characteristics of the information society is the emergence of two-dimensional science [Klir, 1985]. According to this paradigm, the first dimension of science is represented by the *classical science*, divided into many disciplines and specializations, each focusing on the study of certain specific types of objects and phenomena. The second dimension of science, which emerged only recently, has a totally different orientation. It concentrates on the study of the various forms of relationship among entities of all kinds rather than the entities themselves. The term *relationship* is used here in a broad sense, encompassing not only the concept of a mathematical relation (a subset of a relevant Cartesian product) but a large class of kindred concepts such as constraint, interaction, coupling, interdependence, linkage, and the like. Since this emerging new dimension of science is oriented toward the study of various general phenomena of systems, such as communication, control, regulation, adaptation, self-organization, self-production, and so on, it is usually referred to as *systems science*.

It is true that a germ of systems science has always been present in science. What is new, however, is the emergence of systems science as a new domain of inquiry. This domain consists of those properties of systems and associated problems that emanate from a taxonomy of systems based on their relational properties rather than on the type of entities involved. As such, systems science is genuinely interdisciplinary across the classical sciences. Consequently, it is quite natural to view it as a second dimension of science.

The use of information theory in systems science, perhaps one of its most significant roles, became recognized considerably later than many of its other applications. The first proponent of the use of both Hartley and Shannon measures of information for dealing with systems problems was W. Ross Ashby. Although his initial ideas along these lines were already presented in the mid-1950s in his classic book [Ashby, 1956], he returned to this subject more forcefully only one decade later [Ashby, 1965] and continued to be strongly interested in it for the rest of his life [Ashby, 1969, 1970, 1972].

As far as fuzzy set theory is concerned, it was conceived, at least to a great extent, within the realm of research on mathematical system theory in the early 1960s, particularly in the context of attempts to develop suitable mathematical tools for dealing with systems of organized complexity (see Sec. 5.8). The following is a relevant quote from an early paper by Zadeh [1962], one of the most active systems theorists at that time and later the founder of the theory of fuzzy sets:

> There are some who feel . . . the fundamental inadequacy of the conventional mathematics—the mathematics of precisely-defined points, functions, sets, probability measures, etc.—for coping with the analysis of biological systems, and that to deal effectively with such systems, we need a radically different kind of mathematics, the mathematics of fuzzy or cloudy quantities which are not describable in terms of probability distributions. Indeed the need for such mathematics is becoming increasingly apparent even in the realm of inanimate systems.

This theme is further developed, from a more recent perspective, in the following quote from a paper by Gaines [1983]:

> Zadeh's (1962) paper was entitled *From circuit theory to system theory* and this is a very significant context for motivating the development of fuzzy sets. In electronic circuits and their applications to computing, communications and control, we find the apex of modern scientific achievement and our greatest technological triumphs. We might also believe circuit theory to be a confirmation of the intrinsic value of the precisiation process in the paradigm of science—circuits do behave with precision and do follow the underlying mathematics with uncanny veracity—we can design in theory and then implement with exactitude. However, it would be more correct to say that circuits can be *made* to behave with precision—electronics is our ultimate artefact designed to enable us to use our scientific and mathematical techniques, not working *because* they are right, but made to work *as if* they are right. It is when we are fooled by the success of our paradigm in predicting and controlling our artefacts into believing that it is also a tool for predicting and controlling the natural world that problems occur. The notions of *state, stability, adaptivity*, and so on, that had served us so well in engineering, when transferred to biological, social and economic systems became themselves suspect.
>
> *Stability* is an intrinsically imprecise concept and when precisely analysed explodes into a richness of definitions necessary to match the variety of the world but far removed from our intuitive concept of a stable system. . . . The notion of a *state*, so clear in systems design, becomes a mathematical artefact when it has to be inferred from system behavior. . . . The notion of *adaptivity* is particularly interesting because it has biological roots and yet has played an important role in circuit and control theory . . . —again the notion explodes with combinatorial complexity when analysed precisely . . . Zadeh's (1962) paper marked a turning point in his own thought processes—from a major involvement during the 1950s with the frontiers of mathematical system theory—culminating in the early 1960s in formal definitions of basic systemic concepts—calling in 1962 for new foundations for these concepts when applied not only to biological but also to inanimate systems—and providing it in 1965 with fuzzy sets theory. To take the new tools we now have and go back to the fundamental system concepts of state, stability, adaptivity, and so on, and give them exact definitions that accurately reflect their intrinsic imprecision—that is still a task for the fuzzy future.

Let us illustrate the broad applicability of fuzzy set theory and information measures in systems science by describing their roles in a package of methodological tools for dealing with the relationship between systems perceived as wholes and their various subsystems (i.e., parts of the wholes). This package has been referred to as *reconstructability analysis* [Klir, 1985; Klir and Way, 1985].

In terms of the hierarchy of epistemological types of systems outlined in Sec. 5.8, reconstructability analysis deals with structure systems, that is, sets of generative systems viewed as subsystems of an overall generative system. Each subsystem is based on a subset of variables of the overall system and, conversely, the overall system consists of all variables of the subsystems.

Two complementary problems are dealt with by reconstructability analysis. In one of them, a structure system is given and we want to derive as much information as possible about the unknown overall system. This problem is usually called the *identification problem*. In the second problem, an overall system is given and we want to break it down into subsystems, as small as possible, which are adequate to reconstruct the overall system to an acceptable degree of approximation. This problem has been referred to as the *reconstruction problem*.

Identification Problem

The identification problem emerges, for example, when it is impossible or impractical to measure simultaneously all variables of the overall system or when the investigator of the overall system is dependent on data collected only for subsets of its variables, often by specialists in the various disciplines of classical science. In such cases the choice of the subsystems is based solely on practical considerations and does not imply that the overall system is adequately represented by the subsystems. The principal aims of the identification problem are to determine the extent to which the subsystems do adequately represent the overall system and to make meaningful inferences, not necessarily precise, about the latter.

The identification problem clearly belongs to the general class of problems of *ampliative reasoning*. As discussed in Sec. 5.9, the proper way of dealing with these problems is to use appropriate *principles of maximum uncertainty*. Thus far, the identification problem has been investigated for systems conceptualized in terms of either probability theory or possibility theory. In these formal frameworks, we use the principles of maximum entropy or maximum U-uncertainty, respectively. The constraints in these optimization problems are expressed by equations that relate the known marginal probabilities or possibilities to their unknown joint counterparts by the appropriate projection rules.

For probabilistic systems, the projection rules take the form

$$^k p(^k x) = \sum_{x > ^k x} p(x), \tag{6.9}$$

where $p(x)$ are the unknown joint probabilities of the overall system ($x \in X$) and $^k p(^k x)$ are the given marginal probabilities of subsystems k (assume $k \in \mathbb{N}_q$ for some q). For possibilistic systems, the projections are expressed by the equations

$$^k r(^k x) = \max_{x > ^k x} r(x), \tag{6.10}$$

where $r(x)$ and $^k r(^k x)$ are possibilistic counterparts of probabilities $p(x)$ and $^k p(^k x)$ in Eq. (6.9), respectively; we assume again that $k \in N_q$ for some q.

To illustrate Eq. (6.9), consider a simple unknown probabilistic system with three variables v_1, v_2, v_3, each with two states, characterized in Table 6.3(a). Assume that the only information about this overall system is given in terms of the subsystems in Table 6.3(b). Then, Eq. (6.9) stands for eight individual equations, one for each of the given marginal probabilities in Table 6.3(b):

$$p_0 + p_1 = .2 \quad \text{(i)} \qquad p_0 + p_4 = .1 \quad \text{(v)}$$

$$p_2 + p_3 = .5 \quad \text{(ii)} \qquad p_1 + p_5 = .3 \quad \text{(vi)}$$

$$p_4 + p_5 = .2 \quad \text{(iii)} \qquad p_2 + p_6 = 0 \quad \text{(vii)}$$

$$p_6 + p_7 = .1 \quad \text{(iv)} \qquad p_3 + p_7 = .6 \quad \text{(viii)}$$

Solutions of these equations, which are restricted by the requirement of probability theory that $p_i \geq 0$ for all $i = 0, 1, \ldots, 7$, are all the joint probability distributions that are implied by the given marginal probability distributions. The solution set is usually called a *reconstruction family*. In our example, we can solve the equations easily as follows.

From (vii) and the requirement that the unknowns be nonnegative, we obtain $p_2 = p_6 = 0$. From this partial result, (ii), (iv), and the nonnegativity requirement, we obtain $p_3 = .5$ and $p_7 = .1$. It remains to determine p_0, p_1, p_4, and p_5 from Eqs. (i), (iii), (v), and (vi). One of these equations is redundant in the sense that it can be derived from the others. For example, when we add (i) and (iii) and substitute for $p_1 + p_5$ from (vi), we obtain (v). This means that we have three linearly independent equations for four unknowns. When one of the unknowns is chosen as a free parameter, the remaining unknowns are uniquely determined by its value. Using, for example, Eqs. (i), (v), (iii) and choosing p_0 as the free pa-

TABLE 6.3. ILLUSTRATION OF THE IDENTIFICATION PROBLEM FOR PROBABILISTIC SYSTEMS

(a)

v_1 v_2 v_3	$p(x)$
$x =$ 0 0 0	$p(000) = p_0$
0 0 1	$p(001) = p_1$
0 1 0	$p(010) = p_2$
0 1 1	$p(011) = p_3$
1 0 0	$p(100) = p_4$
1 0 1	$p(101) = p_5$
1 1 0	$p(110) = p_6$
1 1 1	$p(111) = p_7$

(b)

v_1 v_2	$^1 p(^1 x)$
$^1 x =$ 0 0	.2
0 1	.5
1 0	.2
1 1	.1

v_2 v_3	$^2 p(^2 x)$
$^2 x =$ 0 0	.1
0 1	.3
1 0	0
1 1	.6

rameter, we obtain

$$p_1 = .2 - p_0,$$

$$p_4 = .1 - p_0,$$

$$p_5 = .2 - p_4 = .1 + p_0,$$

respectively. Since p_0 must be nonnegative, it follows from the equation for p_4 that

$$0 \leq p_0 \leq .1.$$

The reconstruction family is thus defined by the following set of probability distributions:

	v_1	v_2	v_3	$p(x)$
$x =$	0	0	0	$p(000) = p_0 \in [0, .1]$
	0	0	1	$p(001) = p_1 = .2 - p_0$
	0	1	0	$p(010) = p_2 = 0$
	0	1	1	$p(011) = p_3 = .5$
	1	0	0	$p(100) = p_4 = .1 - p_0$
	1	0	1	$p(101) = p_5 = .1 + p_0$
	1	1	0	$p(110) = p_6 = 0$
	1	1	1	$p(111) = p_7 = .1$

For more complex structure systems, reconstruction families may be characterized by several free parameters, and these may be mutually interdependent. The solution process then becomes quite involved and the use of a formal method and the computer is unavoidable. A matrix method for this problem was developed by Cavallo and Klir [1981].

For possibilistic systems, Eq. (6.10) can be conveniently expressed in the form

$$^k r(^k x) = \max_{x \in X} \min[r(x), s(x, {}^k x)], \qquad (6.11)$$

where

$$s(x, {}^k x) = \begin{cases} 1 & \text{if } x > {}^k x \\ 0 & \text{otherwise.} \end{cases} \qquad (6.12)$$

This means that the problem can be formulated in terms of the fuzzy relation matrix equation

$$[^1 r, {}^2 r, \ldots, {}^q r] = r \circ S, \qquad (6.13)$$

where r is the unknown joint possibility distribution, $^k r$ ($k \in \mathbb{N}_q$) are the given marginal distributions, and S is a matrix of a crisp relation between the overall states and the relevant substates defined by its characteristic function (6.12). It is obvious that the method for solving fuzzy relation equations described in Sec.

3.8 is directly applicable to solve Eq. (6.13). More details can be found in a paper by Higashi, Klir, and Pittarelli [1984].

Given a structure system, all members of its reconstruction family are hypothetical overall systems implied by the given subsystem information, but it is not known which of them is the actual overall system. In spite of this uncertainty, we must often select one of them as a basis for making a decision and taking an appropriate action. In order to choose the one that maximally recognizes our ignorance in the given situation, we must maximize the appropriate measure of uncertainty.

The problem of estimating a joint probability or possibility distribution from the knowledge of some of its marginal distributions can thus be characterized as the problem of determining such a distribution for which the Shannon entropy or U-uncertainty, respectively, reaches its maximum within the domain of the reconstruction family. If the given marginal distributions are consistent, this maximum is unique and can be obtained by a special and very simple procedure, without resorting to general optimization methods. Let us describe the procedure for probabilistic systems first.

The procedure to be described is applicable only to structure systems that contain no loops of the following kind but can be extended to any structure system in the manner explained later. Let

$$^1V, {}^2V, \ldots, {}^tV$$

be a sequence of sets of variables of some of the given subsystems such that

$$^kV \cap {}^{k+1}V \neq \varnothing$$

and $k \geq 3$; then, this sequence represents a loop in the structure system if and only if there exists a variable v of the overall system such that

$$v \begin{cases} \in {}^kV & \text{for } k = 1 \text{ and } k = t \\ \notin {}^kV & \text{for } k \in \{2, \ldots, t - 1\}. \end{cases}$$

Structure systems that contain no loops of this kind are called *loopless structure systems*.

Given a probabilistic structure system that is loopless, each component of the maximum entropy joint probability distribution can be obtained as an arithmetic product of appropriate marginal probabilities. The marginal probabilities may enter the product in any order, but arguments of each of them that are also involved in some previous marginal probabilities in the product must be used as conditions. Some probabilities in each product are thus basic probabilities, whereas others are appropriate conditional probabilities. More specifically, given marginal probability distributions $^1\mathbf{p}, {}^2\mathbf{p}, \ldots, {}^s\mathbf{p}$ on some lower-dimensional Cartesian products of $X_1 \times X_2 \times \cdots \times X_n$, each component $p(x_1, x_2, \ldots, x_n)$ of the maximum entropy distribution $\mathbf{p}$ (assuming $x_i \in X_i$ for each $i \in \mathbb{N}_n$) is defined by the product

$$p(x_1, x_2, \ldots, x_n) = \prod_{j=1}^{s} {}^j\dot{p} \tag{6.14}$$

where $^j\dot{p}$ denotes for each $j \in \mathbb{N}_s$ either the basic probability or an appropriate conditional probability defined by $^j p$ for the relevant subset of arguments $x_1, x_2,$ $\ldots, x_n$ according to the following rule: arguments of $^k p$ from some $k \in \mathbb{N}_{2,s}$ that are also employed in $^1 p, ^2 p, \ldots, ^{k-1} p$ are used as conditions in calculating $^k \dot{p}$. For example, given marginal probability distributions $^1\mathbf{p}, ^2\mathbf{p}, ^3\mathbf{p}$ on $X_1 \times X_2,$ $X_2 \times X_3,$ and $X_3 \times X_4,$ respectively, the maximum entropy joint probabilities can be calculated by the product

$$p(x_1, x_2, x_3, x_4) = {}^1p(x_1, x_2) \cdot {}^2p(x_3 \mid x_2) \cdot {}^3p(x_4 \mid x_3)$$

for each $(x_1, x_2, x_3, x_4) \in X_1 \times X_2 \times X_3 \times X_4$. Similarly, given $^1\mathbf{p}, ^2\mathbf{p}, ^3\mathbf{p}$ on $X_1 \times X_2 \times X_3, X_1 \times X_3 \times X_4,$ and $X_5,$ respectively, we have

$$p(x_1, x_2, x_3, x_4) = {}^1p(x_1, x_2, x_3) \cdot {}^2p(x_4 \mid x_1, x_3) \cdot {}^3p(x_5).$$

The following theorem establishes that the joint probability distribution derived from a set of loopless marginal distributions by Eq. (6.14) is indeed the maximum entropy joint distribution.

Theorem 6.1. Given a family of finite sets $X_1, X_2, \ldots, X_n$ and marginal probability distributions $^1\mathbf{p}, ^2\mathbf{p}, \ldots, ^s\mathbf{p}$ defined on some lower-dimensional Cartesian products of $X_1 \times X_2 \times \cdots \times X_n = X$ that cover all sets in the family and represent a loopless structure system, the joint probability distribution determined by Eq. (6.14) is characterized by the maximum value of Shannon entropy among all joint distributions that satisfy the given marginal distributions.

Proof: Let $\mathbf{p}$ denote the joint probability distribution determined from the given marginal distributions by Eq. (6.14) and let $\mathbf{p}'$ denote any other joint distributions that satisfy the given marginal distributions, that is, any other member of the reconstruction family. Using Gibbs' inequality, we have

$$-\sum_{x \in X} p'(x)\log_2 p'(x) \le -\sum_{x \in X} p'(x)\log_2 p(x).$$

By Eq. (6.14),

$$p(x) = \prod_{j=1}^{s} {}^j\dot{p}({}^j x),$$

where $^j x < x$ and $^j\dot{p}(^j x)$ denotes either the basic probability or appropriate conditional probability as required by Eq. (6.14). Hence,

$$-\sum_{x \in X} p'(x)\log_2 p(x) = -\sum_{x \in X} p'(x)\log_2 \prod_{j=1}^{s} {}^j\dot{p}({}^j x)$$

$$= -\sum_{x \in X} p'(x) \sum_{j=1}^{s} \log_2 {}^j\dot{p}({}^j x).$$

Since $\mathbf{p}'$ is a member of the reconstruction family, the given marginal distributions are its projections and, consequently, the terms in the last expression can be grouped together for each ^jx in a way described by the formula

$$- \sum_{x > ^jx} p'(x) \log_2 {}^j\dot{p}(^jx);$$

this formula can also be written as

$$- {}^j\dot{p}(^jx) \log_2 {}^j\dot{p}(^jx).$$

When all these groups of terms are added, we obtain

$$- \sum_{j=1}^{n} \sum_{^jx} {}^j\dot{p}(^jx) \log_2 {}^j\dot{p}(^jx).$$

We can show now that this expression is exactly the same as the entropy of $\mathbf{p}$; indeed, by repeating the previous arguments, we obtain

$$- \sum_{x \in X} p(x) \log_2 p(x) = - \sum_{x \in X} p(x) \sum_{j=1}^{s} \log_2 {}^j\dot{p}(x),$$

where $^jx < x$, and since $\mathbf{p}$ is a member of the reconstruction family ($^j\mathbf{p}$ are projections of $\mathbf{p}$), we can group the terms in the last expression in the same way as before and derive, eventually, the same final expression. Hence,

$$- \sum_{x \in X} p'(x) \log_2 p(x) = - \sum_{x \in X} p(x) \log_2 p(x)$$

and the inequality from which we started becomes

$$- \sum_{x \in X} p'(x) \log_2 p'(x) \le - \sum_{x \in X} p(x) \log_2 p(x),$$

which proves the theorem.   ■

The procedure for calculating the maximum entropy joint distribution on the basis of Theorem 6.1 is usually called a *join procedure*. The name reflects the fact that the procedure is based on the join operation introduced in Sec. 3.2. To illustrate this procedure, let us calculate the maximum entropy joint distribution for the marginal distributions given in Table 6.3. Following Eq. (6.14), we have

$$p(x_1, x_2, x_3) = {}^1p(x_1, x_2) \cdot {}^2p(x_3 \mid x_2)$$

or, alternatively,

$$p(x_1, x_2, x_3) = {}^2p(x_2, x_3) \cdot {}^2p(x_1 \mid x_2).$$

When we apply either of these equations for all states $(x_1, x_2, x_3) \in \{0, 1\}^3$, we obtain

$$p_0 = .05 \qquad p_4 = .05$$

$$p_1 = .15 \qquad p_5 = .15$$

$$p_2 = 0 \qquad p_6 = 0$$

$$p_3 = .5 \qquad p_7 = .1$$

For example,

$$p_1 = p(001) = {}^1p(00) \cdot {}^2p(1 \mid 0) = .2 \times .75 = .15,$$

where

$$^2p(1 \mid 0) = \frac{{}^2p(01)}{{}^2p(00) + {}^2p(01)} = \frac{.3}{.4} = .75.$$

The fact that the maximum entropy joint distribution is represented by the midpoint of the range [0, .1] for p_0 in this example is a coincidence, not a general property.

For structure systems with loops, the join procedure must be repeated iteratively to converge to the maximum entropy joint distribution. It is known that the convergence is guaranteed in a finite number of steps. More details on this issue can be found elsewhere [Cavallo and Klir, 1982; Klir, 1985].

The join procedure for possibilistic systems is even simpler. It is based on the equation

$$r(x_1, x_2, \ldots, x_n) = \min_{j=1}^{s} {}^j r, \tag{6.15}$$

where $^j r$ are the given marginal possibilities of the appropriate substates of $(x_1, x_2, \ldots, x_n)$. No conditional possibilities are needed. Furthermore, Eq. (6.15) is applicable even for structure systems with loops, and no iterations are needed to obtain the maximum U-uncertainty joint possibility distribution. Proofs of these claims can be found in a paper by Cavallo and Klir [1982] or in a book by Klir [1985].

Reconstruction Problem

Let us now discuss the reconstruction problem—the problem of breaking down a given overall system into subsystems that preserve enough information about the overall system. The principle motivation behind the reconstruction problem is the reduction of the complexity of the system involved. That, in turn, is connected with our ability to comprehend and manage the system. For example, it is easier to monitor several small sets of variables than one large set of variables during a crisis situation when decisions to take appropriate actions must be made quickly. However, practical reasons such as those just mentioned are not the only motivations behind the reconstruction problem. A discovery that a system can be represented by a specific set of subsystems may provide the investigator with some knowledge that is not available, at least explicitly, in the given overall system. For example, the subsystem configuration may provide information about causal relationships, the significance of the individual variables, the strength of dependencies among them, and so on. In general, this additional knowledge may help the investigator to develop a better insight into the nature of the relationship of the variables studied.

The reconstruction problem belongs to the general class of problems of *systems simplification*. Given a set of structure systems that are viewed as admissible simplifications of the given overall system, we need to determine how much information about the overall system is contained in each of them. This can be done by reconstructing a hypothetical overall system from each of the considered structure systems and, then, comparing it with the actual overall system. We must ensure that the reconstruction method used utilizes all but no more information than is available in the structure system. Hence, we must determine the maximum uncertainty reconstruction, as discussed in the context of the identification problem, for each of the considered structure systems. This can be done, for example, by the appropriate join procedure.

For probabilistic systems, the comparison between the reconstructed and actual overall systems is expressed in terms of the Shannon cross-entropy

$$H(\mathbf{p}, \mathbf{p}_s) = \sum p(x)\log_2 \frac{p(x)}{p_s(x)} \,,$$

where $\mathbf{p}$ and $\mathbf{p}_s$ denote probability distributions of the given overall system and of the one reconstructed from a structure systems s, respectively. For possibilistic systems, the comparison is expressed by the simple difference

$$U(\mathbf{r}, \mathbf{r}_s) = U(\mathbf{r}_s) - U(\mathbf{r}),$$

where $\mathbf{r}$ and $\mathbf{r}_s$ denote the given possibility distribution and the one reconstructed from a structure system s, respectively [Higashi and Klir, 1983b].

The reconstruction problem requires also that a search through all admissible structure systems be performed. Since the number of possible structure systems grows extremely rapidly with the number of variables, the problem involves formidable computational issues [Klir, 1985]. These, however, are not of our interest in this book.

6.8 OTHER APPLICATIONS

Most of the major areas of application of fuzzy set theory and information theory are covered in the previous sections of this chapter and in the associated notes. In this section, we outline some additional application areas that are not readily classified under the other headings used in this chapter. We also discuss some of the experimental issues in fuzzy set theory associated with the determination of membership grade functions based on subjective perceptions.

Pattern Recognition and Clustering

One of the most active and promising areas of application for the mathematics of uncertainty has been in the field of pattern recognition and clustering. Pattern recognition encompasses a large variety of approaches and techniques aimed at discovering structures within data in order to recognize patterns and classify objects. The technique of clustering is aimed at recognizing the existence of groups or clusters within a set of data whose members display similarity to one another along some relevant dimensions. The application of fuzzy sets to these endeavors

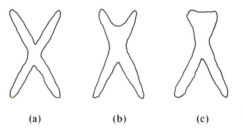

Figure 6.5. Chromosome images: (a) median, (b) submedian, (c) acrocentric.

(a)   (b)   (c)

seems both natural and appropriate due to the fact that many of the categories and patterns we commonly encounter and employ have indistinct or vague boundaries. We are able to recognize similarities and classify patterns such as handwritten characters, faces, and speech, which show a wide range of variations. The importance of pattern recognition extends into areas of image processing and scene analysis, medical diagnosis, speech and handwriting recognition, weather prediction, analysis of aerial photography, and many others. Some of the applications presented in this chapter, in fact, including those in medical diagnosis and meteorology, make use of fuzzy pattern recognition and clustering techniques.

Many different methods are used for pattern recognition, and their selection depends primarily on the specific application. Lee [1975] describes one straightforward method of examining the shape of chromosomes in order to classify them into the three categories pictured in Fig. 6.5. As can be seen from this figure, the classification scheme is based on the ratio of the length of the arms of the chromosome to its total body length. It is difficult to identify sharp boundaries between these three types. Therefore, Lee implements a technique of fuzzy pattern recognition, which compares the angles and arm lengths of the chromosome with those labeled in the idealized skeleton in Fig. 6.6. A test of symmetry is made first, in which a *symmetric* chromosome is defined as one with angles $A_{2i-1} = A_{2i}$ (for $1 \leq i \leq 4$) and sides $a_1 = a_2$, $a_3 = a_4$. The membership grade of each chromosome x in the fuzzy set S of symmetric chromosomes is calculated by

$$\mu_S(x) = 1 - \frac{1}{720°} \sum_{i=1}^{n} |A_{2i-1} - A_{2i}|.$$

Here, the most asymmetric chromosome receives a grade of 0 and the most symmetric, a grade of 1. The further classification of the chromosome into fuzzy categories of *median*, M, *submedian*, SM, and *acrocentric*, AC, make use of the degree of symmetry along with the ratios of arms' lengths. The membership grade of the chomosome x in the fuzzy set M is given by

$$\mu_M(x) = \mu_S(x) \left[1 - \frac{|a_1 - a_4| + |a_2 - a_3|}{a_T} \right],$$

where $a_T = a_1 + a_2 + a_3 + a_4 + a_5$. The membership grade in the fuzzy set SM is given by

$$\mu_{SM}(x) = \mu_S(x) \left[1 - \frac{a_{SM}}{2a_T} \right],$$

where

$$a_{SM} = \min[(\,|\,a_1 - 2a_4\,| + |\,a_2 - 2a_3\,|\,),\,(\,|\,2a_1 - a_4\,| + |\,2a_2 - a_3\,|\,)].$$

Finally, the membership grade in fuzzy set AC is given by

$$\mu_{AC}(x) = \mu_S(x)\left[1 - \frac{a_{AC}}{4a_T}\right],$$

where

$$a_{AC} = \min[(\,|\,a_1 - 4a_4\,| + |\,a_2 - 4a_3\,|\,),\,(\,|\,4a_1 - a_4\,| + |\,4a_2 - a_3\,|\,)].$$

Each chromosome can be classified as *approximately median*, *approximately sub-median*, or *approximately acrocentric* by calculating each of the three membership grades in M, SM, and AC; the category in which the chromosome attains the maximum value is chosen if that value is sufficiently large. If the maximum value falls below some designated threshold, then the image is rejected from all three classes. If the maximum is not unique, then the classification is based on a priority defined on the three classes. As can be seen from this example, geometric similarity is conveniently represented in terms of membership grades in the interval [0, 1]. The further advantages of this type of shape-oriented classification scheme lie in the fact that the method is insensitive to rotation, translation, expansion, or contraction of the chromosome image; these factors do affect the way in which humans are able to classify images.

Available books [Bezdek, 1981; Kandel, 1982; Majumder, 1986] extensively cover the applications of fuzzy set theory within the areas of pattern recognition and clustering. Additional discussions of these applications may be found in Lee and Fu [1972] and Ruspini [1982]; Watanabe [1985] discusses the uses of the minimum entropy principle in pattern recognition. Because a full coverage of this application area is provided by these references, we do not explore the area further in this text.

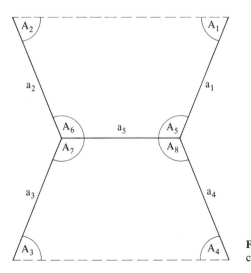

Figure 6.6. Idealized pattern for chromosome classification.

Fuzzy Data

One of the fundamental issues associated with applications of fuzzy set theory is the determination of membership grade functions. In many applications, including some of those outlined in this chapter, relevant membership grade functions are derived by specific and justifiable procedures from data and from various other entities involved in the application. For example, in the climatological study described in Sec. 6.2, a membership function of a fuzzy relation is derived from records of annual precipitation levels; similarly, in one medical diagnosis model discussed in Sec. 6.4, a fuzzy relation characterizing medical knowledge is obtained from previous records of symptoms and diseases of patients by applying a maximum nonspecificity principle.

In some cases of data gathering, it is desirable to define states of a variable as labels of fuzzy sets via an appropriate *fuzzy observation channel*. Each observation expressed in terms of a variable with n states is then an n-tuple of membership grades, one for each fuzzy set recognized on the observed attribute. Variables defined in this way are called *fuzzy variables*. For example, states of a fuzzy variable that characterize the unemployment level in a country may represent fuzzy sets whose linguistic labels are *extremely high*, *very high*, *high*, *moderate*, *low*, *very low*, and *negligible*. Membership functions of fuzzy sets corresponding to these linguistic labels may, for example, be elicited from a group of experts by an appropriate method for scaling their perceptions of concepts expressed by these labels. Various methods have been developed in mathematical psychology for this purpose; one of them is briefly described in this section.

Table 6.4 illustrates the difference between crisp and fuzzy data for two variables, each with three states (values, labels) 0, 1, 2. Observations are distin-

TABLE 6.4.   COMPARISON OF CRISP AND FUZZY DATA

(a) Crisp data

$t =$	1	2	3	4	5	6	7	8	. . .
$v_1(t) =$	1	2	1	1	2	2	0	1	. . .
$v_2(t) =$	1	1	0	2	2	2	2	0	. . .

(b) Comparable fuzzy data

$t =$		1	2	3	4	5	6	7	8	. . .
	0	.1	0	0	0	0	0	.6	.1	. . .
$v_1(t) =$ 1	1	.9	.4	.7	.5	.2	0	.3	.9	. . .
	2	0	.6	.3	.5	.8	1	.1	0	. . .
	0	0	0	.8	.1	0	0	.1	.7	. . .
$v_2(t) =$ 1	1	.7	1	.2	.3	.4	.2	.4	.2	. . .
	2	.3	0	0	.6	.6	.8	.5	.1	. . .

guished by values of the parameter t ($t \in \mathbb{N}$); values of t may represent instances in time, individuals of a population, and the like. In the crisp data (Table 6.4(a)), each observation is a pair of states (values, labels), one for each variable. In the comparable fuzzy data (Table 6.4(b)), each observation consists of two triples of numbers, one triple for each variable. The numbers represent the degrees of membership of the observed value of each attribute in the three fuzzy sets corresponding to the linguistic labels 0, 1, 2 (standing for, as an example, low, medium, and high, respectively). Let us denote these membership degrees by $\mu_{i,s_i}(t)$, where i, s_i, and t identify the variable, the state (linguistic label) of the variable, and the value of the parameter (a specific observation), respectively.

Let us look now at the problem of estimating probabilities of the overall system from fuzzy data. For crisp data, this is done in terms of frequencies of states of the overall system. To calculate similar frequencies from fuzzy data, it is reasonable to require that membership degrees of all states add to one for each variable and each observation, that is,

$$\sum_{s_i} \mu_{i,s_i}(t) = 1$$

for each i and t. If the data given in terms of membership degrees $\mu'_{i,s_i}(t)$ are not normalized in this sense, they can be easily normalized by the formula

$$\mu_{i,s_i}(t) = \frac{\mu'_{i,s_i}(t)}{\sum_{s_i} \mu'_{i,s_i}(t)} .$$

Observe that fuzzy data in Table 6.4 are normalized.

In calculating frequencies of overall states from fuzzy data, the membership degrees pertaining to states of individual variables must be properly aggregated into membership degrees of the overall states. Let S_i denote the state set of variable v_i (assume $i \in \mathbb{N}_n$) and let $\mathbf{s} \in S_1 \times S_2 \times \cdots \times S_n = S$ denote an overall state of the variables involved. Then, we may define

$$\mu_s(t) = \prod_{i \in \mathbb{N}_n} \mu_{i,s_i}(t), \tag{6.16}$$

where $s_i < \mathbf{s}$, $s_i \in S_i$ for all $i \in \mathbb{N}_n$. That is, for each value of t (each observation), the membership degree of the n-tuple $\mathbf{s} = (s_1, s_2, \ldots, s_n)$ is calculated by taking the product of the membership degrees of its components $s_1, s_2, \ldots, s_n$. This is justified by the following properties of Eq. (6.16):

1. For each t, $\mu_s(t) = 0$ if and only if $\mu_{i,s_i}(t) = 0$ for at least one component s_i (i.e., Axiom h1 for aggregation operations is satisfied).
2. For each t, $\mu_s(t) = 1$ if and only if $\mu_{i,s_i}(t) = 1$ for all s_i (i.e., Axiom h1 for aggregation operations is satisfied).
3. For each t, μ_s is a monotonic nondecreasing, continuous, and symmetric function (Axioms h2–h4 for aggregation operations); moreover, if all arguments of μ_s are fixed except one, μ_s is linearly proportional to the value of the free argument.

4. For each t, $\mu_s(t)$ is normalized, that is,

$$\sum_{s \in S} \mu_s(t) = 1.$$

This means that the total (aggregated) membership degree for each observation is equal to 1, even though it is not allocated to one overall state (as in the case of crisp data) but distributed among all overall states proportionally to the membership degrees of their components.

Assuming that $\mu_s(t)$ is calculated by Eq. (6.16), the frequency $N(s)$ of each overall state s ($s \in S$) can then be calculated, in a very natural way, by the formula

$$N(s) = \sum_t \mu_s(t), \tag{6.17}$$

where the sum is taken over all observations included in the given fuzzy data. Since values of $N(s)$ need not be whole numbers, it is better to call them *pseudo-frequencies*.

Probabilities $p(s)$ are usually estimated on the basis of frequencies $N(s)$, $s \in S$. The simplest way is to estimate them in terms of relative frequencies or, when derived from fuzzy data, pseudo-frequencies; then, we have

$$p(s) = \frac{N(s)}{\sum\limits_{x \in S} N(x)}, \tag{6.18}$$

where $N(s)$ is given by Eq. (6.17).

To illustrate the difference between the calculations of frequencies for crisp data and of pseudo-frequencies from fuzzy data, let us use the small segments of data given in Table 6.4. The calculations are summarized in Table 6.5(a) and (b) for crisp and fuzzy sets, respectively. Each observation in the fuzzy data is described by the membership grade function on the set $\{0, 1, 2\}^2$ of all overall states described by Eq. (6.16); for easy comparison, each observation in the crisp data is described by a characteristic function on $\{0, 1, 2\}^2$. For example, for $t = 8$ of the fuzzy data, Eq. (6.16) yields

$$\mu_{00}(8) = \mu_{1,0}(8) \cdot \mu_{2,0}(8) = .1 \times .7 = .07,$$

$$\mu_{01}(8) = \mu_{1,0}(8) \cdot \mu_{2,1}(8) = .1 \times .2 = .02,$$

$$\mu_{02}(8) = \mu_{1,0}(8) \cdot \mu_{2,2}(8) = .1 \times .1 = .01,$$

$$\mu_{10}(8) = \mu_{1,1}(8) \cdot \mu_{2,0}(8) = .9 \times .7 = .63.$$

$$\vdots$$

Pseudo-frequencies $N(s)$ are then calculated by Eq. (6.17). In terms of Table 6.5(b), this amounts to calculating the sum in each row. For example,

$$N(12) = \sum_{t=1}^{8} \mu_{12}(t) = .27 + 0 + 0 + .3 + .12 + 0 + .15 + .09 = .93.$$

TABLE 6.5. ILLUSTRATION OF PROBABILITY OR POSSIBILITY DISTRIBUTION ESTIMATES DERIVED FROM CRISP OR FUZZY DATA

(a) Crisp data

					t							
v_1	v_2	1	2	3	4	5	6	7	8	$N(s)$	$p(s)$	$r(s)$
$s =$ 0	0	0	0	0	0	0	0	0	0	0	0	0
0	1	0	0	0	0	0	0	0	0	0	0	0
0	2	0	0	0	0	0	0	1	0	1	$\frac{1}{8}$	.5
1	0	0	0	1	0	0	0	0	1	2	$\frac{1}{4}$	1
1	1	1	0	0	0	0	0	0	0	1	$\frac{1}{8}$	.5
1	2	0	0	0	1	0	0	0	0	1	$\frac{1}{8}$	.5
2	0	0	0	0	0	0	0	0	0	0	0	0
2	1	0	1	0	0	0	0	0	0	1	$\frac{1}{8}$	.5
2	2	0	0	0	0	1	1	0	0	2	$\frac{1}{4}$	1

(b) Fuzzy data

					t							
v_1	v_2	1	2	3	4	5	6	7	8	$N(s)$	$p(s)$	$r(s)$
$s =$ 0	0	0	0	0	0	0	0	.06	.07	.13	.02	.08
0	1	.07	0	0	0	0	0	.24	.02	.33	.04	.19
0	2	.03	0	0	0	0	0	.30	.01	.34	.04	.20
1	0	0	0	.56	.05	0	0	.03	.63	1.27	.16	.75
1	1	.63	.4	.14	.15	.08	0	.12	.18	1.70	.21	1
1	2	.27	0	0	.3	.12	0	.15	.09	.93	.12	.55
2	0	0	0	.24	.05	0	0	.01	0	.30	.04	.15
2	1	0	.6	.06	.15	.32	.2	.04	0	1.37	.17	.81
2	2	0	0	0	.3	.48	.8	.05	0	1.63	.20	.96

Probabilities $p(s)$, when estimated in terms of relative frequencies, are calculated by Eq. (6.18). The sum of all pseudo-frequencies in our example is 8 and, hence,

$$p(s) = \frac{N(s)}{8}.$$

For example,

$$p(12) = \frac{.93}{8} = .12.$$

Frequencies or pseudo-frequencies may also be used for estimating possibilities of the overall states. One formula, which is a possibilistic counterpart of Eq. (6.18), is based on a simple normalization of the frequency or pseudo-frequency distribution $(N(s) \mid s \in S)$ to obtain a normalized possibility distribution:

$$r(s) = \frac{N(s)}{\max_{x \in S} N(x)}. \tag{6.19}$$

Possibility distributions based on this formula are also given in Table 6.5. For example, the maximum frequency for the fuzzy data in $N(11) = 1.7$ and, hence,

$$r(21) = \frac{N(21)}{1.7} = \frac{1.37}{1.7} = .81.$$

Formula (6.19) may also be generalized to

$$r(\mathbf{s}) = \left[\frac{N(\mathbf{s})}{\max_{\mathbf{x} \in X} N(\mathbf{x})} \right]^w, \tag{6.20}$$

where $w \in [0, 1]$ is a parameter whose value can be adjusted according to the characteristics of each application. Observe that for $w = 0$, Eq. (6.20) yields a crisp possibility distribution in which

$$r(\mathbf{s}) = \begin{cases} 1 & \text{when } N(\mathbf{s}) \neq 0 \\ 0 & \text{when } N(\mathbf{s}) = 0; \end{cases}$$

for $w = 1$, we obtain Eq. (6.19); and for $w = \infty$, Eq. (6.20) yields another crisp possibility distribution in which

$$r(\mathbf{s}) = \begin{cases} 1 & \text{when } N(\mathbf{s}) = \max_{\mathbf{x} \in S} N(\mathbf{x}) \\ 0 & \text{otherwise.} \end{cases}$$

Thus far, we have assumed that fuzzy data are based on observation channels that are explicitly defined. In some application areas, however, the observation channels may be represented implicitly by a human observer, usually an expert in the field of inquiry. For example, in studying animal behavior, ethologists use methods of observation that disturb the natural habitat of the animals investigated as little as possible. One of the methods used in studying groups of animals is to make motion-picture films and from these to determine relevant behavior sequences. For each specific kind of animal, some significant postures and movements are usually recognized. Ethologists often specify them by characteristic pictures supplemented by verbal descriptions. For instance, principal postures of herring gulls are pictorially specified in Fig. 6.7 and are given suggestive names such as "rest," "facing away," "choking," and so on. Each of them is also described verbally; for example: "choking begins with bending down over the nest (or any depression in the ground similar to a nest, such as a human footprint in sand on a beach), followed by a rhythmic up-and-down movement of the head." Clearly, in this case the observation channels cannot be defined mathematically. They are represented by the ethologist's comprehension (based on training and experience) of the combination of pictorial and verbal characterizations of the various postures and movements, which are genuinely fuzzy. The advantage of fuzzy data in an area such as ethology is that it allows the ethologist to be "honest" by permitting the expression of any uncertainty in deciding which posture or movement is in fact being observed in each particular observation. That is, although less precise than crisp data, fuzzy data are in fact a more accurate representation of reality.

Figure 6.7.   Pictorial definitions of typical postures of herring gulls in ethological studies.

Membership Functions Based on Subjective Perceptions

In many applications, membership functions of fuzzy sets are based on subjective perceptions rather than on data or other objective entities involved. Included in this category are fuzzy observation channels, which are based on subjective perceptions of specific linguistic labels by relevant experts.

The problem of assigning numbers to subjective perceptions is dealt with by mathematical psychology and involves the utilization of various principles and methods of the theory of measurement. Many methods are now available for this purpose, but it is beyond the scope of this book to cover them adequately. How-

ever, to illustrate the issues involved, we describe two relevant experiments performed by Norwich and Turksen [1982a–c, 1984].

The experiments to be described are based on the assumption that the interval scale is employed. Two techniques, referred to as direct rating and reverse
rating, were used; these are described later. Experiments were performed with
thirty subjects. Membership functions were constructed for each subject by ran-
domly generating different stimuli (each one at least nine times) and then averaging
the subject's responses. The following is a brief characterization of each of the
two experiments:

1. The first experiment involves the subjective perception of each participant
   of the notion of tall persons. It uses a life-sized wooden figure of adjustable
   height.
2. The second experiment involves the notion of aesthetically pleasing houses
   among one-story houses of fixed width and variable heights. A cardboard
   model of a house is used, which consists of a chimney and a triangular roof
   sitting on a rectangle of width 2 inches and of height adjustable from 0 inches
   to 34 inches.

Membership ratings in both experiments are indicated with a movable
pointer along a horizontal line segment. The right-hand end of this segment corresponds to the membership degree of those persons who are definitely tall (or

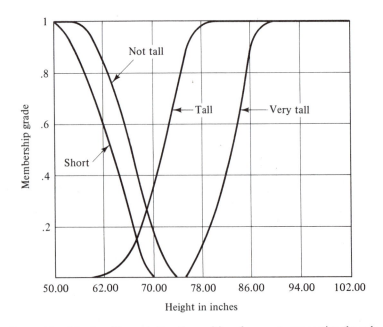

Figure 6.8. Membership grade functions of four fuzzy sets expressing the subjective perceptions of a particular subject related to the concept of tall persons
(adopted from Norwich and Turksen [1984]).

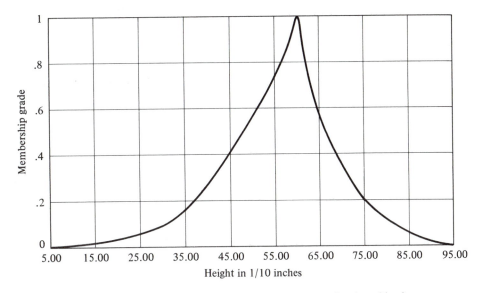

Figure 6.9. Membership grade function of the fuzzy set expressing the subjective perception of a particular subject of aesthetically pleasing houses (adopted from Norwich and Turksen [1984]).

to those houses that are definitely pleasing), and the left-hand end corresponds to those persons felt to be definitely not tall (or to those houses that are definitely not pleasing). The distance at which the pointer is placed between these two endpoints is interpreted as the strength of agreement or truth of the classification of each person (or house) as tall (or pleasing). In the direct rating procedure, the subject is presented with a random series of persons (houses) and asked to use this method of indicating membership degree to rate each one as tall (or pleasing). In the reverse rating procedure, the same method of depicting membership degree is used to give the subject a random membership rating; the subject is then presented with an ordered series of persons (houses) and asked to select the one person (house) best seeming to correspond to the indicated degree of membership in the category of tall persons (or pleasing houses).

An example of the membership functions obtained for the fuzzy sets *tall*, *very tall*, *not tall*, and *short* persons, as perceived by a particular subject, is shown in Fig. 6.8. The plots in this figure represent the mean response to nine stimuli based on the direct rating procedure. An example of a membership function for the fuzzy set of pleasing houses, again representing the mean response of a particular subject to nine stimuli based on the direct rating procedure is shown in Fig. 6.9.

These as well as other experiments of this kind have shown that people do not perceive the same concepts equally. In general, however, graphs of the resulting membership functions have the same shapes and their differences are modest. To summarize perceptions of some concept by a group of people (such as experts in the related field), some aggregation is needed of the various membership functions obtained. Alternatively, the results may be summarized in terms of an appropriate interval-valued fuzzy set.

It is also well known that the membership functions representing the perception of a concept by an individual may depend significantly on the context. Thus, the membership function corresponding to the concept of *small* changes drastically when the reference of the concept changes from birds to elephants; similarly, the concept of *red* varies as the context changes from apples to wines. It has been shown that, in general, people perform some sort of normalization of the scale representing the concept involved and then map this normalized membership function into the range of each particular context. We must admit, however, that our understanding of how the human mind actually operates in this respect is very rudimentary and inconclusive.

NOTES

6.1. Several authors [Gatlin, 1972; Bandler and Kohout, 1980c; Vallet, LeGuyader and Moulin, 1981; Wenstop, 1979] have explored the use of mathematical representations of uncertainty and information within the natural, life, and social sciences. Within the area of psychology and cognitive science, applications have included concept and knowledge formation and manipulation [Oden, 1979; Kochen, 1975, 1982; Oda, Shimomura, and Womack, 1980], memory and learning [Wagner, 1981; Kokawa, Nakamura, Oda, 1975; and Rocha, 1982], and behavior modification [Cervin and Jain, 1982]. Attneave [1959] and Quastler [1955] present the applications of classical information theory to psychology. Applications in the area of sociology have been explored by Aulin [1982], Niedenthal and Cantor [1984], Webber [1979], and Bouchon [1981]. Within the field of economics, applications of information theory are discussed by Theil [1967] and Georgescu-Roegen [1971], and the applications of fuzzy set theory are described by Chen, Lee and Yu [1983]. Additional applications of fuzzy set theory have been made in the area of ecology [Bosserman and Ragade, 1981; and Giering and Kandel, 1983], geography [Gale, 1972], meteorology [Zhang and Chen, 1984], and biology [da Rocha, 1982]. Yu [1976] presents applications of information theory in the field of optics.

6.2. Takagi and Sugeno [1984] discuss the construction of fuzzy control rules from operator actions; Sugeno and Takagi [1983] discuss the derivation of these control rules from fuzzy models. Problems of stability analysis, the selection of appropriate control parameters, and the properties of fuzzy control systems have been discussed by Braae and Rutherford [1979], Kickert and Mamdani [1978], Czogała and Pedrycz [1982], and Burdzy and Kiszka [1983].

The type of linguistic rule described in Sec. 6.3 has formed the basis for the design of several different prototype fuzzy controllers. These include controllers for industrial processes such as a cement kiln [Holmblad and Østergaard, 1982], heat exchange process [Østergaard, 1977], steam engine [Mamdani and Assilian, 1975; King and Mamdani, 1977], as well as others [Sugeno and Kang, 1986; Zeising, Wagenknecht, and Hartmann, 1984]. In addition, fuzzy controllers have been created for the control of a model car [Sugeno and Nishida, 1985], the control of automobile speed [Murakami, 1984], traffic junctions [Mamdani and Pappis, 1977; Nakatsuyama, Nagahashi, and Nishizuka, 1984], robot control [Lakov, 1985; Uragami, Mizumoto, and Tanaka, 1976], and aircraft autopilot control [Larkin, 1985]. Surveys of the field of fuzzy control and overviews of the basics of fuzzy controllers are presented by Tong [1980], Mamdani, Østergaard, and Lembessis [1983], Sugeno [1985b], Mamdani and Sembi [1980], and Mamdani [1976]. Several papers presenting fuzzy control applications and developments have been collected by Sugeno [1985a]. This collection also contains a survey of the state of the field as well as an extensive annotated bibliography.

6.3. Fuzzy sets have been applied in various other areas of engineering such as civil engineering [Blockely, 1982; Hara, 1982], architecture and building damage assessment [Oguntade and Gero, 1981; Ishizuka, Fu, and Yao, 1982; Fu, Ishizuka, and Yao, 1982], switching circuits [Kandel and Lee, 1979], and robotics [Goguen, 1975]. Applications of information theory to engineering are discussed by Bell [1953], by Hyvärinen [1968], and by Reza [1961].

6.4. The use of belief functions in a linguistic model of diagnosis is presented by Gouvernet, Ayme, and Sanchez [1982]. Sanchez, Gouvernet, Bartolin, and Vovan [1982], Vila and Delgado [1983], Lesmo, Saitta, and Torasso [1982], Soula and Sanchez [1982], and Umeyama [1986] discuss some uses of possibility theory and fuzzy logic in modeling the diagnosis process. Within the more restricted area of electrocardiological (ECG) diagnosis, Kerre [1982] and Bortolan and Degani [1984] employ fuzzy methods of decision theory and pattern recognition. Other applications within the area of medicine include the use of fuzzy linguistic variables for questionnaires investigating the relation between social stresses, psychological factors, and the incidence of coronary disease [Saitta and Torasso, 1981], the incorporation of fuzziness in an expert system dealing with the treatment of diabetes [Buisson, Farrey, and Prade, 1985], the use of fuzzy linguistic descriptions of symptoms for the purpose of evaluating different treatment procedures [Oguntade and Beaumont, 1982], and the use of a fuzzy control algorithm for the movement of prosthetic limbs [Saridis and Stephanou, 1977].

6.5. The field of management, including the areas of decision making, operations research, and risk analysis, has been an active area for application of fuzzy set theory and the theory of evidence. Within the field of management, applications of fuzzy set theory have been made for inventory scheduling [Sommer, 1981], personnel management [Ollero and Freire, 1981], new-product development [Nojiri, 1982], investment [Hammerbacher and Yager, 1981], public policy [Esogbue, 1981, 1982; Bye, Biscoe, and Hennessey, 1985], planning [Terano, Sugeno, and Tsukamoto, 1984], management decision-support expert systems [Kacprzyk and Yager, 1985a], and systems management [Fiksel, 1981]. Applications of fuzzy sets in the area of risk analysis have been presented in a paper by Bruce and Kandel [1983] and in a book by Schmucker [1983]. Applications of fuzzy sets in operations research have been explored by Dubois [1983], Prade [1980], and in the book by Zimmerman [1985]. Surveys of the applications of fuzzy set theory to decision making can be found in books by Kickert [1978] and by Kacprzyk [1983].

6.6. Fuzzy mathematical programming is discussed by Zimmermann [1978b], Sakawa [1983], Luhandjula [1983, 1986], Hannan [1981], and Chanas [1983].

6.7. Applications of fuzzy set theory to theories of collective decision making have been presented by Dimitrov [1983], Buckley [1984], Bezdek, Spillman, and Spillman [1978, 1979], Nurmi [1981], and Nojiri [1979].

6.8. The applications of fuzzy sets in the study of decision making extend into the area of multistage or dynamic decision making [Kacprzyk, 1982]. In addition, fuzzy multiple-objective decision making models have been presented by Yager [1977, 1978] and Sukawa and Yano [1985]. Fuzzy decision trees are discussed by Adamo [1980]. The use of fuzzy set theory for the measurement of tranquility and anxiety in decision making is discussed by Yager [1982]. Dubois and Prade [1982d] explore the use of fuzzy numbers in theories of decision making. Other papers discussing fuzzy and possibilistic decision making include those by Yager [1980g], Wierzchon [1982b], Fung and Fu [1975], Enta [1982], Dubois and Prade [1979b], Dompere [1982], Borisov and Krumberg [1983], Van Laarhoven and Pedrycz [1983], and Orlovsky [1978]. Applications of information theory to decision making are presented by Smith [1974] and by Hallefjord and Jornsten [1986].

6.9. A fuzzy database design that accommodates imprecison is presented by Umano [1982, 1984]. This design allows domain values to be specified as possibility distributions and defines relations between these. Zemankova and Kandel [1985] also describe the uses of possibility theory in a fuzzy relational database design. Prade [1984] discusses the uses of possibility theory in the handling of incomplete information. A study of the psychological structures of concept similarity and their application to question-answering information systems is presented by Nakamura and Iwai [1982]. Fuzzy applications in the design of document-retrieval systems have been presented by Kohout, Keravnou, and Bandler [1984] and Zenner, DeCaluwe, and Kerre [1985]. These latter authors also address issues of implementation efficiency, particularly with regard to retrieval time. A discussion of the justifications and advantages of applying fuzzy set theory to information retrieval system design is presented by Radecki [1983] and a mathematical analysis of this type of application is given by Buell [1982].

6.10. A great deal of study has been made of the use of different methods of approximate reasoning for expert systems. Discussions of some of these can be found in Dutta [1985], Baldwin [1985], Bandler and Kohout [1985], Freska [1982], Hajek [1985], Kacprzyk and Yager [1985b], Yager [1985b, 1985c], and Mamdani and Efstathiou [1984]. Whalen and Schott [1985a] performed an interesting comparative study of 11 different logics for approximate reasoning in a small prototype expert system. An overview of the history and implications of fuzzy logic in expert systems is presented by Gaines and Shaw [1985]. Lesmo, Saitta, and Torasso [1983] discuss an approach to incorporating a learning capability for fuzzy production rules. Zemankova-Leech and Kandel [1984] explore the closely connected design issues between fuzzy relational databases and fuzzy expert systems. A great deal of the current interest in expert systems also stems from military applications; the use of fuzzy set theory in this type of expert system is discussed by Dockery [1982]. A book on the applications of fuzzy set theory in expert systems has been written by Negoita [1985]. It contains some extensive notes annotating additional selected readings in the area. A large number of papers dealing with applications of fuzzy logic and the theory of evidence to expert systems have been collected by Gupta, Kandel, Bandler, and Kiszka [1985]. This collection contains papers discussing theoretical issues in approximate reasoning, their application to expert system design, and some actual expert systems that have been implemented using these methods of managing uncertainty.

6.11. The notion of fuzzy algorithms was proposed by Zadeh [1968b] and some possibilities for fuzzy programming languages are discussed by Prade [1983].

6.12. Fuzzy set theory has been found applicable in the area of data display [Benson, 1982]. Here, the graphical display of data that is imprecise or incomplete is approached with the use of fuzzy techniques, which allow a visual representation of subjective categories. This representation is achieved by color scales depicting the variations in fuzzy set membership grades.

6.13. The application of fuzzy set theory to curve fitting is discussed by Chang and Pavlidis [1979]. De Mori [1983] describes the utility of fuzzy algorithms in modeling speech. The uses of information theory in the examination of perception is presented in a book by Moles [1966]. Further discussions and references to additional applications of fuzzy set theory can be found in the book by Dubois and Prade [1980a]; references to other applications of information theory may be found in the book by Guiasu [1977].

6.14. The development of fuzzy set theory has been closely interrelated with mathematical systems theory. This relationship is certainly prominent in the work by Zadeh [1965b, 1971c, 1973, 1974, 1982]. Fuzzy systems are covered in a precise mathematical form in a book by Negoita and Ralescu [1975]; they are covered in a more popular way in a book by Negoita [1981]. Fuzzy-state automata were introduced by Dal Cin [1975] and further studied by Kandel and Lee [1979]. Fuzzy relation equations have been applied to various problems of systems analysis by Pedrycz [1981, 1983c, 1984, 1985] and Higashi and Klir [1984b].

6.15. Reconstructability analysis of both possibilistic and probabilistic systems is covered in the overall context of systems problems solving in a book by Klir [1985]. This book also contains more detailed coverage of fuzzy data in the sense outlined in Sec. 6.8.

6.16. Dubois and Prade [1983a] developed the following well-justified formula connecting probabilities to possibilities:

$$r(x) = \sum_{y \in X} \min[p(x), p(y)].$$

For the inverse formula, let us assume that the universal set $X = \{x_1, x_2, \ldots, x_n\}$ is ordered according to decreasing values of possibilities, that is, $r(x_i) \geq r(x_j)$ for each $i < j$. Then,

$$p(x_i) = \sum_{j=i}^{n} \frac{1}{j} [r(x_j) - r(x_{j+1})],$$

where $r(x_{j+1}) = 0$.

6.17. Applications of both the maximum and minimum entropy principles have been extensive and broad; they have also been very successful. This area is well covered in two publications by Ronald Christensen [1981b, 1986], which perhaps give the most sophisticated uses of these principles. The use of these principles in the broad context of systems problem solving is demonstrated by Klir [1985]. The fundamental connection between the minimum entropy principle and pattern recognition has been established by Watanabe [1985].

6.18. Experimental work oriented toward the determination of membership functions reflecting subjective perceptions in various applications and contexts is far behind the great theoretical advances in fuzzy set theory. Nevertheless, some important results have already been obtained on the experimental side. In addition to the work by Norwich and Turksen [1984] which is briefly outlined in Sec. 6.8, other experimental developments have been reported by Devi and Sarma [1985], Hersh, Caramaza, and Brownell [1979], Hinde [1983], Kempton [1984], Nowakowska [1977, 1979, 1983], and Zimmermann [1978a].

6.19. Ashby's ideas of using information theory in dealing with systems problems have been advanced by various researchers, most notably by Conant [1969, 1974, 1976, 1979], Broekstra [1980, 1981], Klir [1985], and Krippendorff [1982].

6.20. A thought-provoking collection of short articles, comments, and book reviews that discuss the relevance of fuzzy set theory to various applications was published in *Human Systems Management* [1984], Vol. 4, pp. 301–324.

A

Uniqueness of Uncertainty

Measures

A.1. SHANNON ENTROPY

Uniqueness Theorem of Shannon Entropy. If a function $H:\mathcal{P} \to [0, \infty)$ satisfies the requirements of continuity, weak additivity (Eq. (5.40)), monotonicity, branching (Eq. (5.41)), and normalization (as defined in Sec. 5.3), then

$$H(p_1, p_2, \ldots, p_n) = - \sum_{i=1}^{n} p_i \log_2 p_i$$

for all $n \in \mathbb{N}$.

Proof: (i) First, we prove the proposition $f(n^k) = kf(n)$ for all $n, k \in \mathbb{N}$ by induction, where

$$f(n) = H\left(\frac{1}{n}, \frac{1}{n}, \ldots, \frac{1}{n}\right)$$

is the same function that is used in the definition of weak additivity. For $k = 1$, the proposition is trivially true. By the axiom of weak additivity, we have

$$f(n^{k+1}) = f(n^k \cdot n) = f(n^k) + f(n).$$

Assume the proposition is true for some $k \in \mathbb{N}$. Then,

$$f(n^{k+1}) = f(n^k) + f(n) = kf(n) + f(n)$$
$$= (k + 1)f(n),$$

which demonstrates that the proposition is true for all $k \in \mathbb{N}$.

(ii) Next we can prove that $f(n) = \log_2 n$. This proof is identical with the proof of Theorem 5.1, provided that we replace function I (denoting the Hartley information in Theorem 5.1) with our function f. Therefore, we do not repeat the proof here.

(iii) We prove now that $H(p, 1 - p) = -p \log_2 p - (1 - p)\log_2(1 - p)$ for rational p. Let $p = r/s$, where $r, s \in \mathbb{N}$. Then,

$$f(s) = H\left(\underbrace{\frac{1}{s}, \frac{1}{s}, \ldots, \frac{1}{s}}_{r}, \underbrace{\frac{1}{s}, \frac{1}{s}, \ldots, \frac{1}{s}}_{s-r}\right)$$

$$= H\left(\frac{r}{s}, \frac{s-r}{s}\right) + \frac{r}{s} f(r) + \frac{s-r}{s} f(s - r)$$

by the branching axiom. By (ii) and the definition of p, we obtain

$$\log_2 s = H(p, 1 - p) + p \log_2 r + (1 - p)\log_2(s - r).$$

Solving this equation for $H(p, 1 - p)$ results in

$$H(p, 1 - p) = -p \log_2 r + \log_2 s - (1 - p)\log_2(s - r)$$

$$= -p \log_2 r + p \log_2 s - p \log_2 s + \log_2 s$$

$$- (1 - p)\log_2(s - r)$$

$$= -p \log_2 \frac{r}{s} - (1 - p)\log_2 \frac{s - r}{s}$$

$$= -p \log_2 p - (1 - p)\log_2(1 - p).$$

(iv) We extend now (iii) to all real numbers $p \in [0, 1]$ with the help of the continuity axiom. Let p be any real number in the interval $[0, 1]$ and let p' be a series of rational numbers that approach p. Then,

$$H(p, 1 - p) = \lim_{p' \to p} (p', 1 - p')$$

by the continuity axiom. Moreover,

$$\lim_{p' \to p} H(p', 1 - p') = \lim_{p' \to p} [-p'\log_2 p' - (1 - p')\log_2(1 - p')]$$

$$= -p \log_2 p - (1 - p)\log_2(1 - p)$$

since all the functions on the right-hand side of the equation are continuous.

(v) We conclude the proof now by showing that

$$H(p_1, p_2, \ldots, p_n) = - \sum_{i=1}^{n} p_i \log_2 p_i.$$

This is accomplished by induction on n. The result is proved in (ii) and (iv) for n = 1, 2, respectively. For $n \geq 3$, we may use the branching axiom to obtain

$$H(p_1, p_2, \ldots, p_n) = H(p_A, p_n) + p_A H\left(\frac{p_1}{p_A}, \frac{p_2}{p_A}, \ldots, \frac{p_{n-1}}{p_A}\right) + p_n H\left(\frac{p_n}{p_n}\right),$$

where $p_A = p_1 + p_2 + \cdots + p_{n-1}$. Since $H(p_n/p_n) = H(1) = 0$ by (ii), we have

$$H(p_1, p_2, \ldots, p_n) = H(p_A, p_n) + p_A H\left(\frac{p_1}{p_A}, \frac{p_2}{p_A}, \ldots, \frac{p_{n-1}}{p_A}\right).$$

By (iv) and assuming the proposition to be proved is true for $n - 1$, we may rewrite this equation as

$$
\begin{aligned}
H(p_1, p_2, \ldots, p_n) &= -p_A\log_2 p_A - p_n\log_2 p_n - p_A \sum_{i=1}^{n-1} \frac{p_i}{p_A} \log_2 \frac{p_i}{p_A} \\
&= -p_A\log_2 p_A - p_n\log_2 p_n - \sum_{i=1}^{n-1} p_i\log_2 \frac{p_i}{p_A} \\
&= -p_A\log_2 p_A - p_n\log_2 p_n - \sum_{i=1}^{n-1} p_i\log_2 p_i + \sum_{i=1}^{n-1} p_i\log_2 p_A \\
&= -p_A\log_2 p_A - p_n\log_2 p_n - \sum_{i=1}^{n-1} p_i\log_2 p_i + p_A\log_2 p_A \\
&= -\sum_{i=1}^{n} p_i\log_2 p_i. \quad \blacksquare
\end{aligned}
$$

A.2. U-UNCERTAINTY

The uniqueness proof is divided into four parts, each covered in one section of this appendix: a derivation of the branching axiom in terms of basic assignments and its generalizations; a determination of function U for crisp possibility distributions; a determination of function U for two-value possibility distributions; and the uniqueness proof of the U-uncertainty for general possibility distributions.

Generalized Branching Property

First, we determine the form of the branching axiom expressed in terms of basic assignments. This is a subject of the following theorem.

Theorem 1.   The branching axiom expressed by Eq. (5.78) has the following form in terms of basic distributions:

$$U_m(\mu_1, \mu_2, \ldots, \mu_n) = U_m(\mu_1, \mu_2, \ldots, \mu_{k-3}, \mu_{k-2} + \mu_{k-1}, 0, \mu_k, \mu_{k+1}, \ldots, \mu_n)$$

$$+ (\mu_{k-2} + \mu_{k-1}) U_m \bigg(\underbrace{0, 0, \ldots, 0}_{k-3}, \frac{\mu_{k-2}}{\mu_{k-2} + \mu_{k-1}}, \frac{\mu_{k-1}}{\mu_{k-2} + \mu_{k-1}}, \underbrace{0, 0, \ldots, 0}_{n-k+1} \bigg)$$

$$- (\mu_{k-2} + \mu_{k-1})(\underbrace{0, 0, \ldots, 0}_{k-3}, 1, \underbrace{0, 0, \ldots, 0}_{n-k+2}), \quad (1)$$

where $k \in \mathbb{N}_{3,n+1}$.*

     Proof: Since Eq. (1), formed in terms of basic distributions, is claimed to be a counterpart of Eq. (5.78), which is based on possibility distributions, it is required that $t(\rho_1, \rho_2, \ldots, \rho_n) = (\mu_1, \mu_2, \ldots, \mu_n)$. Let $({}^1\mu_1, {}^1\mu_2, \ldots, {}^1\mu_n)$ denote the unknown basic distribution in the first term on the right-hand side of Eq. (1). This distribution must correspond to the possibility distribution in the first term on the right-hand side of Eq. (5.78). By Eq. (4.36), we obtain:

$$^1\mu_i = \rho_i - \rho_{i+1} = \mu_i \quad \text{for all } i \in \mathbb{N} \text{ except } i = k-2, k-1,$$

$$^1\mu_{k-1} = \rho_k - \rho_k = 0,$$

and

$$^1\mu_{k-2} = \rho_{k-2} - \rho_k = (\rho_{k-2} - \rho_{k-1}) + (\rho_{k-1} - \rho_k) = \mu_{k-2} + \mu_{k-1}.$$

     Let $({}^2\mu_1, {}^2\mu_2, \ldots, {}^2\mu_n)$ denote the unknown basic distribution in the second term on the right-hand side of Eq. (1). Expressing its connection with the possibility distribution in the corresponding term of Eq. (5.78) by Eq. (4.36), we obtain:

$$^2\mu_i = \rho_i - \rho_{i+1} = 0 \quad \text{for all } i \in \mathbb{N}_n \text{ except } i = k-2, k-1,$$

$$^2\mu_{k-2} = 1 - \frac{\rho_{k-1} - \rho_k}{\rho_{k-2} - \rho_k} = \frac{\rho_{k-2} - \rho_{k-1}}{\rho_{k-2} - \rho_k} = \frac{\mu_{k-2}}{\mu_{k-2} + \mu_{k-1}},$$

and

$$^2\mu_{k-1} = \frac{\rho_{k-1} - \rho_k}{\rho_{k-2} - \rho_k} = \frac{\mu_{k-1}}{\mu_{k-2} + \mu_{k-1}}.$$

     Similarly, for the basic distribution $({}^3\mu_1, {}^3\mu_2, \ldots, {}^3\mu_n)$ in the last term of Eq. (1), we obtain:

$$^3\mu_i = \rho_i - \rho_{i+1} = 0 \quad \text{for all } i \in \mathbb{N}_n \text{ except } i = k-2,$$

and

$$^3\mu_{k-2} = 1 - 0 = 1.$$

---

     * When $k = 3$, the sequence $\mu_1, \mu_2, \ldots, \mu_{k-3}$ in the first term on the right-hand side of (1) is vacuous and, consequently, $\mu_{k-2} + \mu_{k-1}$ becomes the first component in the basic distribution.

The weight by which the uncertainties in the last two terms of Eq. (1) are multiplied must be derived from the corresponding weight $\rho_{k-2} - \rho_k$ in Eq. (5.78). As already shown, $\rho_{k-2} - \rho_k$ corresponds to $\mu_{k-2} + \mu_{k-1}$. This concludes the proof. ■

Assume now that in formalizing the branching axiom the distinction between ρ_k and ρ_{k+1} in $\mathbf{r}$ is ignored by *replacing the lower possibility value ρ_{k+1} with the higher value $\rho_k(k \in N_{n-1})$*. Then we obtain an alternative form of the axiom:

$$U(\rho_1, \rho_2, \ldots, \rho_n) = U(\rho_1, \rho_2, \ldots, \rho_{k-1}, \rho_k, \rho_k, \rho_{k+2}, \ldots, \rho_n)$$

$$+ (\rho_k - \rho_{k+2})U\left(\underbrace{1, 1, \ldots, 1,}_{k} \frac{\rho_{k+1} - \rho_{k+2}}{\rho_k - \rho_{k+2}}, \underbrace{0, 0, \ldots, 0}_{n - k - 1}\right)$$

$$- (\rho_k - \rho_{k+2})U(\underbrace{1, 1, \ldots, 1,}_{k + 1} \underbrace{0, 0, \ldots, 0}_{n - k - 1}). \tag{2}$$

It can easily be demonstrated, following the line of reasoning in the proof of Theorem 1, that the branching axiom expressed by Eq. (2) has the following form in terms of basic distributions:

$$U_m(\mu_1, \mu_2, \ldots, \mu_n) = U_m(\mu_1, \mu_2, \ldots, \mu_{k-1}, 0, \mu_k + \mu_{k+1}, \mu_{k+2}, \ldots, \mu_n)$$

$$+ (\mu_k + \mu_{k+1})U_m\left(\underbrace{0, 0, \ldots, 0,}_{k - 1} \frac{\mu_k}{\mu_k + \mu_{k+1}}, \frac{\mu_{k+1}}{\mu_k + \mu_{k+1}}, \underbrace{0, 0, \ldots, 0}_{n - k - 1}\right)$$

$$- (\mu_k + \mu_{k+1})U_m(\underbrace{0, 0, \ldots, 0,}_{k} \underbrace{1, 0, 0, \ldots, 0}_{n - k - 1}). \tag{3}$$

Let Eqs. (5.78) and (1) be referred to as the *first forms* of the branching axiom and Eqs. (2) and (3) as its *second forms*.

The first generalized form of the branching property involves the grouping of q neighboring values ($q \geq 2$) of the given basic distribution (rather than two values, as in the branching axioms) by replacing higher values with a single lower value. The form is expressed by the following theorem; the proof of the theorem is done by an induction on q.

Theorem 2.   For all $k \in N_{2,n}$ and all $q \in N_{2,n}$ such that $q < k$, the following equation holds:*

---

* When $q = k - 1$, the sequence $\mu_1, \mu_2, \ldots, \mu_{k-q-1}$ in the first term on the right-hand side of (4) is vacuous and, consequently, $s_1(q)$ becomes the first component in the basic distribution.

$$U_m(\mu_1, \mu_2, \ldots, \mu_n) = U_m(\mu_1, \mu_2, \ldots, \underbrace{\mu_{k-q-1}, s_1(q), 0, 0, \ldots, 0,}_{q-1}$$

$$\mu_k, \mu_{k+1}, \ldots, \mu_n)$$

$$+ s_1(q)U_m\left(\underbrace{0, 0, \ldots, 0,}_{k-q-1} \frac{\mu_{k-q}}{s_1(q)}, \frac{\mu_{k-q+1}}{s_1(q)}, \ldots, \frac{\mu_{k-1}}{s_1(q)}, \underbrace{0, 0, \ldots, 0}_{n-k+1}\right) \tag{4}$$

$$- s_1(q)U_m(\underbrace{0, 0, \ldots, 0,}_{k-q-1} 1, \underbrace{0, 0, \ldots, 0}_{n-k+q}),$$

where

$$s_1(q) = \sum_{i=0}^{q-1} \mu_{k-q+i}. \tag{5}$$

Proof: By Theorem 1, Eq. (4) is true for $q = 2$ and any $k > q$. Assume that the theorem is true for $q - 1$. Then, the following equation holds for any arbitrary $k > q - 1$:

$$U_m(\mu_1, \mu_2, \ldots, \mu_{k-3}, \mu_{k-2} + \mu_{k-1}, 0, \mu_k, \mu_{k+1}, \ldots, \mu_n)$$

$$= U_m(\mu_1, \mu_2, \ldots, \underbrace{\mu_{k-q-1}, s_1(q), 0, 0, \ldots, 0,}_{q-1} \mu_k, \mu_{k+1}, \ldots, \mu_n)$$

$$+ s_1(q)U_m\left(\underbrace{0, 0, \ldots, 0,}_{k-q-1} \frac{\mu_{k-q}}{s_1(q)}, \frac{\mu_{k-q+1}}{s_1(q)}, \ldots, \frac{\mu_{k-3}}{s_1(q)}, \frac{\mu_{k-2} + \mu_{k-1}}{s_1(q)},\right.$$

$$\left.\underbrace{0, 0, \ldots, 0}_{n-k+2}\right) - s_1(q)U_m(\underbrace{0, 0, \ldots, 0,}_{k-q-1} 1, \underbrace{0, 0, \ldots, 0}_{n-k+q}). \tag{6}$$

This equation is obtained by applying Eq. (4) to $q - 1$ and the special basic distribution, which is constructed from the same components $\mu_1, \mu_2, \ldots, \mu_n$ that are used in Eq. (4).

When applying the branching axiom in the form of Eq. (1) to another special basic distribution

$$\left(\underbrace{0, 0, \ldots, 0,}_{k-q-1} \frac{\mu_{k-q}}{s_1(q)}, \frac{\mu_{k-q+1}}{s_1(q)}, \ldots, \frac{\mu_{k-1}}{s_1(q)}, \underbrace{0, 0, \ldots, 0}_{n-k+1}\right)$$

and multiplying the resulting equation by $s_1(q)$, we obtain:

$$s_1(q)U_m\left(\underbrace{0, 0, \ldots, 0}_{k-q-1}, \frac{\mu_{k-q}}{s_1(q)}, \frac{\mu_{k-q+1}}{s_1(q)}, \ldots, \frac{\mu_{k-1}}{s_1(q)}, \underbrace{0, 0, \ldots, 0}_{n-k+1}\right)$$

$$= s_1(q)U_m\left(\underbrace{0, 0, \ldots, 0}_{k-q-1}, \frac{\mu_{k-q}}{s_1(q)}, \frac{\mu_{k-q+1}}{s_1(q)}, \ldots, \frac{\mu_{k-3}}{s_1(q)},\right.$$

$$\left.\frac{\mu_{k-2} + \mu_{k-1}}{s_1(q)}, \underbrace{0, 0, \ldots, 0}_{n-k+2}\right)$$

$$+ (\mu_{k-2} + \mu_{k-1})U_m\left(\underbrace{0, 0, \ldots, 0}_{k-3}, \frac{\mu_{k-2}}{\mu_{k-2} + \mu_{k-1}}, \frac{\mu_{k-1}}{\mu_{k-2} + \mu_{k-1}},\right.$$

$$\left.\underbrace{0, 0, \ldots, 0}_{n-k+1}\right) - (\mu_{k-2} + \mu_{k-1})U_m(\underbrace{0, 0, \ldots, 0}_{k-3}, 1, \underbrace{0, 0, \ldots, 0}_{n-k+2}). \qquad (7)$$

We now perform two substitutions: (i) we express the second term in Eq. (1) explicitly and substitute it for the first term in Eq. (6), and (ii) we express the second term in Eq. (7) explicitly and substitute it for the third term in Eq. (6). The result is the following equation:

$$U_m(\mu_1, \mu_2, \ldots, \mu_n) - (\mu_{k-2} + \mu_{k-1})U_m\left(\underbrace{0, 0, \ldots, 0}_{k-3}, \frac{\mu_{k-2}}{\mu_{k-2} + \mu_{k-1}},\right.$$

$$\left.\frac{\mu_{k-1}}{\mu_{k-2} + \mu_{k-1}}, \underbrace{0, 0, \ldots, 0}_{n-k+1}\right)$$

$$+ (\mu_{k-2} + \mu_{k-1})U_m(\underbrace{0, 0, \ldots, 0}_{k-3}, 1, \underbrace{0, 0, \ldots, 0}_{n-k+2})$$

$$= U_m(\mu_1, \mu_2, \ldots, \mu_{k-q-1}, s_1(q), \underbrace{0, 0, \ldots, 0}_{q-1}, \mu_k, \mu_{k+1}, \ldots, \mu_n)$$

$$+ s_1(q)U_m \left(\underbrace{0, 0, \ldots, 0,}_{k-q-1} \frac{\mu_{k-q}}{s_1(q)}, \frac{\mu_{k-q+1}}{s_1(q)}, \ldots, \frac{\mu_{k-1}}{s_1(q)}, \underbrace{0, 0, \ldots, 0}_{n-k+1} \right)$$

$$- (\mu_{k-2} + \mu_{k+1})U_m \left(\underbrace{0, 0, \ldots, 0,}_{k-3} \frac{\mu_{k-2}}{\mu_{k-2} + \mu_{k-1}}, \frac{\mu_{k-1}}{\mu_{k-2} + \mu_{k-1}}, \right.$$

$$\left. \underbrace{0, 0, \ldots, 0}_{n-k+1} \right) + (\mu_{k-2} + \mu_{k-1})U_m(\underbrace{0, 0, \ldots, 0,}_{k-3} 1, \underbrace{0, 0, \ldots, 0}_{n-k+2})$$

$$- s_1(q)U_m(\underbrace{0, 0, \ldots, 0,}_{k-q-1} 1, \underbrace{0, 0, \ldots, 0}_{n-k+q}).$$

Cancellation of common terms on both sides of this equation yields Eq. (4). This concludes the proof. ∎

The second generalized form of the branching property can be derived in a similar way by using the second form of the branching axiom, expressed by Eq. (3):

$$U_m(\mu_1, \mu_2, \ldots, \mu_n) = U_m(\mu_1, \mu_2, \ldots, \mu_{k-1}, \underbrace{0, 0, \ldots, 0,}_{q-1}$$

$$s_2(q), \mu_{k+q}, \mu_{k+q+1}, \ldots, \mu_n)$$

$$+ s_2(q)U_m \left(\underbrace{0, 0, \ldots, 0,}_{k-1} \frac{\mu_k}{s_2(q)}, \frac{\mu_{k+1}}{s_2(q)}, \ldots, \frac{\mu_{k+q-1}}{s_2(q)}, \underbrace{0, 0, \ldots, 0}_{n-k-q+1} \right)$$

$$- s_2(q)U_m(\underbrace{0, 0, \ldots, 0,}_{k+q-2} 1, \underbrace{0, 0, \ldots, 0}_{n-k-q+1}), \tag{8}$$

where

$$s_2(q) = \sum_{i=0}^{q-1} \mu_{k+i}. \tag{9}$$

For the sake of completeness, let us also describe both generalized forms of the branching property in terms of possibility distributions. They can be easily derived from their counterparts in the domain of basic distributions by Eqs. (4.36) and (4.35). The first form, which corresponds to Eq. (4), is expressed by the following equations:

$$U(\rho_1, \rho_2, \ldots, \rho_n) = U(\rho_1, \rho_2, \ldots, \underbrace{\rho_{k-q}, \rho_k, \ldots, \rho_k}_{q}, \rho_{k+1}, \rho_{k+2}, \ldots, \rho_n)$$

$$+ (\rho_{k-q} - \rho_k)U\left(\underbrace{1, 1, \ldots, 1}_{k-q}, \frac{\rho_{k-q+1} - \rho_k}{\rho_{k-q} - \rho_k}, \frac{\rho_{k-q+2} - \rho_k}{\rho_{k-q} - \rho_k}, \ldots, \frac{\rho_{k-1} - \rho_k}{\rho_{k-q} - \rho_k}, \underbrace{0, 0, \ldots, 0}_{n-k+1}\right)$$

$$- (\rho_{k-q} - \rho_k)U(\underbrace{1, 1, \ldots, 1}_{k-q}, \underbrace{0, 0, \ldots, 0}_{n-k+q}). \tag{10}$$

The second form, which corresponds to Eq. (8), is expressed as follows:

$$U(\rho_1, \rho_2, \ldots, \rho_n) = U(\rho_1, \rho_2, \ldots, \underbrace{\rho_{k-1}, \rho_k, \ldots, \rho_k}_{q}, \rho_{k+q}, \rho_{k+q+1}, \ldots, \rho_n)$$

$$+ (\rho_k - \rho_{k+q})U\left(\underbrace{1, 1, \ldots, 1}_{k}, \frac{\rho_{k+1} - \rho_{k+q}}{\rho_k - \rho_{k+q}}, \frac{\rho_{k+2} - \rho_{k+q}}{\rho_k - \rho_{k+q}}, \ldots, \frac{\rho_{k+q-1} - \rho_{k-q}}{\rho_k - \rho_{k+q}}, \underbrace{0, 0, \ldots, 0}_{n-k-q+1}\right)$$

$$- (\rho_k - \rho_{k+q})U(\underbrace{1, 1, \ldots, 1}_{k+q-1}, \underbrace{0, 0, \ldots, 0}_{n-k-q+1}). \tag{11}$$

Uncertainty Measure for Crisp Possibility Distributions

A crisp possibility distribution $\mathbf{r}_c$ represents uniquely a crisp set. When ordered, any crisp possibility distribution defined on a set with n elements has the form

$$\mathbf{r}_c = (\underbrace{1, 1, \ldots, 1}_{a}, \underbrace{0, 0, \ldots, 0}_{n-a}).$$

Due to the expansibility axiom,

$$U(\mathbf{r}_c) = U(\underbrace{1, 1, \ldots, 1}_{a});$$

that is, U becomes identical to the function I defined in Sec. 5.3. Hence, by Theorem 5.1, we have

$$I(a) = \log_2 a. \tag{12}$$

Let us use this result to express the first form of the generalized branching property (Eq. (4)) in a shorthand notation:

$$U_m(\mathbf{m}) = U_m(^g\mathbf{m}_1) + s_1(q)U_m(^l\mathbf{m}_1) - s_1(q)\log_2(k - q), \tag{13}$$

where $^g\mathbf{m}_1$ and $^l\mathbf{m}_1$ denote the global and local basic distributions in Eq. (4), respectively, and $s_1(q)$ is given by Eq. (5). Similarly, let

$$U_m(\mathbf{m}) = U_m(^g\mathbf{m}_2) + s_2(q)U_m(^l\mathbf{m}_2) - s_2(q)\log_2(k + q - 1) \qquad (14)$$

be a shorthand expression of the second form of the generalized branching property (Eq. (8)), where $^g\mathbf{m}_2$ and $^l\mathbf{m}_2$ denote the global and local basic distributions in Eq. (8), respectively, and $s_2(q)$ is given by Eq. (9). These simplified expressions of the two forms of the branching property are used later.

Uniqueness of the U-uncertainty for Two-Value Possibility Distributions

The purpose of this section is to prove the uniqueness of the U-uncertainty for a special class of possibility distributions. These distributions contain only two possibility values in their components, 1 and some value α less than 1 but greater than 0. More formally, a possibility distribution $\mathbf{r} = (\rho_i \mid i \in \mathbb{N}_n)$ is called a *two-value possibility distribution with u leading* 1s, $u \in \mathbb{N}_{n-1}$, and $n - u$ α's, $0 < \alpha < 1$, if

$$\rho_i = \begin{cases} 1 & \text{for } i \in \mathbb{N}_u \\ \alpha & \text{for } i \in \mathbb{N}_{u+1,n}, \quad 0 < \alpha < 1. \end{cases}$$

The following theorem relates a two-value possibility distribution with a special basic distribution.

Theorem 3. Any two-value possibility distribution of length n and with u leading 1s and $n - u$ α's corresponds to a basic distribution $\mathbf{m} = (\mu_i \mid i \in \mathbb{N}_n)$ such that

$$\mu_i = \begin{cases} \alpha & \text{for } i = n, \quad 0 < \alpha < 1 \\ 1 - \alpha & \text{for } i = u \\ 0 & \text{for all } i \text{ not equal } n \text{ or } u. \end{cases}$$

Such a basic distribution is called a two-value basic distribution with one nonzero component at u.

Proof. The theorem follows immediately from the conversion rule between possibility distributions and basic distributions expressed by Eq. (4.36). ∎

Example 1

The possibility distribution $\mathbf{r} = (1, 1, 1, .3, .3)$ is a two-value possibility distribution with three leading 1s and corresponds to the two-value basic distribution with one nonzero component at three, $\mathbf{m} = (0, 0, .7, 0, .3)$.

The next two theorems develop the form of a joint possibility distribution associated with an arbitrary number of two-value marginal possibility distributions.

Theorem 4. The joint possibility distribution associated with two ordered marginal possibility distributions of length n that are noninteractive and two-value with u leading ones ($u \in \mathbb{N}_{n-1}$) and $n - u$ α's ($0 < \alpha < 1$) is a two-value possibility distribution with u^2 leading 1s and $n^2 - u^2$ α's corresponding to a two-value basic distribution at u^2.

Proof. Let the two marginal possibility distributions in this theorem be denoted by ${}^1\mathbf{r} = ({}^1\rho_i \mid i \in \mathbb{N}_n)$ and ${}^2\mathbf{r} = ({}^2\rho_i \mid i \in \mathbb{N}_n)$. Let $\mathbf{r} = (\rho_{11}, \rho_{12}, \ldots, \rho_{1n}, \rho_{21}, \ldots, \rho_{nn})$ denote the joint possibility distribution, where $\rho_{ij} = \min({}^1\rho_i, {}^2\rho_j)$ for all $i, j \in \mathbb{N}_n$. Observe that, in general, $\mathbf{r}$ is not ordered.

Since ${}^1\rho_i, {}^2\rho_j \in \{1, \alpha\}$ for all $i, j \in \mathbb{N}_n$ and the min operator chooses either ${}^1\rho_1$ or ${}^2\rho_j$, $\rho_{ij} \in \{1, \alpha\}$ for all $i, j \in \mathbb{N}_n$. Moreover, since ${}^1\rho_i = {}^2\rho_i$ for all $i \in \mathbb{N}_n$, $\rho_{ij} = 1$ when $i, j \in \mathbb{N}_r$ and $\rho_{ij} = \alpha$ when $i, j \in \mathbb{N}_{n-r+1,n}$. Hence, when $\mathbf{r}$ is ordered, a two-value possibility distribution with u^2 leading ones and $n^2 - u^2$ α's results.

By Theorem 3, a two-value basic distribution with one nonzero component at u^2, is associated with a two-value possibility distribution with u^2 leading ones and $n^2 - u^2$ α's. ∎

The result in Theorem 4 can be extended for any p ordered marginal distributions of length n, all noninteractive with each other and all two-value with u leading ones ($u \in \mathbb{N}_{n-1}$) and $n - u$ α's ($0 < \alpha < 1$). The joint possibility distribution is a two-value possibility distribution with u^p leading ones and $n^p - u^p$ α's, which corresponds to a two-value basic distribution with $1 - \alpha$ at u^p.

Theorem 5.   Assume that the domain of function U (or alternatively U_m) that satisfies the axioms stated in Sec. 5.6 is restricted to the domain of all two-value possibility distributions (or, alternatively, all two-value basic distributions). Then, for any two-value possibility distribution $\mathbf{r}$ with $u(u \in \mathbb{N}_{n-1})$ leading ones and $n - u$ α's ($0 < \alpha < 1$),

$$U(\mathbf{r}) = U_m(\mathbf{m}) = (1 - \alpha)\log_2 u + \alpha \log_2 n \qquad (15)$$

Proof. Consider the function $U_m(0, 0, \ldots, 0, \underbrace{}_{u^2 - 1} 1 - \alpha, 0, 0, \ldots, 0, \underbrace{}_{n^2 - u^2 - 1} \alpha)$,

where the basic distribution $(0, 0, \ldots, 0, \underbrace{}_{u^2 - 1} 1 - \alpha, 0, 0, \ldots, 0, \underbrace{}_{n^2 - u^2 - 1} \alpha)$ is a joint

distribution associated with two arbitrary marginal distributions from Theorem 4. When applying the first form of the generalized branching property, expressed by Eq. (13), to this function with $k = u^2 + 1$ and $q = u^2 - u + 1$, we obtain

$$U_m(0, 0, \ldots, 0, \underbrace{}_{u^2 - 1} 1 - \alpha, 0, 0, \ldots, 0, \underbrace{}_{n^2 - u^2 - 1} \alpha) = U_m(0, 0, \ldots, 0, \underbrace{}_{u - 1}$$

$$1 - \alpha, 0, 0, \ldots, 0, \underbrace{}_{n^2 - u - 1} \alpha)$$

$$+ (1 - \alpha)U_m(0, 0, \ldots, 0, \underbrace{}_{u^2 - 1} 1, 0, 0, \ldots, 0, \underbrace{}_{n^2 - u^2}) - (1 - \alpha)\log_2(u). \qquad (16)$$

When applying the first form of the generalized branching property to the function
$U_m(0, 0, \ldots, \underbrace{0, 1 - \alpha, 0, 0}_{u-1}, \ldots, \underbrace{0, \alpha}_{n^2 - u - 1})$ with $k = n^2 + 1$ and $q = n^2 - n + 1$,

we obtain

$U_m(0, 0, \ldots, \underbrace{0, 1 - \alpha, 0, 0, \ldots, 0, \alpha}_{u-1 \qquad\qquad n^2 - u - 1})$

$= U_m(0, 0, \ldots, \underbrace{0, 1 - \alpha, 0, 0, \ldots, 0, \alpha, 0, 0, \ldots, 0}_{u-1 \qquad\qquad n-u-1 \qquad\qquad n^2-n})$

$+ \alpha U_m(0, 0, \ldots, \underbrace{0, 1}_{n^2-1}) - \alpha \log_2 n.$ \hfill (17)

Substituting Eq. (17) for the second term in Eq. (16) and observing that

$U_m(0, 0, \ldots, \underbrace{0, 1, 0, 0, \ldots, 0}_{u^2-1 \qquad\qquad n^2-u^2}) = \log_2 u^2$

in Eq. (16) and that

$U_m(\underbrace{0, 0, \ldots, 0, 1}_{n^2-1}) = \log_2 n^2$

in Eq. (17), we get

$U_m(0, 0, \ldots, \underbrace{0, 1 - \alpha, 0, 0, \ldots, 0, \alpha}_{u^2-1 \qquad\qquad n^2-u^2-1})$

$= U_m(0, 0, \ldots, \underbrace{0, 1 - \alpha, 0, 0, \ldots, 0, \alpha, 0, 0, \ldots, 0}_{u-1 \qquad\qquad n-u-1 \qquad\qquad n^2-n})$

$+ \alpha \log_2 n^2 - \alpha \log_2 n + (1 - \alpha)\log_2 u^2 - (1 - \alpha)\log_2 u.$ \hfill (18)

By the additivity property,

$U_m(0, 0, \ldots, \underbrace{0, 1 - \alpha, 0, 0, \ldots, 0, \alpha}_{u^2-1 \qquad\qquad n^2-u^2-1})$

$= 2U_m(0, 0, \ldots, \underbrace{0, 1 - \alpha, 0, 0, \ldots, 0, \alpha}_{u-1 \qquad\qquad n-u-1})$ \hfill (19)

and by the expansibility axiom,

$$U_m(\underbrace{0, 0, \ldots, 0,}_{u - 1} 1 - \alpha, \underbrace{0, 0, \ldots, 0,}_{n - u - 1} \alpha, \underbrace{0, 0, \ldots, 0)}_{n^2 - n}$$

$$= U_m(\underbrace{0, 0, \ldots, 0,}_{u - 1} 1 - \alpha, \underbrace{0, 0, \ldots, 0,}_{n - u - 1} \alpha).$$

From Eqs. (18) and (19) we obtain

$$U_m(\underbrace{0, 0, \ldots, 0,}_{u - 1} 1 - \alpha, \underbrace{0, 0, \ldots, 0,}_{n - u - 1} \alpha)$$

$$= \alpha \log_2 n^2 - \alpha \log_2 n + (1 - \alpha)\log_2 u^2 - (1 - \alpha)\log_2 u. \quad (20)$$

Next we observe that $\alpha \log_2 n^2 - \alpha \log_2 n = 2\alpha \log_2 n - \alpha \log_2 n = \alpha(p - 1)\log_2 n$ and $(1 - \alpha)\log_2 u^2 - (1 - \alpha)\log_2 u = (1 - \alpha)\log_2 u$.

Using these expressions to make appropriate replacements in Eq. (20) gives us the desired form

$$U_m(\underbrace{0, 0, \ldots, 0,}_{u - 1} 1 - \alpha, \underbrace{0, 0, \ldots, 0,}_{n - u - 1} \alpha) = (1 - \alpha)\log_2 u + \alpha \log_2 n. \quad \blacksquare$$

Uniqueness Theorem

In this section, we utilize all previous results regarding the uniqueness of the U-uncertainty for special possibility or basic distributions to prove its uniqueness for arbitrary possibility or basic distributions.

Theorem 6 (*Uniqueness Theorem*).   The only function U (or, alternatively, U_m) that satisfies the axioms stated in Section 5.6 is the U-uncertainty, that is,

$$U(\mathbf{r}) = U_m(\mathbf{m}) = \sum_{i=1}^{n} \mu_i \log_2 i, \quad (21)$$

where $\mathbf{r}$ is an arbitrary ordered possibility distribution and $\mathbf{m} = t(\mathbf{r})$.

Proof. Consider a basic distribution of a specific length $n(n \in \mathbb{N} - \{1\})$ with w nonzero components ($w \in \mathbb{N}_{2,n}$), say $m = (\mu_i \mid i \in \mathbb{N}_n)$. For $w = 2$, the form of Eq. (21) follows from Theorem 5. Assume that Eq. (21) is true for $w - 1$. Applying the first form of the branching axiom (Eq. (1)), we obtain:

$$U_m(\mu_1, \mu_2, \ldots, \mu_n) = U_m(\mu_1, \mu_2, \ldots, \mu_{k-3}, \mu_{k-2} + \mu_{k-1},$$

$$0, \mu_k, \mu_{k+1}, \ldots, \mu_n)$$

$$+ (\mu_{k-2} + \mu_{k-1}) U_m\left(\underbrace{0, 0, \ldots, 0,}_{k - 3} \frac{\mu_{k-2}}{\mu_{k-2} + \mu_{k-1}}, \frac{\mu_{k-1}}{\mu_{k-2} + \mu_{k-1}}, \underbrace{0, 0, \ldots, 0}_{n - k + 1}\right)$$

$$- (\mu_{k-2} + \mu_{k-1})\log_2(k - 2). \quad (22)$$

The distribution associated with the second term of this equation has $w - 1$ nonzero components. Hence, by the assumption that Eq. (21) is true for $w - 1$, we obtain:

$$U_m(\mu_1, \mu_2, \ldots, \mu_{k-3}, \mu_{k-2} + \mu_{k-1}, 0, \mu_k, \mu_{k+1}, \ldots, \mu_n)$$

$$= \sum_{i=1}^{k-3} \mu_i \log_2 i + (\mu_{k-2} + \mu_{k-1})\log_2(k - 2) + \sum_{i=k}^{n} \mu_i \log_2 i. \quad (23)$$

By the expansibility axiom and Theorem 5, Eq. (15) with $u = k - 2$ and $n = k + 1$, we get,

$$U_m\left(\underbrace{0, 0, \ldots, 0,}_{k-3} \frac{\mu_{k-2}}{\mu_{k-2} + \mu_{k-1}}, \frac{\mu_{k-1}}{\mu_{k-2} + \mu_{k-1}}, \underbrace{0, 0, \ldots, 0}_{n-k+1}\right)$$

$$= U_m\left(\underbrace{0, 0, \ldots, 0,}_{k-3} \frac{\mu_{k-2}}{\mu_{k-2} + \mu_{k-1}}, \frac{\mu_{k-1}}{\mu_{k-2} + \mu_{k-1}}\right)$$

$$= \frac{\mu_{k-2}}{\mu_{k-2} + \mu_{k-1}} \log_2 k - 2 + \frac{\mu_{k-1}}{\mu_{k-2} + \mu_{k-1}} \log_2(k - 1). \quad (24)$$

Substituting for the second and third term in Eq. (22) from Eqs. (23) and (24), respectively, we obtain

$$U_m(\mu_1, \mu_2, \ldots, \mu_n) = \sum_{i=1}^{k-3} \mu_i \log_2 i + (\mu_{k-2} + \mu_{k-1})\log_2(k - 2)$$

$$+ \sum_{i=k}^{n} \log_2 i + \mu_{k-2}\log_2(k - 2) + \mu_{k-1}\log_2(k - 1)$$

$$- (\mu_{k-2} + \mu_{k-1})\log_2(k - 2)$$

$$= \sum_{i=1}^{n} \mu_i \log_2 i. \quad \blacksquare$$

Note that a similar proof can be made using the second form of the generalized branching axiom (Eq. (3)) instead of the first form. Therefore, either of these forms may be adopted as the branching axiom for Theorem 6.

A.3. GENERAL MEASURE OF NONSPECIFICITY

Uniqueness theorem of the measure *V* of nonspecificity.   If a function

$$V : \mathcal{M} \rightarrow [0, \infty),$$

where $\mathcal{M}$ is the set of all basic assignments, satisfies the requirements of symmetry, additivity, subadditivity, branching, and normalization (i.e., the axiomatic requirements V1–V5 defined in Sec. 5.6), then

$$V(m) = \sum_{A \in \mathcal{F}} m(A)\log_2(A),$$

where $\mathcal{F}$ denotes the set of focal elements of m.

Proof: To facilitate the proof, let us introduce a convenient notation. Let $(^{\alpha}A, {}^{\beta}B, {}^{\gamma}C, \ldots)$ denote a basic assignment such that $m(A) = \alpha$, $m(B) = \beta$, $m(C) = \gamma, \ldots$, where $A, B, C, \ldots$ are the focal elements of m. When we are concerned only with the cardinalities of the focal elements, the basic assignment may be written in the form $(^{\alpha}a, {}^{\beta}b, {}^{\gamma}c, \ldots)$, where $a, b, c, \ldots$ are positive integers that stand for the cardinalities of the focal elements; occasionally, when $\alpha = 1$, we write (a) instead of (^{1}a). For the sake of simplicity, let us refer to the value $m(A)$ of a basic assignment m as the *weight* of A.

(i) First, we prove that $V(^{1}a) = \log_2 a$. For convenience, let $W(a) = V(^{1}a)$; that is, W is a nonnegative real-valued function on $\mathbb{N}$. From additivity (Axiom V2), we have

$$W(ab) = W(a) + W(b). \tag{a}$$

When $a = b = 1$, $W(1) = W(1) + W(1)$ and, consequently, $W(1) = 0$. To show that $W(a) \le W(a + 1)$, let us take set A of $a(a + 1)$ elements and assume that it is a subset of a Cartesian product $X \times Y$. By symmetry (Axiom V1), $V(A)$ does not depend on how the elements of A are arranged. Let us consider two possible arrangements. In the first arrangements, A is viewed as a rectangle $a \times (a + 1)$, where $\times$ denotes here a Cartesian product of sets with cardinalities a and $a + 1$. Then

$$V(A) = W(a) + W(a + 1) \tag{b}$$

by additivity. In the second arrangement, we view A as a subset of a square $(a + 1) \times (a + 1)$ such that at least one diagonal of the square is fully covered. In this case, the projections onto both X and Y have the same cardinality $a + 1$. Hence, by subadditivity (Axiom V3), we have

$$V(A) \le W(a + 1) + W(a + 1). \tag{c}$$

It follows immediately from (b) and (c) that

$$W(a) \le W(a + 1).$$

Function W is thus monotonic nondecreasing. Since it is also additive in the sense of (a) and normalized by Axiom V5, it is equivalent to the Hartley information I. Hence, by Theorem 5.1,

$$W(a) = V(^{1}a) = \log_2 a. \tag{d}$$

(ii) We prove now that $V(^{\alpha}1, {}^{\beta}1, {}^{\gamma}1, \ldots) = 0$. First, we show that the equality holds only for two focal elements, that is, we show that $V(^{\alpha}1, {}^{\beta}1) = 0$. Consider a set A of a^2 elements for some sufficiently large, temporarily fixed a. Let $B =$

$A \times (^\alpha 1, {}^\beta 1)$. Then, $V(A) = 2 \log_2 a$ and

$$V(B) = 2 \log_2 a + V(^\alpha 1, {}^\beta 1) \tag{e}$$

by Eq. (d) and additivity. Set B can be considered as consisting of two squares $a \times a$ with weights α and β. Let us now place both of these squares into a larger square $(a + 1) \times (a + 1)$ in such a way that at least one of the diagonals of the larger square is totally covered by elements of the smaller squares and, at the same time, the $a \times a$ squares do not completely overlap. Clearly, both projections of this arrangement of set B have the same cardinality $a + 1$. Hence,

$$V(B) \le V(a + 1) + V(a + 1)$$

by subadditivity. Substituting for $V(B)$ from Eq. (e) and applying Eq. (d), we obtain

$$V(^\alpha 1, {}^\beta 1) + 2 \log_2 a \le 2 \log_2(a + 1)$$

or, alternatively,

$$V(^\alpha 1, {}^\beta 1) \le 2[\log_2(a + 1) - \log_2 a].$$

This inequality must be satisfied for all $a \in \mathbb{N}$. When a goes to infinity, the left-hand side of the inequality remains constant, whereas its right-hand side converges to zero. Hence,

$$V(^\alpha 1, {}^\beta 1) = 0.$$

By repeating the same argument for $(^\alpha 1, {}^\beta 1, {}^\gamma 1, \ldots)$ with n focal elements, we readily obtain

$$V(^\alpha 1, {}^\beta 1, {}^\gamma 1, \ldots) \le n[\log_2(a + 1) - \log_2 a]$$

and, again, by allowing a to go to infinity, we obtain

$$V(^\alpha 1, {}^\beta 1, {}^\gamma 1, \ldots) = 0. \tag{f}$$

Applying this result and additivity of V, we also have for an arbitrary basic assignment m the following:

$$V(m \cdot (^\alpha 1, {}^\beta 1, {}^\gamma 1, \ldots)) = V(m) + V(^\alpha 1, {}^\beta 1, {}^\gamma 1, \ldots)$$

$$= V(m). \tag{g}$$

This means that we can replicate all focal elements of m the same number of times, splitting their weights in a fixed proportion $\alpha, \beta, \gamma, \ldots$, and the value of V does not change.

(iii) Next, we prove that $V(m_1) = V(m_2)$, where $m_1 = (^\alpha a, {}^{\beta+\gamma} b)$ and $m_2 = (^\alpha a, {}^\beta b, {}^\gamma b)$. Since the actual weights are not essential in this proof, we omit them for the sake of simplicity; if desirable, they can be easily reconstructed. The proof is accomplished by showing that $V(m_1) \le V(m_2)$ and, at the same time, $V(m_2) \le V(m_1)$. To demonstrate the first inequality, let us view the focal elements of m_1 and m_2 as collections of intervals of lengths a, b, and a, b, b respectively. Furthermore, let us place both intervals of m_1 side by side, the first and second interval of m_2 side by side, and the third interval of m_2 above the second interval of m_2. According to this arrangement, the two projections of m_2 consist of m_1 and a pair of singletons with appropriate weights, say $(^x1, {}^y1)$. It follows then from the subadditivity that

$$V(m_2) \leq V(m_1) + V(^x1, {}^y1).$$

The last term in this inequality is 0 by Eq. (f) and, consequently, $V(m_2) \leq V(m_1)$.

To prove the opposite inequality, let $m_3 = (^\beta1, {}^\gamma1)$ so that $m_1 \cdot m_3$ assigns the same weights to the b's as does m_2. We select a sufficiently large integer n, temporarily fixed. Let s and s' denote squares $n \times n$ and $(n + 1) \times (n + 1)$, respectively. We can view $m_1 \cdot m_3 \cdot s$ as a collection of four parallelepipeds, two with edges a, n, n, and the other two with edges b, n, n. Similarly, $m_2 \cdot s'$ can be viewed as a collection of three parallelepipeds, one with edges $a, n + 1, n + 1$, and two with edges $b, n + 1, n + 1$. We now place two blocks with edges a, n, n of $m_1 \cdot m_3 \cdot s$ inside one block with edges $a, n + 1, n + 1$ of m_2 so that they cover the main diagonal. Furthermore, we place two blocks with edges b, n, n of $m_1 \cdot m_3 \cdot s$ inside the separate blocks with edges $b, n + 1, n + 1$ of $m_2 \cdot s'$ so that they again cover the diagonals. Using additivity and Eq. (d), the construction results in the equation

$$V(m_1 \cdot m_3 \cdot s) = V(m_1) + V(^\beta1, {}^\gamma1) + 2 \log_2 n$$

$$= V(m_1) + 2 \log_2 n.$$

Using subadditivity and the projections of the construction, we obtain

$$V(m_1 \cdot m_3 \cdot s) \leq V(m_2) + V(n + 1) + V(n + 1).$$

Hence,

$$V(m_1) + 2 \log_2 n \leq V(m_2) + 2 \log_2(n + 1)$$

and

$$V(m_2) - V(m_1) \geq 2[\log_2 n - \log_2(n + 1)].$$

For n going to infinity, the right-hand side of this inequality converges to 0 and, consequently,

$$V(m_2) - V(m_1) \geq 0$$

or

$$V(m_2) \geq V(m_1).$$

The proof can easily be extended to the case of a general basic assignment in which any given cardinality may repeat more than twice and the number of different cardinalities is arbitrary.

(iv) Repeatedly applying the branching property, we obtain

$$V(^\alpha a, {}^\beta b, {}^\gamma c, \ldots) = V(^\alpha a, {}^{x_1}1, {}^{x_2}1, \ldots) + V(^\beta b, {}^{y_1}1, {}^{y_2}1, \ldots) + \cdots$$

According to (iii), we can combine the singletons so that the proof of the theorem reduces to the determination of $V({}^{\alpha}a, {}^{1-\alpha}1)$ for arbitrary a and α. Moreover,

$$V({}^{1}a) = V({}^{1/2}a, {}^{1/2}a)$$

and, by the branching axiom,

$$V({}^{1/2}a, {}^{1/2}a) = 2V({}^{1/2}a, {}^{1/2}1).$$

Hence,

$$V({}^{1}a) = 2V({}^{1/2}a, {}^{1/2}1).$$

Similarly,

$$V({}^{1}a) = 3V({}^{1/3}a, {}^{2/3}1) = 4V({}^{1/4}a, {}^{3/4}1) = \cdots$$

Using Eq. (d), we obtain for each $t = 1/n$ the equation

$$V({}^{t}a, {}^{1-t}1) = t \log_2 a.$$

This formula can easily be shown to hold for any rational t. Since an arbitrary real number can be approximated by a monotonic sequence of rational numbers, property (iii) implies that the formula holds also for any $t \in [0, 1]$. This concludes the proof. ∎

B

GLOSSARY OF SYMBOLS

GENERAL SYMBOLS

$\{x, y, \ldots\}$	Set of elements $x, y, \ldots$
$\{x \mid p(x)\}$	Set determined by property p
$(x_1, x_2, \ldots, x_n)$	n-tuple
$[x_{i,j}]$	Matrix
$[a, b]$	Closed interval of real numbers between a and b
$[a, b)$	Interval of real numbers closed in a and open in b
$[a, \infty)$	Set of real numbers greater than or equal to a
X	Universal set (universe of discourse)
$A, B, C, \ldots$	Arbitrary sets (crisp or fuzzy)
$x \in X$	Set membership
$A = B$	Set equality
$A \neq B$	Set inequality
$A - B$	Set difference
$A \subseteq B$	Set inclusion
$A \subset B$	Proper set inclusion ($A \neq B$)
$\varnothing$	Empty set
$\mathscr{P}(X)$	Set of all crisp subsets of X (power set)
$\mathscr{\tilde{P}}(X)$	Set of all fuzzy subsets of X
$\mid A \mid$	Cardinality of set A
$A \cap B$	Set intersection
$A \cup B$	Set union
$A \times B$	Cartesian product of sets A and B
A^2	Cartesian product $A \times A$
$[a, b]^2$	Cartesian product $[a, b] \times [a, b]$
$X \rightarrow Y$	Function from X into Y
$R(X, Y)$	Relation on $X \times Y$
$R \circ Q$	Max-min composition of binary fuzzy relations R and Q
$R \odot Q$	Max-product composition of binary fuzzy relations R and Q

$R * Q$	Join of binary fuzzy relations R and Q
R^{-1}	Inverse of a binary fuzzy relation
$[R \downarrow X]$	Projection of relation R with respect to variables in set X
$[R \uparrow X - Y]$	Cylindric extension of relation R with respect to variables in set $X - Y$
$<$	Less than
$\leq$	Less than or equal to (also used for a partial ordering)
$\prec$	Subsequence (substate) relation or, alternatively, sharper than
$x \wedge y$	Meet (greatest lower bound) in a lattice
$x \vee y$	Join (least upper bound) in a lattice
$x \mid y$	x given y
$x \Rightarrow y$	x implies y
$x \Leftrightarrow y$	x if and only if y
$\forall$	For all
$\exists$	There exists at least one
$\sum$	Summation
$\prod$	Product
$\max(x_1, x_2, \ldots, x_n)$	Maximum of $x_1, x_2, \ldots, x_n$
$\min(x_1, x_2, \ldots, x_n)$	Minimum of $x_1, x_2, \ldots, x_n$
i, j, k	Arbitrary identifiers (indices)
I, J, K	General sets of identifiers (indices)
$\mathbb{N}$	Set of positive integers
$\mathbb{N}_0$	Set of nonnegative integers
$\mathbb{N}_n$	Set $\{1, 2, \ldots, n\}$
$\mathbb{N}_{n,m}$	Set $\{n, n + 1, \ldots, m\}$
$\mathbb{R}$	Set of all real numbers
$\mathbb{R}^+$	Set of all nonnegative real numbers
$n!$	n factorial $(= n(n - 1) \cdots 1)$
$\binom{n}{r}$	Combinatorial number $n!/(n - r)!r!$

SPECIAL SYMBOLS*

A_α	α-cut of fuzzy set A
Bel	Belief measure
c	Operation of fuzzy complement
C	Measure of confusion
$C(A)$	Fuzzy complement of A obtained by c
E	Measure of dissonance
e_c	Equilibrium of operation c (of fuzzy complement)
f	Measure of fuzziness
g	Fuzzy measure
h	Aggregation operation
H	Shannon entropy
I	Hartley information
i	Operation of fuzzy intersection
L_n	Łukasiewicz n-valued logic
p	Probability distribution function
P	Probability measure
Pl	Plausibility measure
r	Possibility distribution function

* Only those special symbols that are used repeatedly are included here.

R_T	Transitive closure of relation R
S	Solution set (in various contexts)
supp A	Support of fuzzy set A
u	Operation of fuzzy union
U	U-uncertainty (possibilistic measure of nonspecificity)
V	Measure of nonspecificity
$\mathcal{F}$	Set of focal elements (in evidence theory)
$\mathcal{R}$	Set of all possibility distributions
$^n\mathcal{R}$	Set of all possibility distributions of length n
$\mathcal{M}$	Set of all basic distributions (corresponding to possibility distributions)
$^n\mathcal{M}$	Set of all basic distributions of length n
$\mathcal{P}$	Set of all probability distributions (also the set of all vectors p in fuzzy relation equation)
$^n\mathcal{P}$	set of all probability distributions of length n
α	number in [0, 1] employed to define an α-cut
δ	Absolute values of local differences of membership grades
η	Necessity measure
μ_A	Membership grade function of fuzzy set A
μ_i	Component of a basic distribution
π	Possibility measure
Λ_A	Level set of fuzzy set A

REFERENCES

Abramson, N. [1963], *Information Theory and Coding*. McGraw-Hill, New York.

Aczél, J. [1966], *Lectures on Functional Equations and Their Applications*. Academic Press, New York.

Aczél, J. and **Z. Daróczy** [1975], *On Measures of Information and Their Characterizations*. Academic Press, New York.

Aczél, J.S., B. Forte, and **C.T. Ng** [1974], "Why the Shannon and Hartley entropies are 'natural.'" *Advances in Applied Probability*, **6**, pp. 131–146.

Adamo, J.M. [1980], "Fuzzy decision trees." *Fuzzy Sets and Systems*, **4**, pp. 207–219.

Adlassnig, K.-P. [1982], "A survey on medical diagnosis and fuzzy subsets." In: Gupta and Sanchez [1982b], pp. 203–217.

Adlassnig, K.-P. [1986], "Fuzzy set theory in medical diagnosis." *IEEE Trans. on Systems, Man, and Cybernetics*, **SMC-16**, pp. 260–265.

Adlassnig, K.-P. and **G. Kolarz** [1982], "CADIAG-2: computer-assisted medical diagnosis using fuzzy subsets." In: Gupta and Sanchez [1982b], pp. 219–247.

Albert, P. [1978], "The algebra of fuzzy logic." *Fuzzy Sets and Systems*, **1**, pp. 203–230.

Alsina, C., E. Trillas, and **L. Valverde** [1983], "On some logical connectives for fuzzy set theory." *J. of Math. Analysis and Applications*, **93**, pp. 15–26.

Ash, R.B. [1965], *Information Theory*. Interscience, New York.

Ashby, W.R. [1956], *An Introduction to Cybernetics*. John Wiley, New York.

Ashby, W.R. [1964], "Constraint analysis of many-dimensional relations." *General Systems Yearbook*, **9**, pp. 99–105.

Ashby, W.R. [1965], "Measuring the internal informational exchange in a system." *Cybernetica*, **1**, pp. 5–22.

Ashby, W.R. [1968], "Some consequences of Bremermann's limit for information-processing systems." In: *Cybernetic Problems in Bionics*, edited by H. Oestreicher and D. Moore, Gordon and Breach, New York, pp. 69–76.

Ashby, W.R. [1969], "Two tables of identities governing information flows within large systems." *ASC Communications*, **1**, pp. 3–8.

Ashby, W.R. [1970], "Information flows within co-ordinated systems." In: *Progress in Cybernetics*, Vol. 1, edited by J. Rose, Gordon and Breach, London, pp. 57–64.

Ashby, W.R. [1972], "Systems and their informational measures." In: *Trends in General Systems Theory*, edited by G.J. Klir, Wiley-Interscience, New York, pp. 78–97.

Ashby, W.R. [1973], "Some peculiarities of complex systems." *Cybernetic Medicine*, **9**, pp. 1–7.

Atlan, H. [1983], "Information theory." In: *Cybernetics: Theory and Applications*, edited by R. Trappl, Hemisphere, Washington, D.C., pp. 9–41.

Attneave, F. [1959], *Applications of Information Theory to Psychology*. Holt, New York.

Aulin, A. [1982], *The Cybernetic Laws of Social Progress: Towards a Critical Social Philosophy and a Criticism of Marxism*. Pergamon Press, Oxford.

Avgers, T.G. [1983], "Axiomatic derivation of the mutual information principle as a method of inductive inference." *Kybernetes*, **12**, pp. 107–113.

Bacon, G. [1981], "Distinction between several subsets of fuzzy measures." *Fuzzy Sets and Systems*, **5**, pp. 291–305.

Baldwin, J.F. [1979a], "A model of fuzzy reasoning through multi-valued logic and set theory." *Intern. J. of Man-Machine Studies*, **11**, pp. 351–380.

Baldwin, J.F. [1979b], "A new approach to approximate reasoning using a fuzzy logic." *Fuzzy Sets and Systems*, **2**, pp. 309–325.

Baldwin, J.F. [1979c], "Fuzzy logic and its application to fuzzy reasoning." In: Gupta, Ragade, and Yager [1979], pp. 93–115.

Baldwin, J.F. [1985], "Fuzzy sets and expert systems." *Information Sciences*, **36**, pp. 123–156.

Baldwin, J.F. and **N.C.F. Guild** [1980a], "Feasible algorithms for approximate reasoning using fuzzy logic." *Fuzzy Sets and Systems*, **3**, pp. 225–251.

Baldwin, J.F. and **N.C.F. Guild** [1980b], "The resolution of two paradoxes by approximate reasoning using a fuzzy logic." *Synthese*, **44**, pp. 397–420.

Baldwin, J.F. and **B.W. Pilsworth** [1980], "Axiomatic approach to implication for approximate reasoning with fuzzy logic." *Fuzzy Sets and Systems*, **3**, pp. 193–219.

Ballmer, T.T. and **M. Pinkal**, eds. [1983], *Approaching Vagueness*. North-Holland, Amsterdam and New York.

Bandler, W. and **L.J. Kohout** [1980a], "Fuzzy power set and fuzzy implication operators." *Fuzzy Sets and Systems*, **4**, pp. 13–30.

Bandler, W. and **L.J. Kohout** [1980b], "Semantics of implication operators and fuzzy relational products." *Intern. J. of Man-Machine Studies*, **12**, pp. 89–116.

Bandler, W. and **L.J. Kohout** [1980c], "Relational products as a tool for analysis and synthesis of the behavior of complex natural and artificial systems." In: Wang and Chang [1980], pp. 341–367.

Bandler, W. and **L.J. Kohout** [1985], "Probabilistic versus fuzzy production rules in expert systems." *Intern. J. of Man-Machine Studies*, **22**, pp. 347–353.

Bandler, W. and **L.J. Kohout** [1986a], "On new types of homomorphisms and congruences for partial algebraic structures and n-ary relations." *Intern. J. of General Systems*, **12**, pp. 149–157.

Bandler, W. and **L.J. Kohout** [1986b], "On the general theory of relational morphisms." *Intern. J. of General Systems*, **13**, pp. 47–68.

Bell, D.A. [1953], *Information Theory and Its Engineering Applications*. Pitman, New York.

Bellman, R. and **M. Giertz** [1973], "On the analytic formalism of the theory of fuzzy sets." *Information Sciences*, **5**, pp. 149–156.

Bellman, R. and **L.A. Zadeh** [1970], "Decision making in a fuzzy environment." *Management Science*, **17**, pp. B-144–B-164.

Benson, W.H. [1982], "An application of fuzzy set theory to data display." In: Yager [1982a], pp. 429–438.

Bezdek, J.C. [1981], *Pattern Recognition with Fuzzy Objective Function Algorithms*. Plenum Press, New York.

Bezdek, J.C., ed. [1985], *Analysis of Fuzzy Information*. CRC Press, Boca Raton, Fla.

Bezdek, J.C., **B. Spillman** and **R. Spillman** [1978], "A fuzzy relations space for group decision theory." *Fuzzy Sets and Systems*, **1**, pp. 255–268.

Bezdek, J.C., **B. Spillman** and **R. Spillman** [1979], "Fuzzy relation spaces for group decision theory: an application." *Fuzzy Sets and Systems*, **2**, pp. 5–14.

Bhattacharya, P. and **N.P. Mukherjee** [1985], "Fuzzy relations and fuzzy groups." *Information Sciences*, **36**, pp. 267–282.

Billingsley, P. [1965], *Ergodic Theory and Information*. John Wiley, New York.

Billingsley, P. [1986], *Probability and Measure* (Second Edition). John Wiley, New York.

Black, M. [1937], "Vagueness: an exercise in logical analysis." *Philosophy of Science*, **4**, pp. 427–455.

Blin, J.M. [1974], "Fuzzy relations in group decision theory." *J. of Cybernetics*, **4**, pp. 17–22.

Blin, J.M. and **A.B. Whinston** [1973], "Fuzzy sets and social choice." *J. of Cybernetics*, **3**, pp. 28–36.

Blockley, D. [1982], "Fuzzy systems in civil engineering." In: Gupta and Sanchez [1982b], pp. 103–115.

Boekee, D.E. and **J.C.A. Van Der Lubbe** [1980], "The R-norm information measure." *Information and Control*, **45**, pp. 136–155.

Boltzmann, L. [1896], *Vorlesungen über Gastheorie*. J.A. Barth, Leipzig.

Bordley, R.F. [1983], "A central principle of science: optimization." *Behavioral Science*, **28**, pp. 53–64.

Borisov, A. and **O. Krumberg** [1983], "A theory of possibility for decision making." *Fuzzy Sets and Systems*, **9**, pp. 13–23.

Bortolan, G. and **R.T. Degani** [1984], "Ranking of fuzzy alternatives in electrocardiography." In: Sanchez [1984a], pp. 409–419.

Bosserman, R. and **R. Ragade** [1981], "Ecosystem analysis using fuzzy set theory." In: Lasker [1981], pp. 3033–3045.

Bouchon, B. [1981], "Fuzzy questionnaires." *Fuzzy Sets and Systems*, **6**, pp. 1–9.

Braae, M. and **D.A. Rutherford** [1979], "Selection of parameters for a fuzzy logic controller." *Fuzzy Sets and Systems*, **2**, pp. 185–199.

Bremermann, H.J. [1962], "Optimization through evolution and recombination." In: *Self-Organizing Systems*, edited by M.C. Yovits et al., Spartan Books, Washington, D.C., pp. 93–106.

Brillouin, L. [1956], *Science and Information Theory*. Academic Press, New York.

Brillouin, L. [1964], *Scientific Uncertainty and Information*. Academic Press, New York.

Broekstra, G. [1980], "On the foundations of GIT (General Information Theory)." *Cybernetics and Systems*, **11**, pp. 143–165.

Broekstra, G. [1981], "C-analysis of C-structures." *Intern. J. of General Systems*, **7**, pp. 33–61.

Brooks, D.R. and **E.O. Wiley** [1986], *Evolution as Entropy*. Univ. of Chicago Press, Chicago.

Bruce, W.S. and **A. Kandel** [1983], "The application of fuzzy set theory to a risk analysis model of computer security." In: Wang [1983], pp. 351–376.

Buckles, B.P. and **F.E. Petry** [1982a], "A fuzzy representation of data for relational databases." *Fuzzy Sets and Systems*, **7**, pp. 213–226.

Buckles, B.P. and **F.E. Petry** [1982b], "Fuzzy databases and their applications." In: Gupta and Sanchez [1982a], pp. 361–371.

Buckles, B.P. and **F.E. Petry** [1983], "Information-theoretical characterization of fuzzy relational databases." *IEEE Trans. on Systems, Man, and Cybernetics*, **SMC-13**, pp. 74–77.

Buckles, B.P. and **F.E. Petry** [1984], "Extension of the fuzzy database with fuzzy arithmetic." In: Sanchez [1984a], pp. 421–426.

Buckley, J.J. [1984], "The multiple judge, multiple criteria ranking problem: a fuzzy set approach." *Fuzzy Sets and Systems*, **13**, pp. 25–37.

Buell, D.A. [1982], "An analysis of some fuzzy subset applications to informational retrieval systems." *Fuzzy Sets and Systems*, **7**, pp. 35–42.

Buisson, J-C., H. Farrey, and **H. Prade** [1985], "The development of a medical expert system and the treatment of imprecision in the framework of possibility theory." *Information Sciences*, **37**, pp. 211–226.

Buoncristiani, J.F. [1983], "Probability on fuzzy sets." *J. of Math. Analysis and Applications*, **96**, pp. 24–41.

Burdzy, K. and **J.B. Kiszka** [1983], "The reproducibility property of fuzzy control systems." *Fuzzy Sets and Systems*, **9**, pp. 161–177.

Butnariu, D. [1978], "Fuzzy games: a description of the concept." *Fuzzy Sets and Systems*, **1**, pp. 181–192.

Bye, B.V., S.D. Biscoe, and **J.C. Hennessey** [1985], "A fuzzy algorithmic approach to the construction of composite indices: an application to a functional limitation index." *Intern. J. General Systems*, **11**, pp. 163–172.

Campbell, J. [1982], *Grammatical Man: Information, Entropy, Language, and Life*. Simon and Schuster, New York.

Cao, H. and **G. Chen** [1983], "Some applications of fuzzy sets to meteorological forecasting." *Fuzzy Sets and Systems*, **9**, pp. 1–12.

Carnap, R. [1977], *Two Essays on Entropy*. University of California Press, Berkeley.

Cavallo, R.E. and **G.J. Klir** [1982], "Reconstruction of possibilistic behavior systems." *Fuzzy Sets and Systems*, **8**, pp. 175–197.

Cervin, V.B. and **R. Jain** [1982], "Towards fuzzy algorithms in behavior modification." In: Gupta and Sanchez [1982b], pp. 291–296.

Chaitin, G.J. [1977], "Algorithmic information theory." *IBM J. of Research and Development*, **21**, pp. 350–359.

Chakravarty, S.R. and **T. Roy** [1985], "Measurement of fuzziness: a general approach." *Theory and Decision*, **19**, pp. 163–169.

Chanas, S. [1983], "The use of parametric programming in fuzzy linear programming." *Fuzzy Sets and Systems*, **11**, pp. 243–251.

Chang, C.L. [1968], "Fuzzy topological spaces." *J. of Math. Analysis and Applications*, **24**, pp. 182–190.

Chang, R.L-P. and **T. Pavlidis** [1979], "Applications of fuzzy sets in curve fitting." *Fuzzy Sets and Systems*, **2**, pp. 67–74.

Chen, G.Q., S.C. Lee, and **E.S.H. Yu** [1983], "Application of fuzzy set theory to economics." In: Wang [1983], pp. 277–305.

Cheng-Zhong, L. [1984], "Reliable solution set of a fuzzy relation equation." *J. of Math. Analysis and Applications*, **2**, pp. 524–532.

Cherry, C. [1957], *On Human Communication.* The M.I.T. Press, Cambridge, Mass.

Choquet, G. [1953], "Theory of capacities." *Ann. Institut Fourier*, **V**, pp. 131–295.

Christensen, R. [1980a], *Entropy Minimax Sourcebook. Vol II. Philosophical Origins.* Entropy, Lincoln, Mass.

Christensen, R. [1980b], *Entropy Minimax Sourcebook. Vol III: Computer Implementation.* Entropy, Lincoln, Mass.

Christensen, R. [1981a], *Entropy Minimax Sourcebook. Vol I: General Description.* Entropy, Lincoln, Mass.

Christensen, R. [1981b], *Entropy Minimax Sourcebook. Vol. IV: Applications.* Entropy, Lincoln, Mass.

Christensen, R. [1982], *Belief and Behavior.* Entropy, Lincoln, Mass.

Christensen, R. [1983], *Multivariate Statistical Modeling.* Entropy, Lincoln, Mass.

Christensen, R. [1984], *Order and Time: A General Theory of Prediction.* Entropy, Lincoln, Mass.

Christensen, R. [1985], "Entropy minimax multivariate statistical modeling—I: theory." *Intern. J. of General Systems*, **11**, pp. 231–277.

Christensen, R. [1986], "Entropy minimax multivariate statistical modeling—II: applications." *Intern. J. of General Systems*, **12**, pp. 193–271.

Civanlar, M.R. and **H.J. Trussell** [1986], "Constructing membership functions using statistical data." *Fuzzy Sets and Systems*, **18**, pp. 1–13.

Conant, R.C. [1969], "The information transfer required in regulatory processes." *IEEE Trans. on Systems Science and Cybernetics*, **SSC-5**, pp. 334–338.

Conant, R.C. [1974], "Information flows in hierarchical systems." *Intern. J. of General Systems*, **1**, pp. 9–18.

Conant, R.C. [1976], "Laws of information which govern systems." *IEEE Trans. on Systems, Man, and Cybernetics*, **SMC-6**, pp. 240–255.

Conant, R.C. [1979], "Communication without a channel." *Intern. J. of General Systems*, **5**, pp. 93–98.

Conant, R., ed. [1981a], *Mechanisms of Intelligence: Ross Ashby's Writings on Cybernetics.* Intersystems, Seaside, Calif.

Conant, R.C. [1981b], "Efficient proofs of identities in N-dimensional information theory." *Cybernetica*, **24**, pp. 191–197.

Cornacchio, J.V. [1977], "System complexity—a bibliography." *Intern. J. of General Systems*, **3**, pp. 267–271.

Csiszár, I. and J. Körner [1981], *Information Theory: Coding Theorems for Discrete Memoryless Systems*. Academic Press, New York.

Czogała, E. and J. Drewniak [1984], "Associative monotonic operations in fuzzy set theory." *Fuzzy Sets and Systems*, **12**, pp. 249–270.

Czogała, E., J. Drewniak, and W. Pedrycz [1982], "Fuzzy relation equations on a finite set," *Fuzzy Sets and Systems*, **7**, pp. 89–101.

Czogała, E. and W. Pedrycz [1981], "On identification in fuzzy systems and its applications in control problems." *Fuzzy Sets and Systems*, **6**, pp. 73–83.

Czogała, E. and W. Pedrycz [1982], "Control problems in fuzzy systems." *Fuzzy Sets and Systems*, **7**, pp. 257–273.

Dal Cin, M. [1975a], "Fuzzy-state automata: their stability and fault tolerance." *Intern. J. of Computer and Information Sciences*, **4**, pp. 63–80.

Dal Cin, M. [1975b], "Modification tolerance of fuzzy-state automata." *Intern. J. of Computer and Information Sciences*, **4**, pp. 81–93.

Dale, A.I. [1980], "Probability, vague statements, and fuzzy sets." *Philosophy of Science*, **47**, pp. 38–55.

Da Rocha, A.F. [1982], "Basic properties of neural circuits." *Fuzzy Sets and Systems*, **7**, pp. 109–121.

Daróczy, Z. [1970], "Generalized information functions." *Information and Control*, **16**, pp. 36–51.

De Fériet, J.K. [1982], "Interpretation of membership functions of fuzzy sets in terms of plausibility and belief." In: Gupta and Sanchez [1982a], pp. 93–98.

De Finetti, B. [1974], *Theory of Probability*. John Wiley, New York.

De Luca, A. and S. Termini [1972], "A definition of a nonprobabilistic entropy in the setting of fuzzy sets theory." *Information and Control*, **20**, pp. 301–312.

De Luca, A. and S. Termini [1974], "Entropy of L-fuzzy sets." *Information and Control*, **24**, pp. 55–73.

De Luca, A. and S. Termini [1977], "On the convergence of entropy measures of a fuzzy set." *Kybernetes*, **6**, pp. 219–227.

De Mori, R. [1983], *Computer Models of Speech Using Fuzzy Algorithms*. Plenum Press, New York.

Dempster, A.P. [1967], "Upper and lower probabilities induced by multivalued mappings." *Annals of Mathematical Statistics*, **38**, pp. 325–339.

Denbigh, K.G. and J.S. Denbigh [1985], *Entropy in Relation to Incomplete Knowledge*. Cambridge Univ. Press, Cambridge.

Devi, B.B. and V.V.S. Sarma [1985], "Estimation of fuzzy memberships from histograms." *Information Sciences*, **35**, pp. 43–59.

Dijkman, J., H. van Haeringen, and S.J. De Lange [1983], "Fuzzy numbers." *J. of Math. Analysis and Applications*, **92**, pp. 302–341.

Dimitrov, V. [1983], "Group choice under fuzzy information." *Fuzzy Sets and Systems*, **9**, pp. 25–39.

Di Nola, A. and S. Sessa [1983], "On the set of solutions of composite fuzzy relation equations." *Fuzzy Sets and Systems*, **9**, pp. 275–285.

Di Nola, A., S. Sessa and W. Pedrycz [1985], "Decomposition problem of fuzzy relations." *Intern. J. of General Systems*, **10**, pp. 123–133.

Di Nola, A., S. Sessa, W. Pedrycz, and M. Higashi [1985], "Minimal and maximal solutions of a decomposition problem of fuzzy relations." *Intern. J. of General Systems*, **11**, pp. 103–116.

Di Nola, A. and **A.G.S. Ventre,** eds. [1986], *The Mathematics of Fuzzy Systems.* TÜV Rheinland, Köln.

Dishkant, H. [1981], "About membership function estimation." *Fuzzy Sets and Systems*, **5**, pp. 141–147.

Dockery, J.T. [1982], "Fuzzy design of military information systems." *Intern. J. of Man-Machine Studies*, **16**, pp. 1–38.

Dockx, S. and **P. Bernays** [1965], *Information and Prediction in Science.* Academic Press, New York.

Dompere, K.K. [1982], "The theory of fuzzy decisions." In: Gupta and Sanchez [1982b], pp. 365–379.

Dombi, J. [1982], "A general class of fuzzy operators, the De Morgan class of fuzzy operators and fuzziness measures induced by fuzzy operators." *Fuzzy Sets and Systems*, **8**, pp. 149–163.

Dretske, F.I. [1981], *Knowledge and the Flow of Information.* The MIT Press, Cambridge, Mass.

Drewniak, J. [1984], "Fuzzy relation equations and inequalities." *Fuzzy Sets and Systems*, **14**, pp. 237–247.

Dubois, D. [1983], "A fuzzy, heuristic, interactive approach to the optimal network problem." In: Wang [1983], pp. 253–276.

Dubois, D. and **H. Prade** [1979a], "Operations in a fuzzy-valued logic." *Information and Control*, **43**, pp. 224–240.

Dubois, D. and **H. Prade** [1979b], "Decision making under fuzziness." In: Gupta, Ragade, and Yager [1979], pp. 279–302.

Dubois, D. and **H. Prade** [1980a], *Fuzzy Sets and Systems: Theory and Applications.* Academic Press, New York.

Dubois, D. and **H. Prade** [1980b], "New results about properties and semantics of fuzzy set-theoretic operators." In: Wang and Chang [1980], pp. 59–75.

Dubois, D. and **H. Prade** [1982a], "A class of fuzzy measures based on triangular norms." *Intern. J. of General Systems*, **8**, pp. 43–61.

Dubois, D. and **H. Prade** [1982b], "Towards fuzzy differential calculus." *Fuzzy Sets and Systems*, **8**, pp. 1–17, 105–116, 225–233.

Dubois, D. and **H. Prade** [1982c], "On several representations of an uncertain body of evidence." In: Gupta and Sanchez [1982a], pp. 167–181.

Dubois, D. and **H. Prade** [1982d], "The use of fuzzy numbers in decision analysis." In: Gupta and Sanchez [1982a], pp. 309–321.

Dubois, D. and **H. Prade** [1983a], "Unfair coins and necessity measures: towards a possibilistic interpretation of histograms." *Fuzzy Sets and Systems*, **10**, pp. 15–20.

Dubois, D. and **H. Prade** [1983b], "Ranking fuzzy numbers in the setting of possibility theory." *Information Sciences*, **30**, pp. 183–224.

Dubois, D. and **H. Prade** [1984a], "Fuzzy logics and the generalized modus ponens revisited." *Cybernetics and Systems*, **15**, pp. 293–331.

Dubois, D. and **H. Prade** [1984b], "Criteria aggregation and ranking of alternatives in the framework of fuzzy set theory." In: Zimmermann, Zadeh, and Gaines [1984], pp. 209–240.

Dubois, D. and **H. Prade** [1985a], "A review of fuzzy set aggregation connectives." *Information Sciences*, **36**, pp. 85–121.

Dubois, D. and **H. Prade** [1985b], "A note on measures of specificity for fuzzy sets." *Intern. J. of General Systems*, **10**, pp. 279–283.

Dubois, D. and **H. Prade** [1985c], "Fuzzy cardinality and the modeling of imprecise quantification." *Fuzzy Sets and Systems*, **16**, pp. 199–230.

Dubois, D. and **H. Prade** [1985d], *Theorie des possibilités*. Masson, Paris (English translation, Plenum Press, New York, 1987).

Dubois, D. and **H. Prade** [1986], "A set-theoretic view of belief functions." *Intern. J. of General Systems*, **12**, pp. 193–226.

Dubois, D. and **H. Prade** [1987], "Properties of measures of information in evidence and possibility theories." *Fuzzy Sets and Systems*, **24**.

Dutta, A. [1985], "Reasoning with imprecise knowledge in expert systems." *Information Sciences*, **37**, pp. 3–24.

Dyckhoff, H. and **W. Pedrycz** [1984], "Generalized means as a model of compensative connectives." *Fuzzy Sets and Systems*, **14**, pp. 143–154.

Emptoz, H. [1981], "Nonprobabilistic entropies and indetermination measures in the setting of fuzzy sets theory." *Fuzzy Sets and Systems*, **5**, pp. 307–317.

Enta, Y. [1982], "Fuzzy decision theory." In: Yager [1982a], pp. 439–449.

Esogbue, A.O. [1981], "A fuzzy set approach to public participation effectiveness measurement in water quality planning." In: Lasker [1981], pp. 3076–3081.

Esogbue, A.O. [1982], "A fuzzy set model for the evaluation of best management practices in nonpoint source water pollution policy formulation." In: Gupta and Sanchez [1981b], pp. 431–436.

Esogbue, A.O. and **R.C. Elder** [1979], "Fuzzy sets and the modelling of physician decision processes: part I: the initial interview-information gathering process." *Fuzzy Sets and Systems*, **2**, pp. 279–291.

Esogbue, A.O. and **R.C. Elder** [1980], "Fuzzy sets and the modelling of physician decision processes: part II: fuzzy diagnosis decision models." *Fuzzy Sets and Systems*, **3**, pp. 1–9.

Esogbue, A.O. and **R.C. Elder** [1983], "Measurement and valuation of a fuzzy mathematical model for medical diagnosis." *Fuzzy Sets and Systems*, **10**, pp. 223–242.

Esteva, F., **E. Trillas**, and **X. Domingo** [1981], "Weak and strong negation function for fuzzy set theory." *Proc. of the Eleventh IEEE Intern. Symp. on Multiple-Valued Logic*, Norman, Oklahoma, pp. 23–27.

Fast, J.D. [1962], *Entropy*. Gordon and Breach. New York.

Feinstein, A. [1958], *Foundations of Information Theory*. McGraw-Hill, New York.

Feller, W. [1950, 1966], *An Introduction to Probability Theory and Its Applications* (2 volumes). John Wiley, New York.

Ferdinand, A.E. [1974], "A theory of system complexity." *Intern. J. of General Systems*, **1**, pp. 19–33.

Fiksel, J. [1981], "Applications of fuzzy set and possibility theory to systems management." In: Lasker [1981], pp. 2966–2973.

Fine, T.L. [1973], *Theories of Probability: An Examination of Foundations*. Academic Press, New York.

Fordon, W.A. and **J.C. Bezdek** [1979], "The application of fuzzy set theory to medical diagnosis." In: Gupta, Ragade, and Yager [1979], pp. 445–461.

Forte, B. [1968], "On the amount of information given by an experiment." In: *Proc. of the Colloquium on Information Theory* (Vol. 1), edited by A. Renyi, Janos Bolyai Mathematical Society, Budapest, pp. 149–166.

Forte, B. [1975], "Why Shannon entropy." *Symposium Mathematica,* **15,** Academic Press, New York, pp. 137–152.

Frank, M.J. [1979], "On the simultaneous associativity of F(x,y) and x + y − F(x,y)." *Aequationes Math.,* **19,** pp. 194–226.

Freska, C. [1982], "Linguistic description of human judgments in expert systems and in the 'soft' sciences." In: Gupta and Sanchez [1982b], pp. 297–305.

Fu, K.S., M. Ishizuka, and **J.T.P. Yao** [1982], "Application of fuzzy sets in earthquake engineering." In: Yager [1982a], pp. 504–523.

Fung, L.W. and **K.S. Fu** [1975], "An axiomatic approach to rational decision making in a fuzzy environment." In: Zadeh, Fu, Tanaka, and Shimura [1975], pp. 227–256.

Gaines, B.R. [1976], "Foundations of fuzzy reasoning." *Intern. J. of Man-Machine Studies,* **8,** pp. 623–668.

Gaines, B.R. [1977], "System identification, approximation and complexity." *Intern. J. of General Systems,* **3,** pp. 145–174.

Gaines, B.R. [1978], "Fuzzy and probability uncertainty logics." *Information and Control,* **38,** pp. 154–169.

Gaines, B.R. [1981], "Logical foundations for database systems." In: Mamdani and Gaines [1981], pp. 289–308.

Gaines, B.R. [1983], "Precise past—fuzzy future." *Intern. J. of Man-Machine Studies,* **19,** pp. 117–134.

Gaines, B.R. and **L.J. Kohout** [1977], "The fuzzy decade: a bibliography of fuzzy systems and closely related topics." *Intern. J. of Man-Machine Studies,* **9,** pp. 1–68.

Gaines, B.R. and **M.L.G. Shaw** [1985], "From fuzzy logic to expert systems." *Information Sciences,* **36,** pp. 5–16.

Gale, S. [1972], "Inexactness, fuzzy sets, and the foundations of behavioral geography." *Geographical Analysis,* **4,** pp. 337–349.

Garey, M.R. and **D.S. Johnson** [1979], *Computers and Intractability: A Guide to the Theory of NP-Completeness.* W.H. Freeman, San Francisco.

Garner, W.R. [1962], *Uncertainty and Structure as Psychological Concepts.* John Wiley, New York.

Gatlin, L.L. [1972], *Information Theory and the Living System.* Columbia University Press, New York.

Georgescu-Roegen, N. [1971], *The Entropy Law and the Economic Process.* Harvard Univ. Press, Cambridge, Mass.

Giering, E.W. and **A. Kandel** [1983], "The application of fuzzy set theory to the modeling of competition in ecological systems." *Fuzzy Sets and Systems,* **9,** pp. 103–127.

Giles, R. [1977], "Łukasiewicz logic and fuzzy set theory." *Intern. J. of Man-Machine Studies,* **8,** pp. 313–327.

Giles, R. [1982], "Foundations for a theory of possibility." In: Gupta and Sanchez [1982a], pp. 183–195.

Goguen, J.A. [1967], "L-fuzzy sets." *J. of Math. Analysis and Applications*, **18**, pp. 145–174.

Goguen, J.A. [1968–69], "The logic of inexact concepts." *Synthese*, **19**, pp. 325–373.

Goguen, J.A. [1975], "On fuzzy robot planning." In: Zadeh, Fu, Tanaka, and Shimura [1975], pp. 429–447.

Gokhale, D.V. and **S. Kullback** [1978], *The Information in Contingency Tables*. Marcel Dekker, New York.

Goldman, S. [1953], *Information Theory*. Dover, New York.

Good, I.J. [1963], "Maximum entropy for hypothesis formulation, especially for multidimensional contingency tables." *The Annals of Mathematical Statistics*, **34**, pp. 911–934.

Goodman, I.R. [1982], "Fuzzy sets as equivalence classes of random sets." In: Yager [1982a], pp. 327–343.

Goodman, I.R. and **H.T. Nguyen** [1985], *Uncertainty Models for Knowledge-Based Systems*. North-Holland, New York.

Gordon, J. and **E.H. Shortliffe** [1985], "A method for managing evidential reasoning in a hierarchical hypothesis space." *Artificial Intelligence*, **26**, pp. 323–357.

Gottwald, S. [1979], "Set theory for fuzzy sets of higher level." *Fuzzy Sets and Systems*, **2**, pp. 125–151.

Gottwald, S. [1980], "Fuzzy propositional logics." *Fuzzy Sets and Systems*, **3**, pp. 181–192.

Gottwald, S. [1985], "Generalized solvability criteria for fuzzy equations." *Fuzzy Sets and Systems*, **17**, pp. 285–296.

Gottwald, S. and **W. Pedrycz** [1986], "Solvability of fuzzy relational equations and manipulation of fuzzy data." *Fuzzy Sets and Systems*, **18**, pp. 45–65.

Gouvernet, J., S. Ayme, and **E. Sanchez** [1982], "Approximate reasoning in medical genetics." In: Yager [1982a], pp. 524–530.

Grad, H. [1961], "The many faces of entropy." *Communications on Pure and Applied Mathematics*, **14**, pp. 323–354.

Guiasu, S. [1971], "Weighted entropy." *Reports on Mathematical Physics*, **2**, pp. 165–179.

Guiasu, S. [1977], *Information Theory and Applications*. McGraw-Hill, New York.

Gupta, M.M., A. Kandel, W. Bandler, and **J.B. Kiszka,** eds. [1985], *Approximate Reasoning in Expert Systems*. North-Holland, New York.

Gupta, M.M., R.R. Martin-Clouaire, and **P.N. Nikiforuk** [1984], "Fuzzy set theory in medical sciences." In: Sanchez [1984a], pp. 29–30.

Gupta, M.M., R.K. Ragade, and **R.R. Yager,** eds. [1979], *Advances in Fuzzy Set Theory and Applications*. North-Holland, New York.

Gupta, M.M., G.N. Saridis, and **B.R. Gaines,** eds. [1977], *Fuzzy Automata and Decision Processes*. North-Holland, New York.

Gupta, M.M. and **E. Sanchez,** eds. [1982a], *Fuzzy Information and Decision Processes*. North-Holland, New York.

Gupta, M.M. and **E. Sanchez,** eds. [1982b], *Approximate Reasoning in Decision Analysis*. North-Holland, New York.

Hacking, I. [1975], *The Emergence of Probability*. Cambridge University Press, Cambridge, Mass.

Hájek, P. [1985], "Combining functions for certainty degrees in consulting systems." *Intern. J. of Man-Machine Studies*, **22**, pp. 59–76.

Hallefjord, Å. and **K. Jörnsten** [1986], "An entropy target-point approach to multiobjective programming." *Intern. J. of Systems Science*, **17**, pp. 639–653.

Hamacher, H. [1978], "Über logische Verknupfungen unscharfer Aussagen und deren Zugehörige Bewertungsfunktionen." In: *Progress in Cybernetics and Systems Research*, Vol. 3, edited by R. Trappl, G.J. Klir, and L. Ricciardi, Hemisphere, Washington, D.C., pp. 276–288.

Hammerbacher, I.M. and **R.R. Yager** [1981], "The personalization of security selection: an application of fuzzy set theory." *Fuzzy Sets and Systems*, **5**, pp. 1–9.

Hanna, J.F. [1969], "Exploration, prediction, description, and information theory." *Synthese*, **20**, pp. 308–334.

Hannan, E.L. [1981], "Linear programming with multiple fuzzy goals." *Fuzzy Sets and Systems*, **6**, pp. 235–248.

Hara, F. [1982], "Fuzzy simulation model of civil evacuation from large-scale earthquake-generated fires." In: Gupta and Sanchez [1982b], pp. 313–326.

Hartley, R.V.L. [1928], "Transmission of information." *The Bell Systems Technical J.*, **7**, pp. 535–563.

Hersh, H.M., A. Caramazza, and **H.H. Brownell** [1979], "Effects of context on fuzzy membership functions." In: Gupta, Ragade, and Yager [1979], pp. 389–408.

Higashi, M., A. Di Nola, S. Sessa, and **W. Pedrycz** [1984], "Ordering fuzzy sets by consensus concept and fuzzy relation equations." *Intern. J. of General Systems*, **10**, pp. 47–56.

Higashi, M. and **G.J. Klir** [1982], "On measures of fuzziness and fuzzy complements." *Intern. J. of General Systems*, **8**, pp. 169–180.

Higashi, M. and **G.J. Klir** [1983a], "Measures of uncertainty and information based on possibility distributions." *Intern. J. of General Systems*, **9**, pp. 43–58.

Higashi, M. and **G.J. Klir** [1983b], "On the notion of distance representing information closeness: possibility and probability distributions." *Intern. J. of General Systems*, **9**, pp. 103–115.

Higashi, M. and **G.J. Klir** [1984a], "Resolution of finite fuzzy relation equations." *Fuzzy Sets and Systems*, **13**, pp. 65–82.

Higashi, M. and **G.J. Klir** [1984b], "Identification of fuzzy relation systems." *IEEE Trans. on Systems, Man, and Cybernetics*, **SMC-14**, pp. 349–355.

Higashi, M., G.J. Klir, and **M.A. Pittarelli** [1984], "Reconstruction families of possibilistic structure systems." *Fuzzy Sets and Systems*, **12**, pp. 37–60.

Hinde, C.J. [1983], "Inference of fuzzy relational tableaux from fuzzy exemplifications." *Fuzzy Sets and Systems*, **11**, pp. 91–101.

Hintikka, J. [1970], "On semantic information." In: Hintikka and Suppes, [1970], pp. 3–27.

Hintikka, J. and **P. Suppes,** eds. [1970], *Information and Inference*. D. Reidel, Dordrecht.

Hisdal, E. [1978], "Conditional possibilities, independence and noninteraction." *Fuzzy Sets and Systems*, **1**, pp. 283–297.

Hisdal, E. [1979], "Possibilistically dependent variables and a general theory of fuzzy sets." In: Gupta, Ragade, and Yager [1979], pp. 215–234.

Hisdal, E. [1980], "Generalized fuzzy set systems and particularization." *Fuzzy Sets and Systems*, **4**, pp. 275–291.

Hisdal, E. [1981], "The IF THEN ELSE statement and interval-valued fuzzy sets of higher type." *Intern. J. of Man-Machine Studies*, **15**, pp. 385–455.

Höhle, U. [1981], "Fuzzy plausibility measures." *Proc. Third Intern. Seminar on Fuzzy Set Theory*, edited by E.P. Klement, Johannes Kepler University, Linz, Austria, pp. 7–30.

Höhle, U. [1982], "Entropy with respect to plausibility measures." *Proc. Twelfth IEEE Intern. Symp. on Multiple-Valued Logic*, Paris, pp. 167–169.

Höhle, U. [1984], "Fuzzy plausibility measures." In: Zimmermann, Zadeh, and Gaines [1984], pp. 83–96.

Höhle, U. and **E.P. Klement** [1984], "Plausibility measures—a general framework for possibility and fuzzy probability measures." In: Skala, Termini, and Trillas [1984], pp. 31–50.

Holmblad, L.P. and **J-J. Østergaard** [1982], "Control of a cement kiln by fuzzy logic." In: Gupta and Sanchez [1982a], pp. 389–399.

Hyvärinen, L.P. [1968], *Information Theory for Systems Engineers*. Springer-Verlag, New York.

Ingels, F.M. [1971], *Information and Coding Theory*. INTEXT Educational Publishers, Scranton, Penn.

Ishizuka, M., K.S. Fu, and **J.T.P. Yao** [1982], "A rule-based inference with fuzzy set for structural damage assessment." In: Gupta and Sanchez [1982b], pp. 261–268.

Islam, S. [1974], "Toward integration of two system theories by Mesarovic and Wymore." *Intern. J. of General Systems*, **1**, pp. 35–40.

Jaynes, E.T. [1968], "Prior probabilities." *IEEE Trans. on Systems Science and Cybernetics*, **SSC-4**, pp. 227–241.

Jaynes, E.T. [1979], "Where do we stand on maximum entropy?" In: Levine and Tribus [1979], pp. 15–118.

Jaynes, E.T. [1982], "On the rationale of maximum entropy methods." *Proc. of IEEE*, **70**, pp. 939–952.

Jaynes, E.T. [1983], *Papers on Probability, Statistics and Statistical Physics* (edited by R.D. Rosenkrantz). D. Reidel, Boston.

Jelinek, F. [1968], *Probabilistic Information Theory: Discrete and Memoryless Models*. McGraw-Hill, New York.

Jones, A., A. Kaufmann, and **H.-J. Zimmermann,** eds. [1986], *Fuzzy Sets Theory and Applications*. D. Reidel, Boston.

Jones, D.S. [1979], *Elementary Information Theory*. Clarendon Press, Oxford.

Jumarie, G. [1979], "The concept of structural entropy and its applications to general systems." *Intern. J. of General Systems*, **5**, pp. 99–120.

Jumarie, G. [1981], "A general paradigm of subjectivity in communication, subjective transinformation and subjective transuncertainty." *J. of Information and Optimization Sciences*, **2**, pp. 273–296.

Jumarie, G. [1983], "Entropy of fuzzy events revisited: an approach to fuzzy transinformation." *Cybernetica*, **26**, pp. 99–116.

Jumarie, G. [1986], *Subjectivity, Information, Systems: Introduction to a Theory of Relativistic Cybernetics*. Gordon and Breach, New York.

Jumarie, G. [1987], "A Minkowskian theory of observation: application to uncertainty and fuzziness." *Fuzzy Sets and Systems*, **24**.

Kacprzyk, J. [1982], "Multistage decision processes in a fuzzy environment: a survey." In: Gupta and Sanchez [1982a], pp. 251–263.

Kacprzyk, J. [1983], *Multistage Decision-Making Under Fuzziness*. Verlag TÜV Rheinland, Köln.

Kacprzyk, J. and **R.R. Yager,** eds. [1985a], *Management Decision-Support Systems, Using Fuzzy Sets and Possibility Theory*. Verlag TÜV Rheinhold, Köln.

Kacprzyk, J. and **R.R. Yager** [1985b], "Emergency-oriented expert systems: a fuzzy approach." *Information Sciences*, **37**, pp. 143–155.

Kaleva, O. and **S. Seikkala** [1984], "On fuzzy metric spaces." *Fuzzy Sets and Systems*, **12**, pp. 215–229.

Kandel, A. [1982], *Fuzzy Techniques in Pattern Recognition*. John Wiley, New York.

Kandel, A. [1986], *Fuzzy Mathematical Techniques with Applications*. Addison-Wesley, Reading, Mass.

Kandel, A. and **A. Lee** [1979], *Fuzzy Switching and Automata: Theory and Applications*. Crane Russak, New York.

Kapur, J.N. [1983], "Twenty-five years of maximum entropy principle." *J. Math. Phys. Sciences*, **17**, pp. 103–156.

Katz, A. [1967], *Principles of Statistical Mechanics: The Information Theory Approach*. W.H. Freeman, San Francisco.

Kaufmann, A. [1975], *Introduction to the Theory of Fuzzy Subsets*. Academic Press, New York.

Kaufmann, A. and **M.M. Gupta** [1985], *Introduction to Fuzzy Arithmetic: Theory and Applications*. Van Nostrand Reinhold, New York.

Kempton, W. [1984], "Interview methods for eliciting fuzzy categories." *Fuzzy Sets and Systems*, **14**, pp. 43–64.

Kerre, E.E. [1982], "The use of fuzzy set theory in electrocardiological diagnostics." In: Gupta and Sanchez [1982b], pp. 277–282.

Khinchin, A.I. [1957], *Mathematical Foundations of Information Theory*. Dover, New York.

Kickert, W.J.M. [1978], *Fuzzy Theories on Decision-Making*. Martinus-Nijhoff, Boston.

Kickert, W.J.M. and **E.H. Mamdani** [1978], "Analysis of a fuzzy logic controller." *Fuzzy Sets and Systems*, **1**, pp. 29–44.

King, P.J. and **E.H. Mamdani** [1977], "The application of fuzzy control systems to industrial processes." In: Gupta, Saridis, and Gaines [1977], pp. 321–330.

Kirschenmann, P.P. [1970], *Information and Reflection*. Humanities Press, New York.

Klement, E.P. [1982], "A theory of fuzzy measures: a survey." In: Gupta and Sanchez [1982a], pp. 59–65.

Klement, E.P. [1984], "Operations on fuzzy sets: an axiomatic approach." *Information Sciences*, **27**, pp. 221–232.

Klir, G.J. [1985], *Architecture of Systems Problem Solving*. Plenum Press, New York.

Klir, G.J. [1987], "Where do we stand on measures of uncertainty, ambiguity, fuzziness, and the like?" *Fuzzy Sets and Systems*, **24**.

Klir, G.J. and **M. Mariano** [1987], "On the uniqueness of possibilistic measure of uncertainty and information." *Fuzzy Sets and Systems*, **24**.

Klir, G.J. and **E.C. Way** [1985], "Reconstructability analysis: aims, results, open problems." *Systems Research*, **2**, pp. 141–163.

Knopfmacher, J. [1975], "On measures of fuzziness." *J. Math. Analysis and Applications,* **49,** pp. 529–534.

Kochen, M. [1975], "Applications of fuzzy sets in psychology." In: Zadeh, Fu, Tanaka, and Shimura [1975], pp. 395–408.

Kochen, M. [1982], "The origin of concepts in nervous systems: can fuzzy set theory clarify the questions?" In: Yager [1982a], pp. 542–550.

Kohout, L.J., E. Keravnou, and **W. Bandler** [1984], "Information retrieval system using fuzzy relational products for thesaurus construction." In: Sanchez [1984a], pp. 7–13.

Kokawa, M.K., K. Nakamura, and **M. Oda** [1975], "Experimental approach to fuzzy simulation of memorizing, forgetting, and inference process." In: Zadeh, Fu, Tanaka, and Shimura [1975], pp. 409–428.

Kolmogorov, A.N. [1950], *Foundations of the Theory of Probability.* Chelsea, New York.

Kolmogorov, A.N. [1965], "Three approaches to the quantitative definition of information." *Problems of Information Transmission,* **1,** pp. 1–7.

Kornwachs, K. and **W. von Lucadou** [1985], "Pragmatic information as a nonclassical concept to describe cognitive processes." *Cognitive Systems,* **1,** pp. 79–94.

Krippendorff, K. [1982], "Q: an interpretation of the information theoretical Q-measure." In: *Progress in Cybernetics and Systems Research,* Vol VIII, edited by R. Trappl, G.J. Klir, and F.R. Pichler, Hemisphere, Washington, D.C., pp. 63–67.

Kullback, S. [1959], *Information Theory and Statistics.* John Wiley, New York. (Reprinted by Dover, New York, 1968.)

Lakoff, G. [1973], "Hedges: a study in meaning criteria and the logic of fuzzy concepts." *J. of Philosophical Logic,* **2,** pp. 458–508.

Lakov, D. [1985], "Adaptive robot under fuzzy control." *Fuzzy Sets and Systems,* **17,** pp. 1–8.

Larkin, L.I. [1985], "A fuzzy logic controller for aircraft flight control." In: Sugeno [1985], pp. 87–103.

Lasker, G.E., ed. [1981], *Applied Systems and Cybernetics. Vol VI: Fuzzy Sets and Fuzzy Systems, Possibility Theory, and Special Topics in Systems Research.* Pergamon Press, New York.

Lee, E.T. [1975], "Shape-oriented chromosome classification." *IEEE Trans. on Systems, Man and Cybernetics,* **SMC-5,** pp. 629–632.

Lee, H.C. and **K.S. Fu** [1972], "A stochastic syntax analysis procedure and its application to pattern classification." *IEEE Trans. on Computers,* **C-21,** pp. 660–666.

Lee, R.C.T. and **C.-L. Chang** [1971], "Some properties of fuzzy logic." *Information and Control,* **19,** pp. 417–431.

Lefkovitch, L.P. [1985], "Entropy and set covering." *Information Sciences,* **36,** pp. 283–294.

Lesmo, L., L. Saitta, and **P. Torasso** [1982], "Learning of fuzzy production rules for medical diagnosis." In: Gupta and Sanchez [1982b], pp. 249–260.

Lesmo, L., L. Saitta, and **P. Torasso** [1983], "Fuzzy production rules: a learning methodology." In: Wang [1983], pp. 181–198.

Levine, R.D. and **M. Tribus,** eds. [1979], *The Maximum Entropy Formalism.* The MIT Press, Cambridge, Mass.

Lipschutz, S. [1968], *Probability.* McGraw-Hill, New York.

Liu, Y.-M. [1985], "Some properties of fuzzy sets." *J. of Math. Analysis and Applications,* **111,** pp. 119–129.

Löfgren, L. [1977], "Complexity of descriptions of systems: a foundational study." *Intern. J. of General Systems,* **3,** pp. 197–214.

Loo, S.G. [1977], "Measures of fuzziness." *Cybernetica,* **20,** pp. 201–210.

Lowen, R. [1976], "Fuzzy topological spaces and fuzzy compactness." *J. of Math. Analysis and Applications,* **56,** pp. 621–633.

Lowen, R. [1978], "On fuzzy complements." *Information Science,* **14,** pp. 107–113.

Lowen, R. [1980], "Convex fuzzy sets." *Fuzzy Sets and Systems,* **3,** pp. 291–310.

Luhandjula, M.K. [1986], "On possibilistic linear programming." *Fuzzy Sets and Systems,* **18,** pp. 15–30.

Luhandjula, M.K. [1983], "Linear programming under randomness and fuzziness." *Fuzzy Sets and Systems,* **10,** pp. 45–55.

MacKay, D.M. [1950], "Quantal aspects of scientific information." *Philosophical Magazine,* **41,** pp. 289–311.

MacKay, D.M. [1969], *Information, Mechanism and Meaning.* The MIT Press, Cambridge, Mass.

Majumder, D.K.D. [1986], *Fuzzy Mathematical Approach to Pattern Recognition.* John Wiley, New York.

Mamdani, E.H. [1976], "Advances in the linguistic synthesis of fuzzy controllers." *Intern. J. of Man-Machine Studies,* **8,** pp. 669–678.

Mamdani, E.H. [1977], "Applications of fuzzy set theory to control systems: a survey." In: Gupta, Saridis, and Gaines [1977], pp. 77–88.

Mamdani, E.H. and **S. Assilian** [1975], "An experiment in linguistic synthesis with a fuzzy logic controller." *Intern. J. of Man-Machine Studies,* **7,** pp. 1–13.

Mamdani, E.H. and **J. Efstathion** [1984], "An analysis of formal logics as inference mechanisms on expert systems." *Intern. J. of Man-Machine Studies,* **21,** pp. 213–227.

Mamdani, E.H. and **B.R. Gaines,** eds. [1981], *Fuzzy Reasoning and its Applications.* Academic Press, London.

Mamdani, E.H., J.-J. Østergaard, and **E. Lembessis** [1983], "Use of fuzzy logic for implementing rule-based control of industrial processes." In: Wang [1983], pp. 307–323.

Mamdani, E.H. and **C.P. Pappis** [1977], "A fuzzy logic controller for a traffic junction." *IEEE Trans. on Systems, Man, and Cybernetics,* **SMC-7,** pp. 707–717.

Mamdani, E.H. and **B.S. Sembi** [1980], "Process control using fuzzy logic." In: Wang and Chang [1980], pp. 249–266.

Mathai, A.M. and **P.N. Rathie** [1975], *Basic Concepts of Information Theory and Statistics.* John Wiley, New York.

McEliece, R.J. [1977], *The Theory of Information and Coding.* Addison-Wesley, Reading, Mass.

Menger, K. [1942], "Statistical metrics." *Proc. Nat. Acad. Sci.,* **28,** pp. 535–537.

Menges, G. [1974], *Information, Inference and Decision.* D. Reidel, Boston.

Mesarovic, M.D. and **Y. Takahara** [1975], *General Systems Theory: Mathematical Foundations.* Academic Press, New York.

Miyakoshi, M. and **M. Shimbo** [1985], "Solutions of composite fuzzy relational equations with triangular norms." *Fuzzy Sets and Systems,* **16,** pp. 53–63.

Mizumoto, M. [1981], "Note on the arithmetic rule by Zadeh for fuzzy conditional inference." *Cybernetics and Systems,* **12,** pp. 247–306.

Mizumoto, M. and **K. Tanaka** [1976], "Some properties of fuzzy sets of type 2." *Information and Control,* **31,** pp. 312–340.

Mizumoto, M. and **K. Tanaka** [1981], "Fuzzy sets of type 2 under algebraic product and algebraic sum." *Fuzzy Sets and Systems,* **5,** pp. 277–290.

Mizumoto, M. and **H.-J. Zimmermann** [1982], "Comparison of fuzzy reasoning methods." *Fuzzy Sets and Systems,* **8,** pp. 253–283.

Moles, A. [1966], *Information Theory and Esthetic Perception.* University of Illinois Press, Urbana, Ill.

Moore, R.E. [1966], *Interval Analysis.* Prentice-Hall, Englewood Cliffs, N.J.

Moore, R.E. [1979], *Methods and Applications of Interval Analysis.* SIAM, Philadelphia, Pa.

Morris, C. [1938], *Foundations of the Theory of Signs* (2 volumes). University of Chicago Press, Chicago, Ill.

Morris, C. [1946], *Signs, Language, and Behavior.* Prentice-Hall, Englewood Cliffs, N.J.

Morris, C. [1964], *Signification and Significance.* M.I.T. Press, Cambridge, Mass.

Mou-Chao, M. and **C. Zhi-Quiang** [1982], "The multistage evaluation method in psychological measurement: an application of fuzzy sets theory to psychology." In: Gupta and Sanchez [1982b], pp. 307–312.

Murakami, S. [1984], "Application of fuzzy controller to automobile speed control system." In: Sanchez [1984a], pp. 43–48.

Nakamura, K. and **S. Iwai** [1982], "A representation of analogical inference by fuzzy sets and its application to information retrieval system." In: Gupta and Sanchez [1982a], pp. 373–386.

Nakatsuyama, M., H. Nagahashi, and **N. Nishizuka** [1984], "Fuzzy logic phase controller for traffic functions in the one-way arterial road." *Proc. IFAC 9th Triennial World Congress,* Pergamon Press, Oxford, pp. 2865–2870.

Nauta, D. [1972], *The Meaning of Information.* Mouton, The Hague.

Negoita, C.V. [1981], *Fuzzy Systems.* Abacus, Tunbridge Wells (U.K.).

Negoita, C.V. [1985], *Expert Systems and Fuzzy Systems.* Benjamin/Cummings, Menlo Park, Calif.

Negoita, C.V. and **D.A. Ralescu** [1975], *Applications of Fuzzy Sets to Systems Analysis.* Birkhäuser, Basel and Stuttgart.

Nguyen, H.T. [1978], "On conditional possibility distributions." *Fuzzy Sets and Systems,* **1,** pp. 299–309.

Nguyen, H.T. [1979], "Toward a calculus of the mathematical notion of possibility." In: Gupta, Ragade, and Yager [1979], pp. 235–246.

Niedenthal, P.M. and **N. Cantor** [1984], "Making use of social prototypes: from fuzzy concepts to firm decisions." *Fuzzy Sets and Systems,* **14,** pp. 5–27.

Nojiri, H. [1979], "A model of fuzzy team decision." *Fuzzy Sets and Systems,* **2,** pp. 201–212.

Nojiri, H. [1982], "A model of the executive's decision processes in new product development." *Fuzzy Sets and Systems,* **7,** pp. 227–241.

Norwich, A.M. and **I.B. Turksen** [1982a], "The fundamental measurement of fuzziness." In: Yager [1982a], pp. 49–60.

Norwich, A.M. and **I.B. Turksen** [1982b], "The construction of membership functions." In: Yager [1982a], pp. 61–67.

Norwich, A.M. and **I.B. Turksen** [1982c], "Meaningfulness in fuzzy set theory." In: Yager [1982a], pp. 68–74.

Norwich, A.M. and **I.B. Turksen** [1984], "A model for the measurement of membership and the consequences of its empirical implementation." *Fuzzy Sets and Systems,* **12,** pp. 1–25.

Novák, V. [1984], "Fuzzy sets—the approximation of semisets." *Fuzzy Sets and Systems,* **14,** pp. 259–272.

Nowakowska, M. [1977], "Methodological problems of measurement of fuzzy concepts in the social sciences." *Behavioral Science,* **22,** pp. 107–115.

Nowakowska, M. [1979], "Fuzzy concepts: their structure and problems of measurement." In: Gupta, Ragade, and Yager [1979], pp. 361–387.

Nowakowska, M. [1983], "Some problems of observability theory and its applications." *Mathematical Social Sciences,* **4,** pp. 1–23.

Nowakowska, M. [1986], *Cognitive Sciences.* Academic Press, New York.

Nurmi, H. [1981], "Approaches to collective decision making with fuzzy preference relations." *Fuzzy Sets and Systems,* **6,** pp. 249–259.

Oblow, E.M. [1987a], "O-theory: a hybrid uncertainty theory." *Intern. J. of General Systems,* **13,** pp. 95–106.

Oblow, E.M. [1987b], "Foundations of O-theory I: the intersection rule." *Intern. J. of General Systems,* **13.**

Oda, M., T. Shimomura, and **B.F. Womack** [1980], "Concept structure and its distortion on the communication and formation process of morality concept." In: Wang and Chang [1980], pp. 369–389.

Oden, G.C. [1979], "Fuzzy propositional approach to psycholinguistic problems: an application of fuzzy set theory in cognitive science." In: Gupta, Ragade, and Yager [1979], pp. 409–420.

Oguntade, O.O. and **P.E. Beaumont** [1982], "Ophthalmological prognosis via fuzzy subsets." *Fuzzy Sets and Systems,* **7,** pp. 123–138.

Oguntade, O.O. and **J.S. Gero** [1981], "Evaluation of architectural design profiles using fuzzy sets." *Fuzzy Sets and Systems,* **5,** pp. 221–234.

Ollero, A. and **E. Freire** [1981], "The structure of relations in personnel management." *Fuzzy Sets and Systems,* **5,** pp. 115–125.

Orlovsky, S.A. [1978], "Decision-making with a fuzzy preference relation." *Fuzzy Sets and Systems,* **1,** pp. 155–167.

Østergaard, J.-J. [1977], "Fuzzy logic control of a heat exchanger process." In: Gupta, Saridis, and Gaines [1977], pp. 285–320.

Ovchinnikov, S.V. [1981], "Innovations in fuzzy set theory." *Proc. of the Eleventh IEEE International Symposium on Multiple-Valued Logic,* Norman, Oklahoma, pp. 226–227.

Ovchinnikov, S.V. [1983], "General negations in fuzzy set theory." *J. of Math. Analysis and Applications,* **92,** pp. 234–239.

Ovchinnikov, S.V. [1984], "Representations of transitive fuzzy relations." In: Skala, Termini, and Trillas [1984], pp. 105–118.

Padet, C. [1985], "Comparative properties of several entropies in informational observation theory." *Kybernetes,* **14,** pp. 17–24.

Pappis, C.P. and **M. Sugeno** [1985], "Fuzzy relational equations and the inverse problem." *Fuzzy Sets and Systems,* **15,** pp. 79–90.

Pawlak, Z. [1984], "Rough classification." *Intern. J. of Man-Machine Studies,* **20,** pp. 469–483.

Pedrycz, W. [1981], "An approach to the analysis of fuzzy systems." *Intern. J. of Control,* **34,** pp. 403–421.

Pedrycz, W. [1983a], "Fuzzy relational equations with generalized connectives and their applications." *Fuzzy Sets and Systems,* **10,** pp. 185–201.

Pedrycz, W. [1983b], "Numerical and applicational aspects of fuzzy relational equations." *Fuzzy Sets and Systems,* **11,** pp. 1–18.

Pedrycz, W. [1983c], "Some applicational aspects of fuzzy relational equations in systems analysis." *Intern. J. of General Systems,* **9,** pp. 125–132.

Pedrycz, W. [1984], "An identification algorithm in fuzzy relational systems." *Fuzzy Sets and Systems,* **13,** pp. 153–167.

Pedrycz, W. [1985], "Structured fuzzy models." *Cybernetics and Systems,* **16,** pp. 103–117.

Prade, H. [1980], "Operations research with fuzzy data." In: Wang and Chang [1980], pp. 155–170.

Prade, H. [1983], "Fuzzy programming: why and how?—some hints and examples." In: Wang [1983], pp. 237–251.

Prade, H. [1984], "Lipski's approach to incomplete information data bases restated and generalized in the setting of Zadeh's possibility theory." *Information Systems,* **9,** pp. 27–42.

Prevot, M. [1981], "Algorithm for the solution of fuzzy relations." *Fuzzy Sets and Systems,* **5,** pp. 319–322.

Puri, M.L. and **D. Ralescu** [1982], "A possibility measure is not a fuzzy measure." *Fuzzy Sets and Systems,* **7,** pp. 311–313.

Quastler, H., ed. [1955], *Information Theory in Psychology.* The Free Press, Glencoe, Ill.

Radecki, T. [1983], "A theoretical background for applying fuzzy set theory in information retrieval." *Fuzzy Sets and Systems,* **10,** pp. 169–183.

Ramer, A. [1987], "Uniqueness of information measure in the theory of evidence." *Fuzzy Sets and Systems,* **24.**

Ramer, A. and **L. Lander** [1987], "Axiomatic characterization of possibilistic measure of uncertainty and information." *Fuzzy Sets and Systems,* **24.**

Reichenbach, H. [1949], *The Theory of Probability.* University of California Press, Berkeley.

Rényi, A. [1960], "On measures of entropy and information." *Proc. 4th Berkeley Symp. on Math. Stat. and Prob.,* University of California Press, Berkeley, pp. 547–561.

Rényi, A. [1970], *Probability Theory.* North-Holland, Amsterdam, (Chapter IX, Introduction to Information Theory, pp. 540–616).

Rescher, N. [1969], *Many-valued Logic.* McGraw-Hill, New York.

Reza, F.M. [1961], *Introduction to Information Theory.* McGraw-Hill, New York.

Rocha, A.F. [1982], "Toward a theoretical and experimental approach of fuzzy learning." In: Gupta and Sanchez [1982], pp. 191–200.

Rosenfeld, A. [1971], "Fuzzy groups." *J. of Math. Analysis and Applications,* **35,** pp. 512–517.

Ruspini, E.H. [1982], "Recent developments in fuzzy clustering." In: Yager [1982a], pp. 133–147.

Saitta, L. and **P. Torasso** [1981], "Fuzzy characterization of coronary disease." *Fuzzy Sets and Systems,* **5,** pp. 245–258.

Sakawa, M. [1983], "Interactive computer programs for fuzzy linear programming with multiple objectives." *Intern. J. of Man-Machine Studies,* **18,** pp. 489–503.

Sakawa, M. and **H. Yano** [1985], "Interactive fuzzy decision-making for multi-objective non-linear programming using reference membership intervals." *Intern. J. of Man-Machine Studies,* **23,** pp. 407–421.

Sanchez, E. [1976], "Resolution of composite fuzzy relation equations." *Information and Control,* **30,** pp. 38–48.

Sanchez, E. [1977], "Solutions in composite fuzzy relation equations." In: Gupta, Saridis, and Gaines [1977], pp. 221–234.

Sanchez, E. [1979a], "Compositions of fuzzy relations." In: Gupta, Ragade, and Yager [1979], pp. 421–433.

Sanchez, E. [1979b], "Medical diagnosis and composite fuzzy relations." In: Gupta, Ragade, and Yager [1979], pp. 437–444.

Sanchez, E., ed. [1984a], *Fuzzy Information, Knowledge Representation and Decision Analysis.* Pergamon Press, Oxford.

Sanchez, E. [1984b], "Solution of fuzzy equations with extended operations." *Fuzzy Sets and Systems,* **12,** pp. 237–248.

Sanchez, E., J. Gouvernet, R. Bartolin, and **L. Vovan** [1982], "Linguistic approach in fuzzy logic of the WHO classifications of dyslipoproteinemias." In: Yager [1982a], pp. 582–588.

Saridis, G.N. and **H.E. Stephanou** [1977], "Fuzzy decision-making in prosthetic devices." In: Gupta, Saridis, and Gaines [1977], pp. 387–402.

Savage, L.J. [1972], *The Foundations of Statistics.* Dover, New York.

Schmucker, K.J. [1983], *Fuzzy Sets, Natural Language Computations, and Risk Analysis.* Computer Science Press, Rockville, Md.

Schweizer, B. and **A. Sklar** [1960], "Statistical metric spaces." *Pacific J. of Mathematics,* **10,** pp. 313–334.

Schweizer, B. and **A. Sklar** [1961], "Associative functions and statistical triangle inequalities." *Publicationes Mathematicae Debrecen,* **8,** pp. 169–186.

Schweizer, B. and **A. Sklar** [1963], "Associative functions and abstract semi-groups." *Publicationes Mathematicae Debrecen,* **10,** pp. 69–81.

Schweizer, B. and **A. Sklar** [1983], *Probability Metric Spaces.* North-Holland, New York.

Sessa, S. [1984], "Some results in the setting of fuzzy relation equations theory." *Fuzzy Sets and Systems,* **14,** pp. 281–297.

Shackle, G.L.S. [1969], *Decision, Order and Time in Human Affairs.* Cambridge University Press, Cambridge.

Shafer, G. [1976a], *A Mathematical Theory of Evidence.* Princeton University Press, Princeton.

Shafer, G. [1976b], "A theory of statistical evidence." In: *Foundations of Probability Theory, Statistical Inference, and Statistical Theories of Science,* Vol. II, edited by W.L. Harper and C.A. Hooker, D. Reidel, Dordrecht-Holland, pp. 365–436.

Shafer, G. [1978], "Non-additive probabilities in the work of Bernoulli and Lambert." *Archive for History of Exact Sciences,* **19,** pp. 309–370.

Shafer, G. [1979], "The allocation of probability." *The Annals of Probability,* **7,** pp. 827–839.

Shafer, G. [1981a], "Jeffrey's rule of conditioning." *Philosophy of Science,* **48,** pp. 337–362.

Shafer, G. [1981b], "Constructive probability." *Synthese,* **48,** pp. 1–60.

Shafer, G. [1982], "Lindley's paradox." *J. of the American Statistical Association,* **77,** pp. 325–351.

Shafer, G. [1985], "Belief functions and possibility measures." In: Bezdek [1985].

Shafer, G. and **A. Tversky** [1985], "Languages and designs for probability judgment." *Cognitive Science,* **9,** pp. 309–339.

Shannon, C.E. [1948], "The mathematical theory of communication." *The Bell System Technical Journal,* **27,** pp. 379–423, 623–656.

Shannon, C.E. and **W. Weaver** [1964], *The Mathematical Theory of Communication.* University of Illinois Press, Urbana, Ill.

Shore, J.E. and **R.W. Johnson** [1980], "Axiomatic derivation of the principle of maximum entropy and the principle of minimum cross-entropy." *IEEE Trans. on Information Theory,* **IT-26,** pp. 26–37.

Silvert, W. [1979], "Symmetric summation: a class of operations on fuzzy sets." *IEEE Trans. on Systems, Man, and Cybernetics,* **SMC-9,** pp. 657–659.

Skala, H.J. [1978], "On many-valued logics, fuzzy sets, fuzzy logics and their applications." *Fuzzy Sets and Systems,* **1,** pp. 129–149.

Skala, H.J., S. Termini, and **E. Trillas,** eds. [1984], *Aspects of Vagueness.* D. Reidel, Dordrecht and Boston.

Smets, P. [1981a], "The degree of belief in a fuzzy event." *Information Sciences,* **25,** pp. 1–19.

Smets, P. [1981b], "Medical diagnosis: fuzzy sets and degrees of belief." *Fuzzy Sets and Systems,* **5,** pp. 259–266.

Smets, P. [1983], "Information content of an evidence." *Intern. J. of Man-Machine Studies,* **19,** pp. 33–43.

Smith, C.R. and **W.T. Grandy,** eds. [1985], *Maximum-Entropy and Bayesian Methods in Inverse Problems.* D. Reidel, Boston.

Smith, S.A. [1974], "A derivation of entropy and the maximum entropy criterion in the context of decision problems." *IEEE Trans. on Systems, Man, and Cybernetics,* **SMC-4,** pp. 157–163.

Sommer, G. [1981], "Fuzzy inventory scheduling." In: Lasker [1981], pp. 3052–3060.

Soula, G. and **E. Sanchez** [1982], "Soft deduction rules in medical diagnostic processes." In: Gupta and Sanchez [1982b], pp. 77–88.

Sugeno, M. [1977], "Fuzzy measures and fuzzy integrals—a survey." In: Gupta, Saridis, and Gaines [1977], pp. 89–102.

Sugeno, M., ed. [1985a], *Industrial Applications of Fuzzy Control.* North-Holland, New York.

Sugeno, M. [1985b], "An introductory survey of fuzzy control." *Information Sciences,* **36,** pp. 59–83.

Sugeno, M. and **G.T. Kang** [1986], "Fuzzy modelling and control of multilayer incinerator." *Fuzzy Sets and Systems,* **18,** pp. 329–346.

Sugeno, M. and **M. Nishida** [1985], "Fuzzy control of model car." *Fuzzy Sets and Systems,* **16,** pp. 103–113.

Sugeno, M. and **T. Takagi** [1983], "A new approach to design of fuzzy controller." In: Wang [1983], pp. 325–334.

Takagi, T. and **M. Sugeno** [1984], "Derivation of fuzzy control rules from human operator's control actions." In: Sanchez [1984a], pp. 55–60.

Tanaka, H., T. Okuda, and **K. Asai** [1979], "Fuzzy information and decision in statistical model." In: Gupta, Ragade, and Yager [1979], pp. 303–320.

Terano, T. and **M. Sugeno** [1975], "Conditional fuzzy measures and their applications." In: Zadeh, Fu, Tanaka, and Shimura [1975], pp. 151–170.

Terano, T., M. Sugeno, and **Y. Tsukamoto** [1984], "Planning in management by fuzzy dynamic programming." In: Sanchez [1984a], pp. 381–386.

Thole, U., H.-J. Zimmermann, and **P. Zysno** [1979], "On the suitability of minimum and product operators for the intersection of fuzzy sets." *Fuzzy Sets and Systems,* **2,** pp. 167–180.

Theil, H. [1967], *Economics and Information Theory.* North-Holland, Amsterdam, and Rand McNally, Chicago.

Togai, M. and **H. Watanabe** [1986], "A VLSI implementation of a fuzzy-inference engine: toward an expert system on a chip." *Information Sciences,* **38,** pp. 147–163.

Tong, R.M. [1979], "The construction and evaluation of fuzzy models." In: Gupta, Ragade, and Yager [1979], pp. 559–576.

Tong, R.M. [1980], "The evaluation of fuzzy models derived from experimental data." *Fuzzy Sets and Systems,* **4,** pp. 1–12.

Tong, R.M. [1984], "A retrospective view of fuzzy control systems." *Fuzzy Sets and Systems,* **14,** pp. 199–210.

Trillas, E., C. Alsina, and **L. Valverde** [1982], "Do we need max, min, l-j in fuzzy set theory?" In: Yager [1982a], pp. 275–297.

Trillas, E. and **T. Riera** [1978], "Entropies of finite fuzzy sets." *Information Sciences,* **15,** pp. 159–168.

Tsichritzis, D. [1971], "Participation measures." *J. of Mathematical Analysis and Applications,* **36,** pp. 60–72.

Turksen, I.B. and **D.D.W. Yao** [1984], "Representations of connectives in fuzzy reasoning: the view through normal form." *IEEE Trans. on Systems, Man, and Cybernetics,* **SMC-14,** pp. 146–151.

Tzannes, N.S. and **J.P. Noonan** [1973], "The mutual information principle and applications." *Information and Control,* **22,** pp. 1–12.

Umano, M. [1982], "Freedom-0: fuzzy database system." In: Gupta and Sanchez [1982a], pp. 339–347.

Umano, M. [1984], "Retrieval from fuzzy database by fuzzy relational algebra." In: Sanchez [1984a], pp. 1–6.

Umeyama, S. [1986], "The complementary process of fuzzy medical diagnosis and its properties." *Information Sciences,* **38,** pp. 229–242.

Uragami, M., M. Mizumoto, and **K. Tanaka** [1976], "Fuzzy robot controls." *J. of Cybernetics,* **6,** pp. 39–64.

Vallet, C.L., H. Le Guyader, and **Th. Moulin** [1981], "Ambiguity and imprecision in arithmetical models of natural systems." In: Lasker [1981], pp. 3070–3075.

Van Der Lubbe, J.C.A. [1984], "A generalized class of certainty and information." *Information Sciences, 32,* pp. 187–215.

Van Emden, M.H. [1971], *An Analysis of Complexity.* Mathematical Centre, Amsterdam.

Van Laarhoven, P.J.M. and **W. Pedrycz** [1983], "A fuzzy extension of Saaty's priority theory." *Fuzzy Sets and Systems, 11,* pp. 229–241.

Vila, M.A. and **M. Delgado** [1983], "On medical diagnosis using possibility measures." *Fuzzy Sets and Systems, 10,* pp. 211–222.

Vopěnka, P. [1979], *Mathematics in the Alternative Set Theory.* Teubner, Leipzig.

Vopěnka, P. and **P. Hájek** [1972], *The Theory of Semisets.* North-Holland, Amsterdam.

Voxman, W. and **R. Goetschel** [1983], "A note on the characterization of the max and min operators." *Information Sciences, 30,* pp. 5–10.

Wagner, W. [1981], "A fuzzy model of concept representation in memory." *Fuzzy Sets and Systems, 6,* pp. 11–26.

Walley, P. and **T.L. Fine** [1979], "Varieties of model (classificatory) and comparative probability." *Synthese, 41,* pp. 321–374.

Walley, P. and **T.L. Fine** [1982], "Toward a frequentist theory of upper and lower probability." *The Annals of Statistics, 10,* pp. 741–761.

Wang, P.P. and **S.K. Chang,** eds. [1980], *Fuzzy Sets: Theory and Applications to Policy Analysis and Information Systems.* Plenum Press, New York.

Wang, P.P., ed. [1983], *Advances in Fuzzy Sets, Possibility Theory, and Applications.* Plenum Press, New York.

Watanabe, S. [1960a], "Information theoretical analysis of multivariate correlations." *IBM J. of Research and Development, 4,* pp. 66–82.

Watanabe, S. [1960b], "Information-theoretical aspects of inductive and deductive inference." *IBM J. of Research and Development, 4,* pp. 208–231.

Watanabe, S. [1969], *Knowing and Guessing.* John Wiley, New York.

Watanabe, S. [1985], *Pattern Recognition: Human and Mechanical.* John Wiley, New York.

Weaver, W. [1968], "Science and complexity." *American Scientist, 36,* pp. 536–544.

Webber, M.J. [1979], *Information Theory and Urban Spatial Structure.* Croom Helm, London.

Weber, S. [1984], "Measures of fuzzy sets and measures of fuzziness." *Fuzzy Sets and Systems, 13,* pp. 247–271.

Wernecke, S.J. and **L.R. D'Addario** [1977], "Maximum entropy image reconstruction." *IEEE Trans. on Computers, C-26,* pp. 351–364.

Wenstop, F. [1979], "Exploring linguistic consequences of assertions on social sciences." In: Gupta, Ragade, and Yager [1979], pp. 501–518.

Whalen, T. and **B. Schott** [1985a], "Alternative logics for approximate reasoning in expert systems: a comparative study." *Intern. J. of Man-Machine Studies, 22,* pp. 327–346.

Whalen, T. and **B. Schott** [1985b], "Goal-directed approximate reasoning in a fuzzy production system." In: Gupta, Kandel, Bandler, and Kiszka [1985], pp. 505–517.

White, R.B. [1979], "The consistency of the axiom of comprehension in the infinite-valued predicate logic of Łukasiewicz." *Journal of Philosophical Logic, 8,* pp. 509–534.

Whittemore, B.J. and **M.C. Yovits** [1974], "The quantification and analysis of information used in decision processes." *Information Sciences, 2,* pp. 171–184.

Wierzchon, S.T. [1982a], "On fuzzy measure and fuzzy integral." In: Gupta and Sanchez [1982a], pp. 79–86.

Wierzchon, S.T.[1982b], "Application of fuzzy decision-making theory to coping with ill-defined problems." *Fuzzy Sets and Systems*, **7**, pp. 1–18.

Wierzchon, S.T. [1983], "An algorithm for identification of fuzzy measure." *Fuzzy Sets and Systems*, **9**, pp. 69–78.

Williams, P.M. [1980], "Bayesian conditionalisation and the principle of minimum information." *British J. for the Philosophy of Science*, **31**, pp. 131–144.

Wolfowitz, J. [1961], *Coding Theorems and Information Theory*. Prentice-Hall, Englewood Cliffs, N.J.

Wong, C.K. [1975], "Fuzzy topology." In: Zadeh, Fu, Tanaka, and Shimura [1975], pp. 171–190.

Wymore, A.W. [1969], *A Mathematical Theory of Systems Engineering: The Elements*. John Wiley, New York.

Wymore, A.W. [1976], *Systems Engineering Methodology for Interdisciplinary Teams*. John Wiley, New York.

Wyner, A.D. [1981], "Fundamental limits in information theory." *Proc. of IEEE*, **69**, pp. 239–251.

Yager, R.R. [1977], "Multiple objective decision-making using fuzzy sets." *Intern. J. of Man-Machine Studies*, **9**, pp. 375–382.

Yager, R.R. [1978], "Fuzzy decision making including unequal objectives." *Fuzzy Sets and Systems*, **1**, pp. 87–95.

Yager, R.R. [1979a], "A measurement-informational discussion of fuzzy union and intersection." *Intern. J. of Man-Machine Studies*, **11**, pp. 189–200.

Yager, R.R. [1979b], "On the measure of fuzziness and negation. Part I: Membership in the unit interval." *Intern. J. of General Systems*, **5**, pp. 221–229.

Yager, R.R. [1980a], "On the measure of fuzziness and negation. Part II: Lattices." *Information and Control*, **44**, pp. 236–260.

Yager, R.R. [1980b], "On a general class of fuzzy connectives." *Fuzzy Sets and Systems*, **4**, pp. 235–242.

Yager, R.R. [1980c], "Fuzzy thinking as quick and efficient." *Cybernetica*, **23**, pp. 265–298.

Yager, R.R. [1980d], "Aspects of possibilistic uncertainty." *Intern. J. of Man-Machine Studies*, **12**, pp. 283–298.

Yager, R.R. [1980e], "A foundation for a theory of possibility." *J. of Cybernetics*, **10**, pp. 177–204.

Yager, R.R. [1980f], "Fuzzy subsets of type II in decisions." *J. of Cybernetics*, **10**, pp. 137–159.

Yager, R.R. [1980g], "Satisfaction and fuzzy decision functions." In: Wang and Chang [1980], pp. 171–194.

Yager, R.R. [1980h], "On modeling interpersonal communication." In: Wang and Chang [1980], pp. 309–320.

Yager, R.R. [1981], "Some properties of fuzzy relationships." *Cybernetics and Systems*, **12**, pp. 123–140.

Yager, R.R., ed. [1982a], *Fuzzy Set and Possibility Theory*. Pergamon Press, Oxford.

Yager, R.R. [1982b], "Some procedures for selecting fuzzy set-theoretic operators." *Intern. J. of General Systems*, **8**, pp. 115–124.

Yager, R.R. [1982c], "Measuring tranquility and anxiety in decision making: an application of fuzzy sets." *Intern. J. of General Systems*, **8**, pp. 139–146.

Yager, R.R. [1982d], "Fuzzy prediction based on regression models." *Information Sciences*, **26**, pp. 45–63.

Yager, R.R. [1982e], "A new approach to the summarization of data." *Information Sciences*, **28**, pp. 69–86.

Yager, R.R. [1982f], "Measures of fuzziness based on t-norms." *Stochastica*, **6**, pp. 207–229.

Yager, R.R. [1983a], "Entropy and specificity in a mathematical theory of evidence." *Intern. J. of General Systems*, **9**, pp. 249–260.

Yager, R.R. [1983b], "On the implication operator in fuzzy logic." *Information Sciences*, **31**, pp. 141–164.

Yager, R.R. [1983c], "Some relationships between possibility, truth and certainty." *Fuzzy Sets and Systems*, **11**, pp. 151–156.

Yager, R.R. [1983d], "Quantifiers in the formulation of multiple objective decision functions." *Information Sciences*, **31**, pp. 107–139.

Yager, R.R. [1984a], "Probabilities from fuzzy observations." *Information Sciences*, **32**, pp. 1–31.

Yager, R.R. [1984b], "Fuzzy subsets with uncertain membership grades." *IEEE Trans. on Systems, Man, and Cybernetics*, **SMC-14**, pp. 271–275.

Yager, R.R. [1984c], "On different classes of linguistic variables defined via fuzzy subsets." *Kybernetes*, **13**, pp. 103–110.

Yager, R.R. [1985a], "On truth functional modification." *Intern. J. of General Systems*, **10**, pp. 105–121.

Yager, R.R. [1985b], "On the relationship of methods of aggregating evidence in expert systems." *Cybernetics and Systems*, **16**, pp. 1–21.

Yager, R.R. [1985c], "Strong truth and rules of inference in fuzzy logic and approximate reasoning." *Cybernetics and Systems*, **16**, pp. 23–63.

Yager, R.R. [1986], "A characterization of the extension principle." *Fuzzy Sets and Systems*, **18**, pp. 205–217.

Yaglom, A.M. and **I.M. Yaglom** [1983], *Probability and Information*. D. Reidel, Boston.

Yeh, R.T. and **S.Y. Bang** [1975], "Fuzzy relations, fuzzy graphs, and their applications." In: Zadeh, Fu, Tanaka, and Shimura [1975], pp. 125–149.

Yu, F.T.S. [1976], *Optics and Information Theory*. John Wiley, New York.

Zadeh, L.A. [1962], "From circuit theory to systems theory." *IRE Proc.*, **50**, pp. 856–865.

Zadeh, L.A. [1965a], "Fuzzy sets." *Information and Control*, **8**, pp. 338–353.

Zadeh, L.A. [1965b], "Fuzzy sets and systems." In: *System Theory*, edited by J. Fox, Polytechnic Press, Brooklyn, N.Y., pp. 29–37.

Zadeh, L.A. [1968a], "Probability measures of fuzzy events." *J. of Math. Analysis and Applications*, **23**, pp. 421–427.

Zadeh, L.A. [1968b], "Fuzzy algorithms." *Information and Control*, **12**, pp. 94–102.

Zadeh, L.A. [1971a], "Similarity relations and fuzzy orderings." *Information Sciences,* **3,** pp. 177–200.

Zadeh, L.A. [1971b], "Quantitative fuzzy semantics." *Information Sciences,* **3,** pp. 159–176.

Zadeh, L.A. [1971c], "Towards a theory of fuzzy systems." In: *Aspects of Networks and Systems Theory,* edited by R.E. Kalman and R.N. DeClairis; Holt, Rinehart & Winston, New York, pp. 469–490.

Zadeh, L.A. [1972], "A fuzzy set interpretation of linguistic hedges." *J. of Cybernetics,* **2,** pp. 4–34.

Zadeh, L.A. [1973], "Outline of a new approach to the analysis of complex systems and decision processes." *IEEE Trans. on Systems, Man, and Cybernetics,* **SMC-1,** pp. 28–44.

Zadeh, L.A. [1974], "A new approach to system analysis." In: *Man and Computer,* edited by M. Marois, North-Holland, Amsterdam, pp. 55–94.

Zadeh, L.A. [1975a], "Calculus of fuzzy restrictions." In: Zadeh, Fu, Tanaka, and Shimura [1975], pp. 1–39.

Zadeh, L.A. [1975b], "The concept of a linguistic variable and its application to approximate reasoning." *Information Sciences,* **8,** pp. 199–249, 301–357; **9,** pp. 43–80.

Zadeh, L.A. [1975c], "Fuzzy logic and approximate reasoning." *Synthese,* **30,** pp. 407–428.

Zadeh, L.A. [1976], "A fuzzy-algorithmic approach to the definition of complex or imprecise concepts." *Intern. J. of Man-Machine Studies,* **8,** pp. 249–291.

Zadeh, L.A. [1977], "Linguistic characterization of preference relations as a basis for choice in social systems." *Erkenntnis,* **11,** pp. 383–410.

Zadeh, L.A. [1978a], "Fuzzy sets as a basis for a theory of possibility." *Fuzzy Sets and Systems,* **1,** pp. 3–28.

Zadeh, L.A. [1978b], "PRUF-a meaning representation language for natural languages." *Intern. J. of Man-Machine Studies,* **10,** pp. 395–460.

Zadeh, L.A. [1979a], "Fuzzy sets and information granularity." In: Gupta, Ragade, and Yager [1979], pp. 3–18.

Zadeh, L.A. [1979b], "On the validity of Dempster's rule of combination of evidence." *Memo UCB/ERL No.* 79/24, University of California, Berkeley.

Zadeh, L.A. [1981], "Possibility theory and soft data analysis." In: *Mathematical Frontiers of the Social and Policy Sciences,* edited by L. Cobb and R.M. Thrall, Westview Press, Boulder, Colorado, pp. 69–129.

Zadeh, L.A. [1982], "Fuzzy systems theory: a framework for the analysis of humanistic systems." In: *Systems Methodology in Social Science Research,* edited by R.E. Cavallo, Kluwer-Nijhoff, Boston and The Hague, pp. 25–41.

Zadeh, L.A. [1983a], "The role of fuzzy logic in the management of uncertainty in expert systems." *Fuzzy Sets and Systems,* **11,** pp. 199–228.

Zadeh, L.A. [1983b], "A computational approach to fuzzy quantifiers in natural languages." *Computers and Mathematics,* **9,** pp. 149–184.

Zadeh, L.A. [1984], "A theory of commonsense knowledge." In: Skala, Termini, and Trillas [1984], pp. 257–296.

Zadeh, L.A. [1985], "Syllogistic reasoning in fuzzy logic and its application to usuality and reasoning with dispositions." *IEEE Trans. on Systems, Man, and Cybernetics,* **SMC-15,** pp. 754–765.

Zadeh, L.A., K.-S. Fu, K. Tanaka, and **M. Shimura,** eds. [1975], *Fuzzy Sets and Their Applications to Cognitive and Decision Processes.* Academic Press, New York.

Zeigler, B.P. [1976], *Theory of Modelling and Simulation.* John Wiley, New York.

Zeigler, B.P. [1984], *Multifacetted Modelling and Discrete Event Simulation.* Academic Press, New York.

Zeising, G., M. Wagenknecht, and **K. Hartmann** [1984], "Synthesis of distillation trains with heat integration by a combined fuzzy and graphical approach." *Fuzzy Sets and Systems,* **12,** pp. 103–115.

Zemankova-Leech, M. and **A. Kandel** [1984], *Fuzzy Relational Data Bases—A Key to Expert Systems.* Verlag TÜV Rheinland, Köln.

Zemankova-Leech, M. and **A. Kandel** [1985], "Implementing imprecision in information systems." *Information Sciences,* **37,** pp. 107–141.

Zenner, R.B.R.C., R.M.M. DeCaluwe, and **E.E. Kerre** [1985], "A new approach to information retrieval systems using fuzzy expressions." *Fuzzy Sets and Systems,* **17,** pp. 9–22.

Zhang, W. and **Y. Chen** [1984], "Mathematical models of multifactorial decisions and weather forecast." In: Sanchez [1984a], pp. 265–269.

Zimmermann, H.-J. [1978a], "Results of empirical studies in fuzzy set theory." In: *Applied General Systems Research,* edited by G.J. Klir, Plenum Press, New York, pp. 303–312.

Zimmermann, H.-J. [1978b], "Fuzzy programming and linear programming with several objective functions." *Fuzzy Sets and Systems,* **1,** pp. 45–55.

Zimmermann, H.-J. [1985], *Fuzzy Set Theory—and Its Applications.* Kluwer-Nijhoff, Boston.

Zimmermann, H.-J., L.A. Zadeh, and **B.R. Gaines,** eds. [1984], *Fuzzy Sets and Decision Analysis.* North-Holland, New York.

Zimmerman, H.-J. and **P. Zysno** [1980], "Latent connectives in human decision making." *Fuzzy Sets and Systems,* **4,** pp. 37–51.

Name Index

Abramson, N., 316
Aczél, J., 223–24, 316
Adamo, J.M., 292, 316
Adlassnig, K.P., 246, 248, 316
Aigner, M., 224
Albert, P., 316
Alsina, C., 62, 315, 336
Apostol, T.M., 43
Aristotle, 27
Asai, K., 336
Ash, R.B., 223–24, 316
Ashby, W.R., 103, 193–94, 210–11, 224, 270, 294, 316–17, 320
Assilian, S., 290, 330
Atlan, H., 317
Attneave, F., 290, 317
Aulin, A., 290, 317
Avgers, T.G., 226, 317
Ayme, S., 291, 325

Bacon, G., 131, 134, 317
Baldwin, J.F., 33, 292, 317
Ballmer, T.T., 317
Bandler, W., 63, 103, 290, 292, 317–18, 325, 329, 337
Bang, S.Y., 103, 339
Bartolin, R., 291, 334
Beaumont, P.E., 291, 332
Bell, D.A., 223, 291, 318
Bellman, R., 62, 255, 257, 318
Benson, W.H., 292, 318

Bernays, P., 322
Bernoulli, N., 214
Bertsekas, D.P., 226
Bezdek, J.C., 252, 281, 291, 318, 324, 335
Bhattacharya, P., 318
Billingsley, P., 223, 318
Biscoe, S.D., 291, 319
Black, M., 32, 318
Blin, J.M., 258, 318
Blockley, D., 291, 318
Bochvar, D.A., 28–29
Boekee, D.E., 224, 318
Boltzmann, L., 167–69, 222, 224, 228, 318
Bordley, R.F., 318
Borisov, A., 292, 318
Bortolan, G., 318
Bosserman, R., 290, 318
Bouchon, B., 290, 318
Braae, M., 290, 318
Bremermann, H.J., 205, 208–11, 230, 316, 318
Brillouin, L., 223, 319
Broekstra, G., 294, 319
Brooks, D.R., 223, 319
Brownell, H.H., 293, 326
Bruce, W.S., 291, 319
Buckles, B.P., 261, 264, 319
Buckley, J.J., 291, 319
Buell, D.A., 292, 319
Buisson, J.-C., 291, 319
Buoncristiani, J.F., 319

Subject Index

(Boldface numbers indicate pages that contain definitions of individual terms.)